HEALTHCARE TECHNOLOGIES SERIES 36

Applications of Machine Learning in Digital Healthcare

IET Book Series on e-Health Technologies

Book Series Editor: Professor Joel J.P.C. Rodrigues, College of Computer Science and Technology, China University of Petroleum (East China), Qingdao, China; Senac Faculty of Ceará, Fortaleza-CE, Brazil and Instituto de Telecomunicações, Portugal

Book Series Advisor: Professor Pranjal Chandra, School of Biochemical Engineering, Indian Institute of Technology (BHU), Varanasi, India

While the demographic shifts in populations display significant socio-economic challenges, they trigger opportunities for innovations in e-Health, m-Health, precision and personalized medicine, robotics, sensing, the Internet of things, cloud computing, big data, software defined networks, and network function virtualization. Their integration is however associated with many technological, ethical, legal, social, and security issues. This book series aims to disseminate recent advances for e-health technologies to improve healthcare and people's wellbeing.

Could you be our next author?

Topics considered include intelligent e-Health systems, electronic health records, ICT-enabled personal health systems, mobile and cloud computing for e-Health, health monitoring, precision and personalized health, robotics for e-Health, security and privacy in e-Health, ambient assisted living, telemedicine, big data and IoT for e-Health, and more.

Proposals for coherently integrated international multi-authored edited or co-authored handbooks and research monographs will be considered for this book series. Each proposal will be reviewed by the book Series Editor with additional external reviews from independent reviewers.

To download our proposal form or find out more information about publishing with us, please visit https://www.theiet.org/publishing/publishing-with-iet-books/.

Please email your completed book proposal for the IET Book Series on e-Health Technologies to: Amber Thomas at athomas@theiet.org or author_support@theiet.org.

Applications of Machine Learning in Digital Healthcare

Edited by
Miguel Hernandez Silveira and Su-Shin Ang

The Institution of Engineering and Technology

Published by The Institution of Engineering and Technology, London, United Kingdom

The Institution of Engineering and Technology is registered as a Charity in England & Wales (no. 211014) and Scotland (no. SC038698).

First published 2022

The Institution of Engineering and Technology
Futures Place
Kings Way, Stevenage
Herts SG1 2UA, United Kingdom

www.theiet.org

British Library Cataloguing in Publication Data
A catalogue record for this product is available from the British Library

ISBN 978-1-83953-335-8 (hardback)
ISBN 978-1-83953-336-5 (PDF)

Typeset in India by MPS Limited

Cover Image: Ying Hiy Phumi Wat'hna/EyeEm via Getty Images

Contents

About the editors

Miguel Hernandez Silveira is the CEO and a principal consultant at Medical Frontier Technology Ltd, UK. He is also CTO of SENTI TECH LTD, UK. He held positions as visiting lecturer at the University of Surrey, UK, and a visiting researcher at Imperial College London, UK. He is also a member of the IET Healthcare Technical Profession Network Committee, and reviewer of IEEE Sensors and IEEE Biomedical Circuits and Systems Journals. His research interests include machine learning, wireless low-power healthcare systems, biomedical sensors, instruments and algorithms, and digital signal processing.

Su-Shin Ang is the CEO and a principal consultant at Medical Frontier Technology Asia Pte Ltd, Singapore. He is a practising engineer, whose passion lies in the application of cutting-edge technology to the improvement of patient care. His research interests include machine learning, healthcare technology, development and deployment of medical devices, and the Internet of medical things.

Chapter 1
Introduction

Miguel Hernandez Silveira[1] *and Su-Shin Ang*[1]

1.1 Why?

Our bodies are definitely the most precious possession of our lives. Most people are concerned about their bodies; and thus spend a great deal of time and resources to maintain it at its best possible condition. As a consequence, the volume of data generated from health monitoring devices is increasing very rapidly, particularly through the use of wearable devices that we use to track our health, fitness, and well-being on daily basis [1]. Increasingly, the use of these informal devices in the outpatient context is being incorporated to provide a more holistic assessment of individuals. In general, these datasets are complex and encapsulate physiological and biomechanical information collected using arrangements of different sensors contained in a single wearable device such as a digital wristwatch [2].

An important reason why such massive and diverse datasets are required to describe, monitor, and diagnose the status or condition of persons is the inherent complexities of our bodies. While this data is the source for comprehensive assessment of people's state of health, the job of analysing and interpreting this data is performed by specialists e.g. clinicians and personal trainers. This is often a challenging job and makes a compelling case for automatic analysis and interpretation. Even partial but effective automation can be of great help to physicians, not only allowing laborious and error prone tasks to be carried out more efficiently and accurately, but also providing specialists with the ability to focus on other tasks that truly require human intervention, judgement, and expertise.

While clinical decision support systems are clearly beneficial, their design and development pose a challenging problem. The latter is associated with the volume and types of data involved, as well as the fact that clinical decisions are often predicated on a multiplicity of factors e.g. physiological conditions, comorbidities, and individuals' demographics. Furthermore, in some cases, the underlying pathophysiology behind the illnesses and conditions are not fully understood, and varies from patient to patient. As such, traditional top-down and rule-based systems are not always effective frameworks for implementing such a tool. What is required are flexible data driven

[1]Medical Frontier Technology Ltd, UK

frameworks that can be applied to different types of data and problems. In simple terms, there is a constant need for intelligent systems capable of generalising and adjusting to changes dictated by new incoming data. The degrees of freedom within the framework should also be adjustable to cope with the varying complexity of the problem.

Why are we writing this book? Simply because we believe Machine Learning (ML) is an important ingredient in the solution as it overcomes many of the limitations of traditional frameworks. Our intention is to demonstrate its potential by showcasing some recent advances. We feel incredibly fortunate that several world class researchers and engineers have agreed to share their work in this book. We hope that it will inspire our reader whatever their background may be.

1.2 How?

In this book, we will start by giving a simple overview of ML i.e. describing some of its fundamental building blocks as well as how to develop algorithms that produce effective predictions. If you are familiar with ML, skip ahead to subsequent chapters.

Further, to illustrate its utility and versatility, we will be looking at the role of ML in fitness and healthcare at different scales. At the macro-level, we look at health records of population segments, in order to guide evidenced-based policy decisions. For individual patients, an interesting guidance technique involving the use of digital holograms to help surgeons in microsurgeries is investigated. At an even lower scale, we will examine the ML techniques that are applied to high dimensional sensor data acquired from subsystems, which includes the thermoregulatory system, brain, and heart. Apart from the obvious health and wellness impact on the general public, these topics have applications in certain occupational contexts e.g. sportsmen/women and professional soldiers.

From our experiences accrued in this journey, we are humbled by the complexity of the problems faced by physicians and the multiplicity of factors that they have to consider in order to provide the right diagnosis and the appropriate course of treatment. Since patient welfare and safety are always of the utmost importance, the risks and benefits brought by any technology will always have to be carefully balanced. In the concluding chapter, we look at the aviation industry in an attempt to learn how this balance is achieved in another safety critical industry, where levels of safety and reliability have been maintained effectively at scale for an extended period of time.

1.3 What is ML?

Often, ML is seen as a miraculous black-box that solves all problems. To a certain extent, it is true as ML is a an extremely powerful and versatile tool that proved to be highly accurate and effective in several application domains. Indeed, its track record includes beating world class players at the game *Go* as well as enabling the autonomous driving of cars.

However, it is important to understand that each box is highly specific to a problem. Moreover, there are many options for filling said box as there are many different types of ML frameworks. Therefore, engineering skills, empirical trial and error are very essential to the design process. Even when a singular framework is suited to several problems, a substantial amount of data, skills, and effort are required to retrain the network in order to retrofit it to the new problem. For the rest of this chapter, rather than giving a general and abstract overview of ML, we will make use of a specific problem as we believe that concrete examples will serve us best in terms of illustrating these concepts.

At the time of writing, the world was in the grip of the COVID-19 pandemic. The virus responsible for the COVID-19 disease is known as the SARS-CoV2 virus, which is genetically similar to the Severe Acute Respiratory Syndrome (SARS) and the Middle East Respiratory Syndrome (MERS) viruses. Symptoms include fever, shortness of breath, fatigue, and sputum production. Since these symptoms are manifested in other types of illnesses, diagnosing a patient via a general consultation is difficult. For this reason, the Reverse Transcription Polymerase Chain Reaction (RT-PCR) is the most commonly used diagnosis tool against the COVID-19 disease and regarded as the gold standard. However, studies have shown that its accuracy is only about 89% [3], so chest X-rays (CXRs) are routinely carried out for symptomatic patients in addition to PCR testing in order to rule out alternative conditions like Lobar collapse, Pneumothorax, or Pleural effusion.

As it is not an easy task, an expert clinical assessment of the CXR is required to help diagnose the COVID-19 disease from normal CXRs as well as other types of conditions. Radiologists can potentially benefit from a clinical decision support tool developed to aid them in diagnosis. We will design a naïve version of a COVID-19 classifier in order to demonstrate important aspects of ML.

1.4 The problem

Let us begin by framing the problem more precisely. As is usually the case for ML-based problems, we start by looking at the data available to us. We have at our disposal an arbitrarily large dataset, consisting of two different types of Posterior–Anterior (PA) CXRs.

- CXRs from symptomatic patients infected by the SARS-CoV2 virus.
- CXRs from healthy patients.

An example of each type is shown in Figure 1.1. The CXR on the right is normal – the ribs and other pertinent structures are clearly defined with minimal occlusions. On the other hand, the CXR on the left is taken from a patient who has been diagnosed with the COVID disease. In contrast with the other CXR, this image is 'blurred' and artefacts known as Ground Glass Opacities (GGOs takes the appearance of powered glass that populate regions of the CXR, obscuring other anatomical structures) are

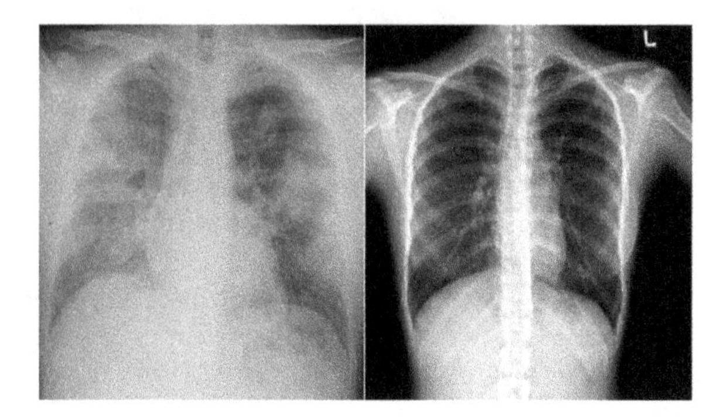

Figure 1.1 Chest X-rays for a COVID-infected individual (left) and a healthy individual (right) [4]

clearly seen throughout the CXR. Note that there are other conditions that may be mistaken for the COVID disease and we will increase the number of categories to allow us to explain certain ideas more clearly in the later sections. For the moment, we will restrict ourselves to only two categories.

We can see each category as a class. Therefore, with this dataset, we will be able to create a classifier that will tell us whether a given CXR falls in one of the two classes. The simplest form of output could be '1' for detected signs of the COVID disease and '0' for a healthy CXR. Instead of adopting this method, we make use of a pair of probabilities p_0, p_1 with the following definitions:

- p_0: the probability that the CXR belongs to a COVID-infected and symptomatic patient given Image I.
- p_1: the probability that the CXR belongs to a healthy individual given Image I.

Under the second scheme, unclear results can be taken into account. For instance, if $p_0 = p_1 = 0.5$, it would mean that there is a high level of uncertainty and other methods should then be applied to ascertain the patient's condition. Conversely, the first scheme does not do so which may then produce a misleading result. Note that both p_0 and p_1 are predicated upon Image I, so they are effectively conditional probabilities and $p_1 = 1 - p_0$. Separately, for the same set of CXRs, we obtain a set of reliable reference values by asking an expert panel to look at each CXR and categorise them into one of the two possible classes; specifically giving a reference value of '1' for a positive diagnosis and '0' otherwise. To maintain a high level of diagnostic accuracy, CXRs will be discarded if consensus cannot be reached among panel members.

In summary, the input will consist of a CXR with assumed dimensions of 512 by 512 pixels and the output will consist of a set of probabilities $\{p_0, p_1\}$ – the conditional probabilities of whether the CXR comes from an infected and symptomatic patient or not, respectively. Each CXR is accompanied by a reference value or label which is used to determine the 'goodness' of the model prediction.

1.5 Gradient descent

Now that the problem as well as its inputs and outputs have been specified, the next step involves the actual internals of the black box. One of its components comprises a mapping function capable of estimating the conditional probability p_0. Since $p_1 = 1 - p_0$, we only have to estimate p_0 for our two-class problem – this is illustrated in Figure 1.2(a). Specifically, we flatten a two-dimensional image into a one-dimensional vector I and feed it to function f. Since we have a set of reference values obtained from the expert panel, we are able to obtain the error e between the function output O and the expected value (refer to Figure 1.2(a)) – an example could be a label of '1' for a positive diagnosis and a model predicted output of 0.7. The weights, $\mathbf{w}$, are parameters of function f and we are able to minimise error e by searching for the optimum value for vector $\mathbf{w}$.

The relationship between the error and weights is shown in the idealised contour chart in Figure 1.2(b) where we adopted only two weights w_0 and w_1 (w_0 and w_1 forms a two-dimensional vector $\mathbf{w} = [w_0, w_1]$, representing a point in the chart) for this particular example. The optimum weight vector $\mathbf{w}$ is represented by the black dot or the 'eye' of the contour graph, and the primary objective is to find this point, starting from an initial point at the boundary of the contour graph. To do so, we can make use of the Batch Gradient Descent (BGD) algorithm, which is shown in Figure 1.3(a). In this case, we consider the averaged squared error (Line 1.1) and we

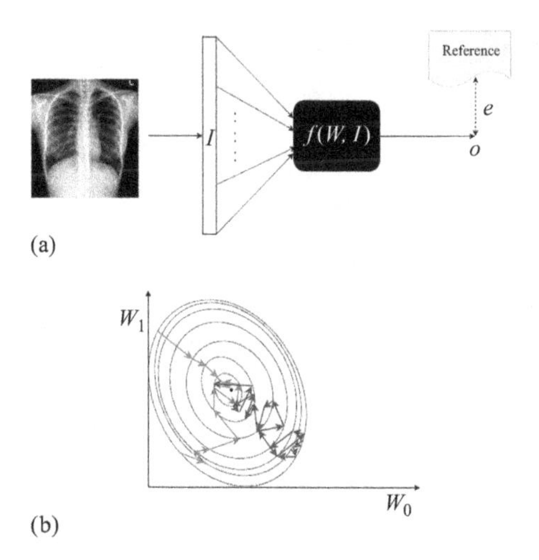

*Figure 1.2 (a) A CXR is flattened into a one-dimensional vector **I** and fed to a black box containing a function f. (b) We would like to find the optimal **w** that will minimise the error e. Red arrows indicate a more direct path taken by batch gradient descent (BGD), whereas green arrows indicate small, convoluted steps taken by Stochastic Gradient Descent algorithm (SGD). Purple arrows indicate the path taken by Minibatch gradient descent.*

Batch GD	**SGD Algorithm**
Inputs: Images ***I***, Reference values ***ref***	Inputs: Images ***I***, Reference values ***ref***
Output: ***w***	Output: ***w***
1.1 $e_j(w) = (f(w, I_j) - ref_j)^2/2M$	1.1 $e_j(w) = (f(w, I_j) - ref_j)^2/2$
1.2 $e_j'(w) = (f(w, I_j) - ref_j)f'(w, I_j)/M$	1.2 $e_j'(w) = (f(w, I_j) - ref_j)f'(w, I_j)$
2. for $i = 1 \ldots N$	***Randomise_Dataset***()
$E = 0$	2. for $i = 1 \ldots N'$
2.1 for $j = 1 \ldots M$	2.1 for $j = 1 \ldots M$
2.1.1 $\boldsymbol{E} = \boldsymbol{E} + \mu e_j'(\boldsymbol{w})$	2.1.1 $\boldsymbol{w} = \boldsymbol{w} + \mu e_j'(\boldsymbol{w})$
2.2 $\boldsymbol{w} = \boldsymbol{w} - \boldsymbol{E}$	
(a)	(b)

Minibatch GD Algorithm

Inputs: Images ***I***, Reference values ***ref***, Batch size m

Output: ***w***

1.1 $e_j(w) = \mathrm{Sum}_{i=1\ldots m}[(f(w, I_j) - ref_j)^2/2m$

1.2 $ej'(w) = \mathrm{Sum}_{i=1\ldots m}[(f(w, I_j) - ref_j)f'(w, I_j)]/m$

Randomise_Dataset()

2. for $i = 1 \ldots N{-}m{+}1$ in steps of m

2.1 $\boldsymbol{E} = \mu e_j'(\boldsymbol{w})$

2.2 $\boldsymbol{w} = \boldsymbol{w} - \boldsymbol{E}$

(c)

Figure 1.3 (a) Batch gradient descent algorithm. (b) Stochastic gradient descent algorithm. (c) Mini Batch gradient descent algorithm. Note that μ refers to the learning rate and it is used to regulate the step size taken in each iteration.

scale it by ½ so that its partial derivative $e_j'(\mathbf{w})$ is simply the averaged error (Line 1.2). At the beginning, the initial solution is likely to be poor, so this will result in a large $e_j'(\mathbf{w})$. We take the average of all derivatives across all M images and reference values, accumulating it in E (Line 2.1.1). We then update the initial solution with E (Line 2.2). This is akin to taking the first step (first red arrow in Figure 1.2(b)) from the initial solution in the contour graph – since the first solution is poor, the corresponding derivative is high and a large initial step is therefore taken. This process is repeated N times, with steps of decreasing size as the algorithm converges towards the solution.

The Batch Gradient Descent (Batch GD) algorithm is named as such because it considers the derivatives or gradients as an entire batch and averages them across all data points before updating the weight $\mathbf{w}$. Doing so makes implementation challenging especially for large values of M (large number of images) because the inner loop 2.1 requires substantial intermediate memory storage. Instead, we consider a simplified algorithm known as the Stochastic Gradient Descent (SGD) that is described in Figure 1.3(b). Rather than averaging across all derivatives in a single batch; we first randomise the dataset, and then work out the derivative for a single data point (or image) and update weight $\mathbf{w}$ immediately, as shown in Loop 2.1 of Figure 1.3(b). In this way, we do not need to wait for all gradients to be computed and averaged before refining $\mathbf{w}$. We see this illustrated by the green path in Figure 1.2(b) which is substantially noisier and more convoluted than the batched method largely because the noise reduction effect via averaging is omitted from the SGD algorithm. The initial data randomisation step is necessary to prevent the algorithm from falling into local minimas – these are minimum points that occur in real-world irregular error surfaces and could 'mislead' the algorithm that it has converged. Therefore, randomisation is essential for reducing this risk.

A typical compromise between SGD and Batch GD algorithms is the Minibatch GD algorithm. The Minibatch GD algorithm behaves in the same manner as the Batch GD except that it operates on a much smaller batch of data points in order to compute a single step for gradient descent. Each batch usually contains only 32, 64, or 128 data points. This compromise will lead to a less noisy path than SGD, without the excessively large memory requirements of Batch GD. Its path is depicted in purple in the contour graph of Figure 1.2(a). Gradient Descent is an important concept as it facilitates an operation known as 'error backpropagation', where the squared error is iteratively worked out for each data point and used to update the weights of the Artificial Neural Network (ANN) in order to improve its performance incrementally. This notion will be made more explicit in the following sections.

1.6 Structural components of the ANN

1.6.1 The fully connected neural network

Having defined the problem and described how we can iteratively improve our model, let us turn our attention to the function f, where we will feed a CXR as an input in order to produce the required probabilities. In this case, we can fill the black box with a Fully Connected Artificial Neural Network (FCN). To understand how an FCN works, we will first talk about its most basic unit – the neuron which is shown in Figure 1.4. This neuron takes in N inputs, $I_0 \ldots I_{N-1}$, and returns a single output O. Each input I_i is multiplied by a corresponding weight w_i and added to the rest of the weighted inputs. Subsequently, an activation function f (note that this function

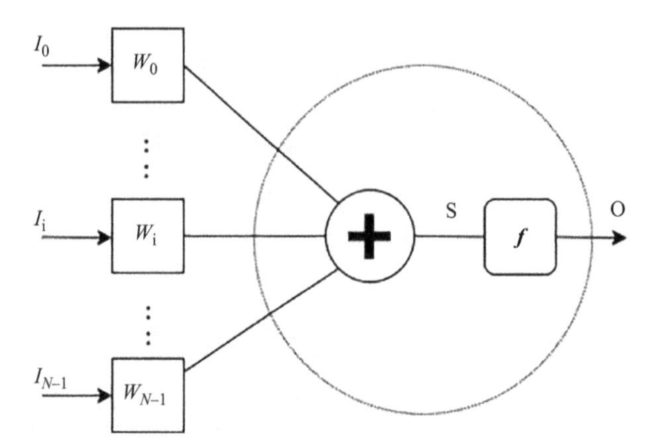

Figure 1.4 A neuron – the most basic component in an artificial neural network (ANN)

is different from the function f in Figure 1.2(a)) is applied to the sum of products in order to produce output O.

These neurons are connected to form a network that is shown in Figure 1.5. The network is composed of three types of layers: Input, Hidden, and Output. The composition of the nodes in the hidden and output layers are shown in Figure 1.4. An input node (yellow) is different as it simply broadcasts the value of the input (current input value is copied to every output).

Some observations can be made about the structure of the FCN.

1. Each column within the network corresponds to a layer. Within a hidden layer, each node is given a unique number with a subscript that corresponds to the layer. For example, 0_0 refers to node *0* located at hidden layer column *0*.
2. Each node in a layer is fully connected to the nodes in the subsequent layer – Input node *0* is connected to all the nodes, $0_0 - 2_0$, in the next layer. Therefore, this network is known as fully connected.
3. The input layer consists of broadcast nodes. For example, for yellow node *0*, all three of its outputs (black arrows) are equal to I_0.
4. As seen in Figure 1.4, each input or connection between two nodes is multiplied with a weight w_{ij} where i represents the destination node and j represents the origin node. For instance, between the input layer and the first hidden layer, w_{01} corresponds to the weight between yellow node *1* and red node 0_0.
5. The set of weights for each connection between the yellow nodes and the first red hidden layer is stored in a matrix $\mathbf{W}_0$. Each element of the matrix is w_{ij} that resides at row i and column j, which in turn represents destination node i and origin node j. The same applies to matrices $\mathbf{W}_1$ and $\mathbf{W}_2$.
6. Each red node output is a function of the weighted sum of inputs and the bias for that particular layer. This is shown specifically for red node output 0_0 in (1.1), and it corresponds to the first stage of Figure 1.4, producing result S.

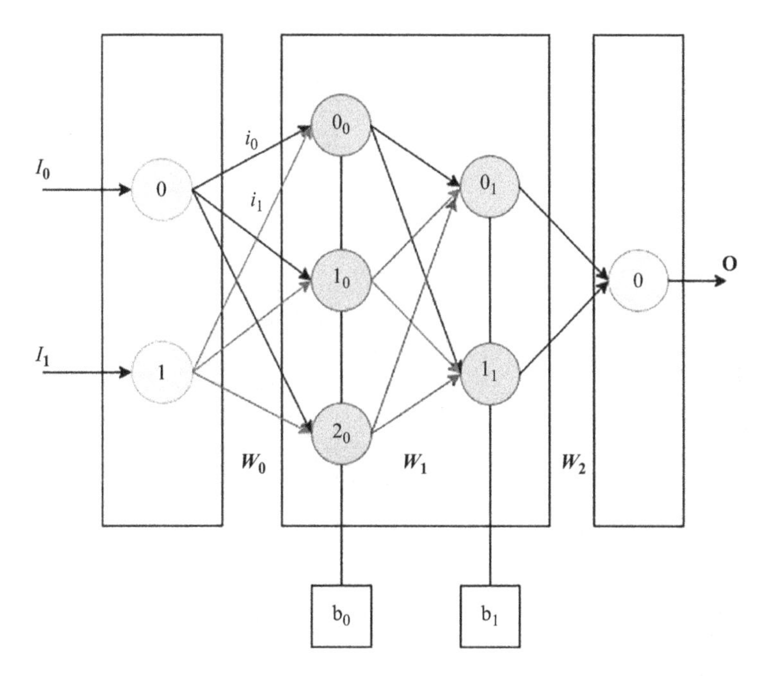

Figure 1.5 ANN. Yellow, red, and blue nodes correspond to neurons in the input, hidden, and output layers, respectively.

7. The second stage involves an activation function f which is described using (1.2). The reason for its inclusion is to create a smoothly varying output that is bounded between 0 and 1 (look at Figure 1.6). Note that there are different types of activation functions. The sigmoid function is defined in (1.2) and seen on the left in Figure 1.6. Another popular activation function is the *ReLU* which is seen on the right of Figure 1.6.
8. Output O of the neuron is broadcasted to two nodes in the next layer, 0_1 and 1_1.
9. The blue node in the output layer is similar to nodes in the hidden layer. Note that there are often multiple output nodes. We simplify it to only one output here since we need only to estimate one probability p_0.

$$S = w_{00}i_0 + w_{01}i_1 + b_0 \tag{1.1}$$

$$O = \frac{1}{1 + e^{-S}} \tag{1.2}$$

In this simple network, there are some parameters that need to be determined, including $\mathbf{W}_0$, $\mathbf{W}_1$, $\mathbf{W}_2$, $\mathbf{b}_0$, and $\mathbf{b}_1$. Random values are usually assigned to them at the start. Subsequently, an iterative process is required to train the network. As part of the Stochastic Gradient Descent (SGD) algorithm, forward propagation step followed by an error backward propagation is required as part of a training step. Training will be further discussed in Section 1.7.1.

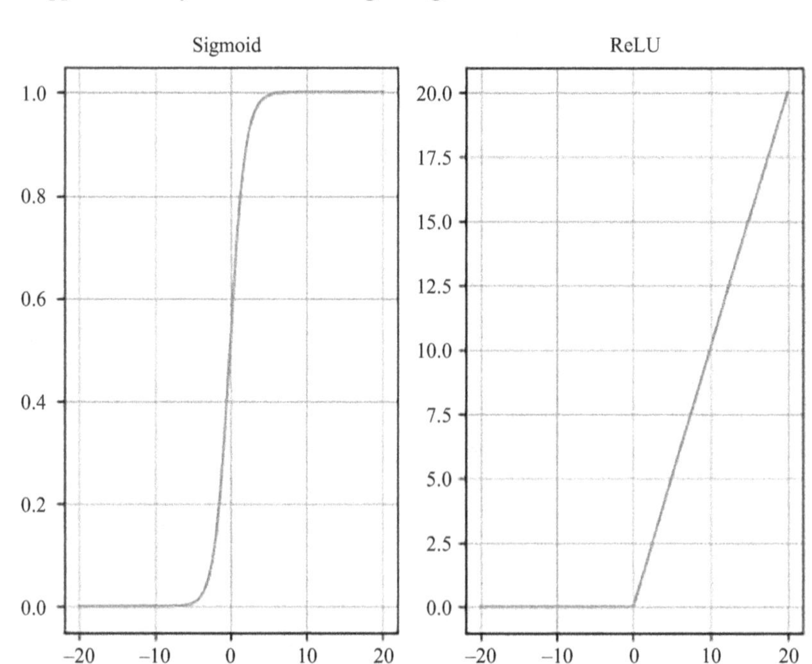

Figure 1.6 Chart of the sigmoid activation function (left). Chart of the ReLU activation function (right).

- *Forward propagation*: Imagine that we have an input vector $\mathbf{I} = [0.3, 0.6]$ and the desired output T is 0.7. Let us make an assumption that the weights and the biases have been initialised to random values. We then work out the result of each of the nodes (all node outputs are stored as they are necessary for back propagation) using (1.1) and (1.2) and propagate them forward till we eventually get to the final output node. Given that the parameters have been randomised, we will most likely get an error. For instance, if the predicted output O is found to be 0.1, we get an error of 0.18 according to (1.3).
- *Backward propagation*: This step is necessary to adjust the parameters, or the weights and biases, in order to improve the ANN's predictive capability. Unlike forward propagation i.e. which starts from the inputs to the output (from left to right in Figure 1.5), we work backwards from the output error e and stop at the input layer i.e. from right to left:

$$e = \frac{1}{2}(T - O)^2 \tag{1.3}$$

Given an error e, we wish to know how much adjustment to apply to each weight. Let us consider the blue neuron in the output layer. A useful parameter is $\partial e/\partial w$, which is the partial derivative of the error with respect to the weight. Effectively, the partial derivative is the gradient in the error surface e defined by the weight space $\mathbf{W}$. We can then use this partial derivative to update the weight using (1.4), where w and μ refer to a particular element of a weight vector $\mathbf{W}$ and the learning rate respectively. The learning rate is a constant that we determine and fix at design time in order to control the size of the update (see Section 1.7.3 regarding the adjustment of learning rate):

$$w = w - \mu \frac{\partial e}{\partial w} \tag{1.4}$$

$$\frac{\partial e}{\partial w} = \frac{\partial S}{\partial w} \frac{\partial O}{\partial S} \frac{\partial e}{\partial O} \tag{1.5}$$

Obtaining $\partial e/\partial w$ is not straightforward. However, we can derive it using the chain rule in (1.5) and a more intuitive illustration is shown in Figure 1.7. The various partial derivatives can be worked out using (1.6)–(1.8). Here, we continue from the previous example, where the target and predicted output values are 0.7 and 0.1 respectively. We use a Sigmoid activation function in this case and its partial derivative is described by (1.7). Finally, since S is the sum of products between the inputs and their corresponding weights, its partial derivative is simply I, as shown in (1.8):

$$\begin{aligned} \frac{\partial e}{\partial O} &= \frac{\partial}{\partial O} \frac{1}{2}(T - O)^2 \\ &= O - T \\ &= 0.1 - 0.7 = -0.6 \end{aligned} \tag{1.6}$$

$$\frac{\partial O}{\partial S} = \frac{\partial}{\partial S} \frac{1}{1 + e^{-S}} = O(1 - O) \tag{1.7}$$

$$\frac{\partial S}{\partial w} = \frac{\partial}{\partial w}(Iw + I^1 w^1) = I \tag{1.8}$$

Intermediate values such as S and I are known from the forward propagation stage or forward pass. Therefore, they can be substituted into (1.7) and (1.8) to obtain the new weights. Let us consider error back propagation in the hidden layer. This is shown in Figure 1.8 where we have three nodes – one node in Hidden layer 1 is connected to two child nodes in Hidden layer 2. In this case, we determine how to

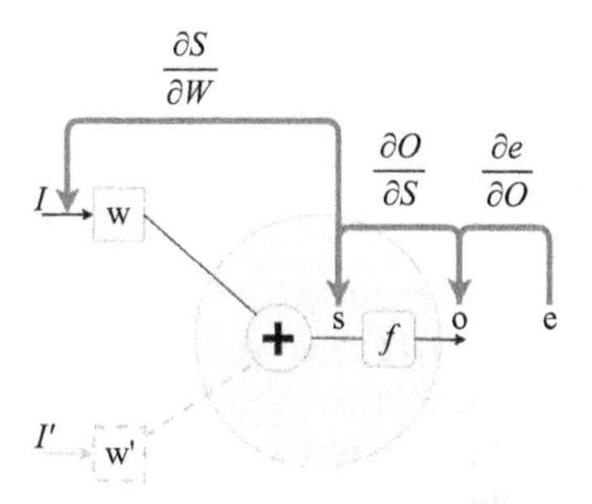

Figure 1.7 Partial derivatives involved in weight update for output layer neuron

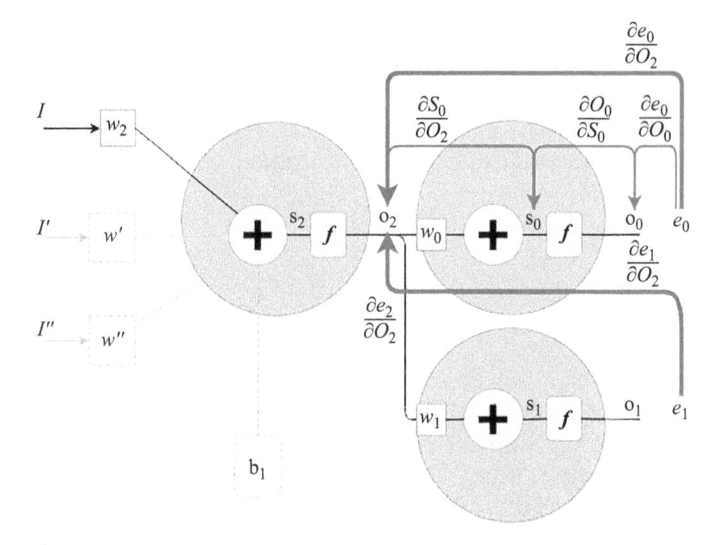

Figure 1.8 Chain rule with partial derivatives in the hidden layer. Parent node on the left, connected to two child nodes on the right.

update weight w_2. To do so, we require the partial derivative $\partial e_2/\partial w_2$ so that we can make use of (1.4) to update w_2.

Since the parent node is connected to only two child nodes in Figure 1.8, errors at the child node outputs O_0 and O_1 will both contribute to the error at output O_2. Therefore, we sum up their contributions to obtain $\partial e_2/\partial w_e$ (abbreviated as δ_2) as shown in (1.9). Subsequently, we make use of the chain rule to obtain (1.10) – this is shown more literally in Figure 1.8. The term on the extreme left, $\partial e_0/\partial O_0$ can be obtained recursively from the previous layer in the same manner that we are obtaining $\partial e_0/\partial O_2$ in (1.10). These are known as the delta errors which we will propagate backwards from layer to layer:

$$\frac{\partial e_2}{\partial O_2} = \frac{\partial e_0}{\partial O_2} + \frac{\partial e_1}{\partial O_2} = \delta_2 \tag{1.9}$$

$$\frac{\partial e_0}{\partial O_2} = \frac{\partial e_0}{\partial O_0}\frac{\partial O_0}{\partial S_0}\frac{\partial S_0}{\partial O_2} \tag{1.10}$$

The remaining two partial derivatives, $\partial O_0/\partial S_0$ and $\partial S_0/\partial O_2$, can be obtained with (1.11) and (1.12). Note that (1.11) involves taking the derivative of the activation function. As discussed earlier, other activation functions apart from the Sigmoid can be used e.g. *ReLU* (chart on the right of Figure 1.6). The prerequisite for a working activation function is that it must be differentiable to allow error backpropagation to function correctly (in this case we used the sigmoid function). The last derivative,

$\partial S_0/\partial O_2$, is simply the weight w_0. Finally, the desired partial derivative, $\partial e_2/\partial w_2$ is obtained using (1.13):

$$\frac{\partial O_0}{\partial S_0} = f'(S_0) = O_0(1 - O_0) \tag{1.11}$$

$$\frac{\partial S_0}{\partial O_2} = \frac{\partial}{\partial O_2}(w_0 O_2) = w_0 \tag{1.12}$$

$$\begin{aligned}\frac{\partial e_2}{\partial w_2} &= \frac{\partial e_2}{\partial O_2}\frac{\partial O_2}{\partial S_2}\frac{\partial S_2}{\partial w_2} \\ &= \delta_0 w_0 O_0(I - O_0)IO_2(I - O_2),\ \delta_0 = \frac{\partial e_0}{\partial O_0}\end{aligned} \tag{1.13}$$

Note that the partial derivative $\partial e_0/\partial O_0$ is represented using a delta δ_0. A similar expression can be found at connection O_2 i.e. δ_2 and the way that it is computed is shown in (1.11). During backpropagation, δ_2 will be used to update the weights in the adjacent layer to the left. Unlike the output layer, we are required to update the bias b_1 as well. To do this, we apply a similar method to weight update, as shown in (1.14). Since $\partial e_2/\partial b_1$ is not immediately known, we make use of the chain rule to decompose it into more manageable bits in (1.15). The resulting expression is almost the same as (1.13) with the exception of the last derivative $\partial S_2/\partial b_1$ in (1.16):

$$b_1 = b_1 + \mu\frac{\partial e_2}{\partial b_1} \tag{1.14}$$

$$\begin{aligned}\frac{\partial e_2}{\partial b_1} &= \frac{\partial e_2}{\partial O_2}\frac{\partial O_2}{\partial S_2}\frac{\partial S_2}{\partial b_1} \\ &= \delta_0 w_0 O_0(1 - O_0)O_2(1 - O_2)\end{aligned} \tag{1.15}$$

$$\begin{aligned}\frac{\partial S_2}{\partial b_1} &= \frac{\partial}{\partial b_1}(Iw_2 + I'w' + I''w'' + b_1) \\ &= 1\end{aligned} \tag{1.16}$$

The operations required to update other weights and biases within the same layer are the same in spite of involving different nodes. By reusing delta error δ of the current layer, the weights for layers towards the left can be similarly updated without the need for parameters from other layers. Consequently, these parameters can be updated layer by layer, from the right to the left, by propagating the error one layer at a time, by working out the required deltas and corresponding updates and then applying these updates to the weights and biases.

Having described an FCN, let us apply it to a CXR. First, the image is flattened i.e. converted into a one-dimensional vector row-wise from the left to the right and from the top to the bottom. Subsequently, we apply this vector to the input layer of the network as seen in Figure 1.9.

The size of the image is 512 by 512 pixels, and the first hidden layer has 10 nodes. It follows that the input layer has 512^2 nodes and matrix $\mathbf{W}_1$ contains $512^2 \times 10$ weights. For a reasonably sized image, this is a massive number of design parameters to consider. Apart from the huge amount of memory required to store these weights, it

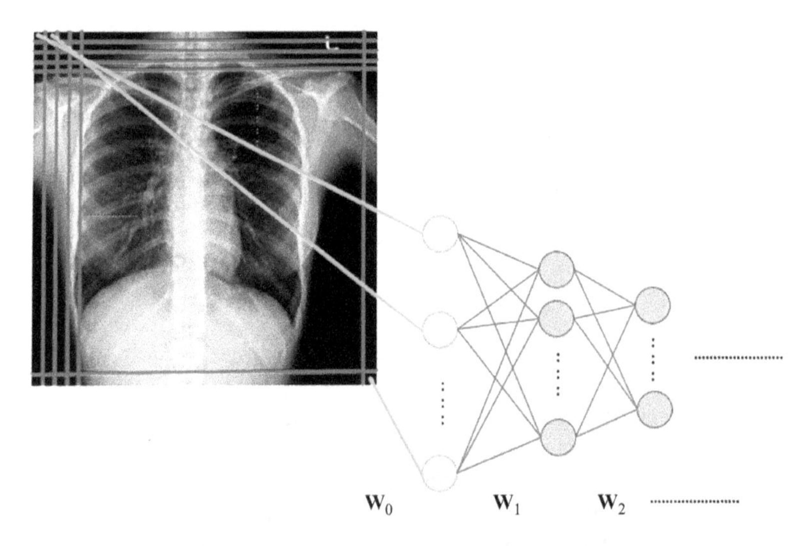

Figure 1.9 Applying an image [4], pixel-by-pixel, to the input layer (yellow) of ANN fully connected network. Matrices $\mathbf{W}_1$ and $\mathbf{W}_2$ contain the weights between subsequent layers; the biases are not illustrated here.

will take a considerable amount of time to compute the best weights for this network. Finally, the number of parameters of the network is likely to exceed the number of data samples available for training; thus, leading to overfitting i.e. inability to generalise when processing new/unseen data.

Due to the expensive computational requirements, an additional feature extraction stage is typically added before the input layer. An example feature could be the proportion of bright pixels within the image. Specifically, this can be worked out using (1.17) where we count the number of pixels with intensities greater than τ. Since the image contains a set of 512^2 pixels belonging to set P, $p \in P$, Function $I(p)$ is used to work out the brightness of the pixel p and function $T_p(C)$ is used as a 'counting function' to work out the number of times when condition C is true over Set P:

$$F_0 = \frac{T_P(I(p) > \tau)}{512^2}, p \in P \tag{1.17}$$

By making use of (1.17), we have reduced the entire Pixel set P to a single number i.e. 512^2 to 1. However, the number of bright pixels as a feature alone is probably not good enough – we are throwing away too much information in the process of collapsing the raw image pixels to one number. We have the option of computing (selecting or extracting) other features and feeding them to the input layer: $F_0, \ldots, F_{N-1}, N \ll 512^2$ such that the problem becomes much more tractable by substantially reducing its dimensionality while capturing the most pertinent information in the form of features. In fact, this popular approach has been widely adopted by the ML community for a long time. However, there are two disadvantages in this approach:

- Features that produce good classification results are often difficult to identify. Also, features that are individually good may not necessarily work well together with others making feature selection a challenging task, as it involves identifying good features as well as finding the right combinations that will work well together. This process is often tedious, time-consuming and expensive.
- Even if an ideal set of features is found, they are usually relevant to one particular problem and feature engineering will have to be redone for other problems.

Whilst connecting a FCN directly to an image is computationally in feasible, we will look at an alternative type of network in Section 1.6.2 that facilitates direct connections with the image while keeping the number of parameters at a manageable level. This is extremely useful as it eliminates the need for feature engineering.

1.6.2 Convolutional neural network

As an alternative to the FCN, we can drastically reduce the computational requirements by using Convolutional Neural Networks (CNNs). CNNs allow a weight sharing scheme where groups of pixels share the same weights. This drastically reduces the number of weights or parameters that need to be found. Since no feature extraction is required, similar CNN structures become directly reusable across many different problems, albeit after some weight adjustments.

Central to the CNN is the filter kernel, which contains filter coefficients or weights. This filter can be seen as a matrix that we slide across the image, left to right and top to bottom or in a zigzag manner as shown in Figure 1.10. At each location, the sum of products is obtained to produce another structure that is known as a feature map – this moving sum is known commonly as a convolution operation. Notice that we have three images stacked on top of one another towards the left in Figure 1.10. This is because colour images have three primary channels: Red, Green, and Blue. Any colour of any intensity can be obtained by adjusting the intensity of each channel, and superimposing these three primary colours. In order for the scheme to work, a stacked filter that is smaller in dimensions than the image is required, and the filter kernel in each layer moves in lock step to obtain the feature map shown on the right. One slice or layer of the filter is seen in Figure 1.10 – this comprises of a two-dimensional matrix (3-by-3 matrix) that operates on each channel and its elements are $w_{00}, \ldots w_{22}$. Like the FCN, these weights are determined using error backpropagation. There are a number of important design parameters relating to the CNN structure.

- *Filter kernel*: This refers to the size of the kernel. In Figure 1.10, we make use of a 3-by-3 filter kernel but filters of other sizes may be used as well.
- *Stride*: When the filter kernel is moved across the image, both horizontally and vertically, the smallest step size is one. However, a step size larger than one is commonly used as well and this is known as the stride (measured in number of pixels).
- *Channels*: In Figure 1.10, three channels are required to represent the intensities of the colour map. The number of channels or the height of the stack is arbitrary for intermediate feature maps.

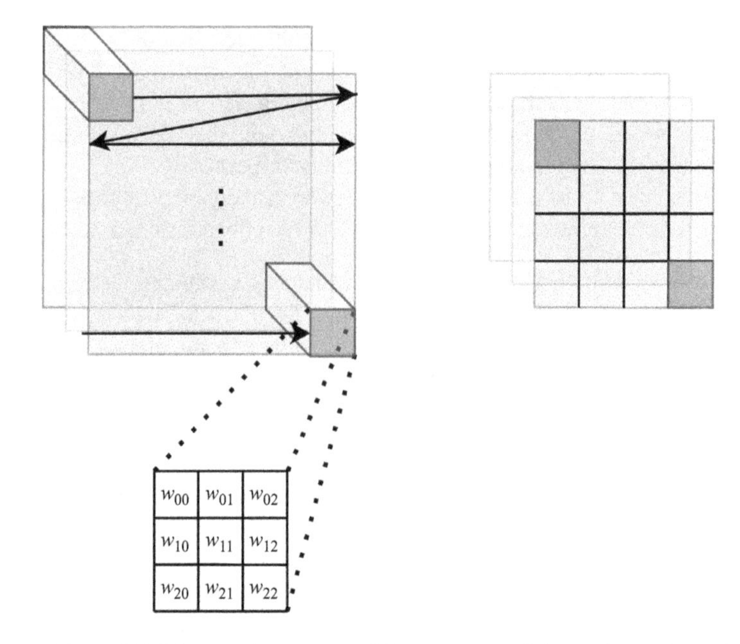

Figure 1.10 A CNN for a colour image. This consists of three channels: red, green, and blue. The filter kernel is three dimensional matrix that slides across the image from left to right, top to bottom. At each location, a sum of products is computed with the 3-by-3 kernel (shaded in blue) to arrive at a corresponding sum on the right (coloured in blue), forming a 3-dimensional structure on the right.

- *Padding*: The pixels nearer to the edges of the image are considered less by the filter relative to those in the middle because there are more overlaps between convolution windows for the pixels in the middle. Padding can be used to alleviate this issue and this involves adding p layers of zeros around the image for a padding of p.

Next, let us describe the way that error backpropagation works with CNNs. We consider the convolution between an image (one channel) and a small 2-by-2 filter as shown in Figure 1.11. The first step of the convolution operation produces O_{00} using (1.18). We can work out the three other convolution outputs in the same manner after moving the filter kernel in a zigzag manner, one step at a time to produce the purple output window shown in Figure 1.11. These partial derivatives will be helpful to adjust the filter weights as part of the backpropagation process:

$$O_{00} = I_{00}w_{00} + I_{01}w_{01} + I_{10}w_{10} + I_{11}w_{11} \tag{1.18}$$

$$O_{01} = I_{01}w_{00} + I_{02}w_{01} + I_{11}w_{10} + I_{12}w_{11} \tag{1.19}$$

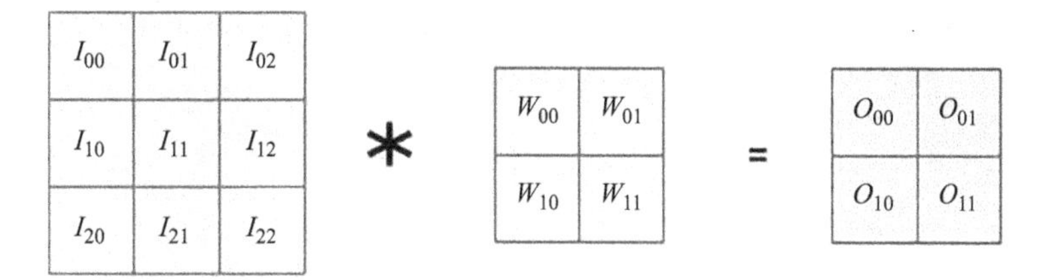

*Figure 1.11 Convolution between Image **I** and filter kernel **W** to produce output **O***

From (1.18), we obtain the partial derivatives (1.20). Our immediate objectives are to find the partial derivatives necessary to update the current filter weights, $\partial e/\partial w_{00}, \ldots, \partial e/\partial w_{11}$, as well as to as to allow the delta errors to be backpropagated through to the earlier layers, $\partial e/\partial I_{00}, \ldots, \partial e/\partial I_{11}$, as shown in (1.21). We make use of the Chain rule to obtain the partial derivatives in (1.22) and (1.23):

$$\frac{\partial O_{00}}{\partial w_{00}} = I_{00}, \frac{\partial O_{00}}{\partial w_{01}} = I_{01}, \frac{\partial O_{00}}{\partial w_{10}} = I_{10}, \frac{\partial O_{00}}{\partial w_{11}} = I_{11} \tag{1.20}$$

$$w_{ij} = w_{ij} - \mu \frac{\partial e}{\partial w_{ij}} \tag{1.21}$$

$$\frac{\partial e}{\partial w_{ij}} = \sum_k \sum_l \frac{\partial e}{\partial O_{kl}} \frac{\partial O_{kl}}{\partial w_{ij}} \tag{1.22}$$

$$\frac{\partial e}{\partial I_{xy}} = \sum_k \sum_l \frac{\partial e}{\partial O_{kl}} \frac{\partial O_{kl}}{\partial I_{xy}} \tag{1.23}$$

Expanding on (1.22), we obtain the partial derivative corresponding to w_{00} in (1.24). The partial derivatives for other weights can be worked out in a similar fashion:

$$\begin{aligned}\frac{\partial e}{\partial w_{00}} &= \frac{\partial e}{\partial O_{00}} \frac{\partial O_{00}}{\partial w_{00}} + \ldots + \frac{\partial e}{\partial O_{11}} \frac{\partial O_{11}}{\partial w_{00}} \\ &= \frac{\partial e}{\partial O_{00}} I_{00} + \frac{\partial e}{\partial O_{01}} I_{01} + \frac{\partial e}{\partial O_{10}} I_{10} + \frac{\partial e}{\partial O_{11}} I_{11}\end{aligned} \tag{1.24}$$

We are still lacking the delta error for this stage, $\partial e/\partial I_{xy}$, as this is necessary for backpropagation to continue to the layer on the upstream (layer to the left). To obtain this set of partial derivatives, we make use of (1.25) and (1.26). Notice that the number of terms on the right-hand side differ for different deltas in (1.25) and (1.26). Why is that the case? To answer that question, we refer back to (1.19), where we find that O_{01} is a function of I_{01}, but not of I_{00}. It follows that $\partial O_{01}/\partial I_{00} = 0$ so we can disregard this term. The same reasoning applies to the partial derivatives with respect to all other inputs (which are not shown here).

$$\frac{\partial e}{\partial I_{00}} = \frac{\partial e}{\partial O_{00}}\frac{\partial O_{00}}{\partial I_{00}} = \frac{\partial e}{\partial O_{00}}w_{00} \tag{1.25}$$

$$\begin{aligned}\frac{\partial e}{\partial I_{01}} &= \frac{\partial e}{\partial O_{00}}\frac{\partial O_{00}}{\partial I_{01}} + \frac{\partial e}{\partial O_{01}}\frac{\partial O_{01}}{\partial I_{01}} \\ &= \frac{\partial e}{\partial O_{00}}w_{01} + \frac{\partial e}{\partial O_{01}}w_{00}\end{aligned} \tag{1.26}$$

To obtain the propagated delta errors or $\partial e/\partial I_{00}, \ldots, \partial e/\partial I_{11}$, it is less clear that the equations, (1.25) and (1.26), involves convolution. Indeed, we need to make modifications to **W** (square matrix containing the filter weights w_{ij}) and $\partial e/\partial O_i$ for convolution to work. Specifically, we get results for (1.25) and (1.26) by first padding **W** with one layer of zeros and subsequently reflecting the delta error matrix both horizontally and vertically. Finally, we convolve these matrices in Figure 1.12.

So far, we have assumed a stride of one pixel and no padding but we can generalise the above approach by assuming an input image I and filter kernel **W** of arbitrary size, which are denoted by N and M, respectively. Furthermore, stride s and padding p are applied to the CNN.

$$\mathbf{dim(O)} = \left\lfloor \frac{N + 2p - M}{s} + 1 \right\rfloor \tag{1.27}$$

$$P[i,j] = P'[\mathbf{dim(W)} - 1 - i, \mathbf{dim(W)} - 1 - j] \tag{1.28}$$

- The width of the output square matrix O is obtained using (1.27).
- The amount of padding required for weight matrix **W** is $\mathbf{dim(W)} - 1$ (blue matrix in Figure 1.13).
- Reflecting the delta error matrix can be generally carried out using (1.28).
- The convolution operations will need to be adjusted accordingly for different strides and padding sizes:

To summarise, we covered the following points relating to CNNs.

1. Using CNNs drastically reduces the number of parameters due to weight sharing in comparison with an FCN.

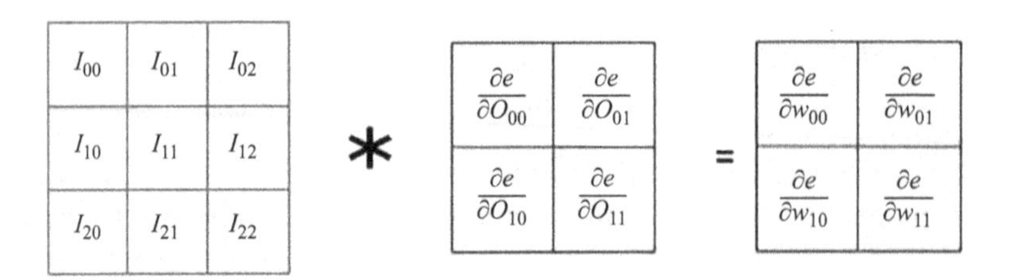

Figure 1.12 Convolution of the Image and the delta errors (from the previous stage) to obtain the partial differentials corresponding to each weight

$$\begin{bmatrix} 0 & 0 & 0 & 0 \\ 0 & w_{00} & w_{01} & 0 \\ 0 & w_{10} & w_{11} & 0 \\ 0 & 0 & 0 & 0 \end{bmatrix} * \begin{bmatrix} \frac{\partial e}{\partial O_{11}} & \frac{\partial e}{\partial O_{10}} \\ \frac{\partial e}{\partial O_{01}} & \frac{\partial e}{\partial O_{00}} \end{bmatrix} = \begin{bmatrix} \frac{\partial e}{\partial I_{00}} & \frac{\partial e}{\partial I_{01}} & \frac{\partial e}{\partial I_{02}} \\ \frac{\partial e}{\partial I_{10}} & \frac{\partial e}{\partial I_{11}} & \frac{\partial e}{\partial I_{12}} \\ \frac{\partial e}{\partial I_{20}} & \frac{\partial e}{\partial I_{21}} & \frac{\partial e}{\partial I_{22}} \end{bmatrix}$$

Figure 1.13 Convolution of the image and the delta errors (from the previous stage) to obtain the partial differentials for weights, which can then be used to update the weights by means of (1.23)

2. CNN involves the use of a stacked two-dimensional filters that is separately convolved with each channel of the image (or feature map) to produce a feature map.
3. Error backpropagation can be used to incrementally improve the weights or coefficients of the filters.
4. The gradients used for weight updates can be obtained through the convolution between the input image and the delta error matrix (errors propagated from the previous layer).
5. The delta errors for the current stage can be obtained by means of convolution between a padded weight matrix and the vertically and horizontally reflected delta error matrix from the previous stage.

1.6.3 Pooling layers

Pooling operations are frequently used to consolidate data and to reduce the feature map resolutions (downsample) in subsequent stages of the network by inserting them between CNN layers. The Maximum (or **Max**) and Average (or **Avg**) are the most commonly used operators in conjunction with convolution to obtain the consolidated feature maps. An example of maximum pooling is seen in Figure 1.14, where **Max** pooling is carried out by convolving a 2-by-2 mask with an input feature map to produce O. The pooling results are shown in (1.29) to (1.32):

$$O_{00} = \mathbf{Max}(I_{00}, I_{01}, I_{10}, I_{11}) = I_{11} = 250 \tag{1.29}$$

$$O_{01} = \mathbf{Max}(I_{01}, I_{02}, I_{11}, I_{12}) = I_{02} = 255 \tag{1.30}$$

$$O_{10} = \mathbf{Max}(I_{10}, I_{11}, I_{20}, I_{21}) = I_{20} = 255 \tag{1.31}$$

$$O_{11} = \mathbf{Max}(I_{11}, I_{12}, I_{21}, I_{22}) = I_{11} = 250 \tag{1.32}$$

Unlike the CNN layer, the **Max** operator does not contain any weights or parameters that need to be 'learnt' (this refers to weights that are updated during gradient descent). It is a non-linear operator that is intrinsically non-differentiable. On the other hand, we can certainly differentiate the result of the **Max** operation. Even though there are no parameters to adjust in the **Max** pooling layer, we need to backpropagate the

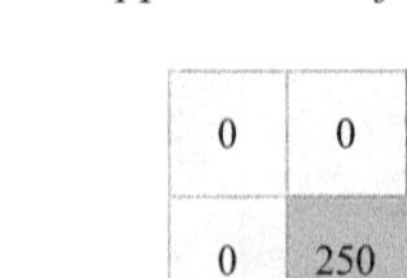

*Figure 1.14 Max pooling layer involving Input feature map **I** and the **Max** operator to obtain **O***

error at the output $\partial e/\partial O$ to get $\partial e/\partial I$, the gradients corresponding to the input feature map (big yellow box). In this particular case, we need only to consider the non-zero entries (small red boxes), or $\partial e/\partial I_{02}, \partial e/\partial I_{11}$, and $\partial e/\partial I_{20}$, since these are the only elements that contribute to Output O. From (1.30), we can deduce that $\partial O_{01}/\partial I_{02} = 1$ and that the partial derivatives involving other elements e.g. I_{11} are zeros. It follows that the gradient w.r.t I_{02} is shown in (1.33). Further, since input I_{11} affects both outputs O_{00} and O_{11}, the gradient $\partial e/\partial I_{11}$ involves the sum of gradients w.r.t. O_{00} and O_{11}, or (1.34). Obtaining gradient $\partial e/\partial I_{20}$ is straightforward since it is only affected by one output O_{10}, as we show in (1.35).

$$\frac{\partial e}{\partial I_{02}} = \frac{\partial e}{\partial O_{01}}\frac{\partial O_{01}}{\partial I_{02}} = \frac{\partial e}{\partial O_{01}} \tag{1.33}$$

$$\frac{\partial e}{\partial I_{11}} = \frac{\partial e}{\partial O_{00}} + \frac{\partial e}{\partial O_{11}} \tag{1.34}$$

$$\frac{\partial e}{\partial I_{20}} = \frac{\partial e}{\partial O_{10}} \tag{1.35}$$

We can generalise error backpropagation for the pooling layer in (1.36) in order to obtain the delta errors or gradients at the input to facilitate backpropagation through the **Max** pooling layer:

$$\frac{\partial e}{\partial I_y} = \sum_z \frac{\partial e}{\partial O_z}\frac{\partial O_z}{\partial I_y} \tag{1.36}$$

$$\frac{\partial O_z}{\partial I_y} = \begin{cases} 1, & \text{if } I_y \in O \\ 0, & \text{otherwise} \end{cases} \tag{1.37}$$

In the case of **Avg** pooling, this has strong similarity with the CNN operation shown in Figure 1.11. Here are crucial differences between them:

1. All of the weights are set to a constant value, $1/N$, where N refers to the size of the filter kernel. In the example above, $N = 4$.
2. There are no parameters to learn since all the weights are assigned to constant values. Therefore, the same procedure can be used for backpropagation as seen in Figure 1.13 to obtain the delta errors $\partial e/\partial I$.

In summary, Pooling is very useful tool for filtering the input feature map and for reducing its dimensionality. Furthermore, error backpropagation is certainly possible through these operators.

1.6.4 The SoftMax function

Taking a step back, the problem at hand is to determine the probability that an input CXR belongs to an infected and symptomatic COVID patient or a problem that involves only two classes. Let us expand the scope of the problem to a 'multi-class' one where each CXR can belong to one of the following categories:

1. Healthy patient.
2. COVID-infected patient.
3. Patient suffering from pneumothorax.
4. Patient suffering from a respiratory condition unrelated to be those described above.

In other words, we wish to estimate the conditional probability that each input CXR will belong to one of the four classes $C_i, 0 \leq i \leq 3$. We create four random variables $Y_i, 0 \leq i \leq 3$ such that $Y_i = C_i$ if the input CXR belongs to class C_i and $Y_i \neq C_i$ otherwise. The objective is to estimate the conditional probabilities that Y_i belongs to class C_i given input x_i (1.38). Equations (1.39) and (1.40) indicate constraints that should be met for the conditional probabilities:

$$P(Y_i = C_i|x_i), 0 \leq i \leq 3 \tag{1.38}$$

$$0 \leq P(Y_i = C_i|x_i) \leq 1, 0 \leq i \leq 3 \tag{1.39}$$

$$\sum_{i=0}^{3} P(Y_i = C_i|x_i) = 1 \tag{1.40}$$

In the context of a Neural Network (ANN), we estimate these probabilities at the last stage. This implies that input **X** refers to the input at the last stage of the ANN rather than the initial input CXR. One implementation option for probability estimation that can work with the network is the Sigmoid function that we saw in Figure 1.6. Since the Sigmoid output is between 0 and 1, it satisfies constraint (1.39). On the other hand, we cannot guarantee that Constraint (1.40) will be satisfied since each Sigmoid functions independently unless the problem only consists of two possible classes. In this case, we can make use of one Sigmoid to obtain the probability of one class and obtain the residual probability for the other class, as shown in (1.41). Note that we make use of the notation P' because the Sigmoid produces an estimate rather than an actual probability value:

$$P'(Y_1 = C_1|x_1) = 1 - P'(Y_0 = C_0|x_0) \tag{1.41}$$

To generalise the solution such that we can solve multi-class problems, we make use of the SoftMax method to estimate the conditional probabilities by means of (1.42). Unlike the Sigmoid, this scheme produces probability estimates that satisfy all of the constraints in (1.38)–(1.40). The SoftMax method is denoted using function f

and it operates on a vector **X**, which in turn contains the individual inputs from the previous ANN stage:

$$P(Y_i = C_i|x_i) \approx f(x_i) = \frac{e^{x_i}}{\sum_{j=0}^{3} e^{x_j}} \tag{1.42}$$

Structurally, the SoftMax function is contained within the last layer of nodes, where the number of nodes corresponds to the number of classes for the problem - this is illustrated in Figure 1.15. How do we backpropagate the error through the SoftMax layer? To do this, we define the cross-entropy loss function, which is described in (1.43). Notice that we have a new parameter t_i. Each element belongs to a reference vector $\mathbf{T} = [t_0, t_1, t_2, t_3]$. Specifically, during the training process, we feed the entire network with a dataset comprising of input images and the known results. For example, an Image I is known to belong to a healthy patient so its corresponding vector $\mathbf{T} = [1, 0, 0, 0]$ – this is known as a one-hot encoding scheme:

$$E = -\sum_{i=0}^{3} t_i \ln y_i \tag{1.43}$$

The cross-entropy function is used to compare the differences between two statistical distributions, derived from random variables T and Y. Large entropy values will imply big differences and vice versa. It follows that the cross-entropy loss (1.43) will be small when applied to the results of an accurate ANN and the reference values. Consequently, the purpose of the training process is to minimise cross entropy loss.

There are no parameters that need to be learnt in SoftMax so we are only required to backpropagate the error from output y_i to input x_i or equivalently to obtain $\partial E/\partial x_i$. In addition, the partial differential $\partial x_j/\partial x_i$ will always be zero except when $i = j$.

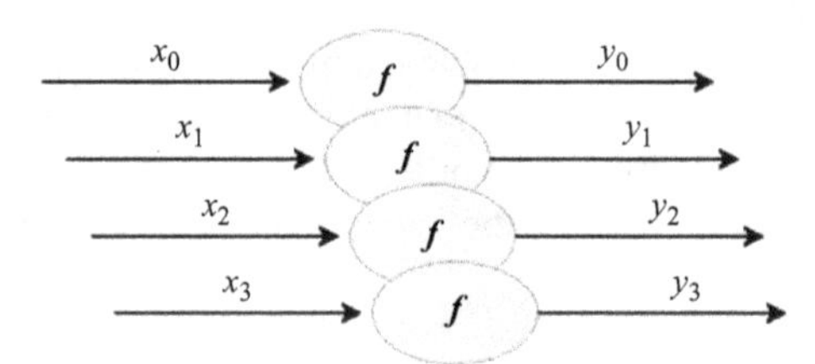

*Figure 1.15 Structural illustration of last layer of the CNN, consisting of a SoftMax function. Input vector **X** contains the results from the previous stage of the ANN, and this returns output **Y** which are the conditional probabilities for each of the four classes described above.*

Consequently, we can replace the differential with a shifted delta function $\delta(i-j)$ to obtain a rather complicated expression (1.44), where $\phi = \sum_j e^{x_j}$:

$$\begin{aligned}
\frac{\partial E}{\partial x_i} &= -\sum_j t_j \frac{\partial}{\partial x_i} \ln y_j \\
&= -\sum_j t_j \frac{\partial}{\partial x_i} \ln \frac{e^{x_j}}{\phi} \\
&= -\sum_j t_j \left(\frac{\partial x_j}{\partial x_i} - \frac{\partial}{\partial x_i} \ln \phi \right) \\
&= -\sum_j t_j \left(\delta(i-j) - \frac{\partial}{\partial x_i} \ln \phi \right),
\end{aligned} \tag{1.44}$$

$$\delta(i-j) = \begin{cases} 1, i = j \\ 0, \text{otherwise} \end{cases}$$

We can further simplify the expression as follows by first obtaining the partial derivative of the logarithmic function (1.45) which in fact reduces to the SoftMax output y_i:

$$\frac{\partial}{\partial x_i} \ln(\phi) = \frac{1}{\phi} \frac{\partial \phi}{\partial x_i} = \frac{1}{\phi} e^{x_i} = y_i \tag{1.45}$$

Finally, we substitute (1.45) back into (1.44). Since x_j is effectively a delta function shifted by j positions, we make use of both bits of information to obtain a rather simple expression (1.46), which is simply the difference between SoftMax output y_i and it's corresponding label t_i. While the derivation for the propagated error is rather involved, it results in a very simple expression which can be trivially implemented. This is an important reason why the SoftMax function is so popular:

$$\begin{aligned}
\frac{\partial E}{\partial x_i} &= \sum_j t_j (y_i - \delta(i-j)) \\
&= y_i \sum_j t_j - \sum_j t_j \delta(i-j), \sum_j t_j = 1 \\
\frac{\partial E}{\partial x_i} &= y_i - t_i
\end{aligned} \tag{1.46}$$

The SoftMax output consists of a vector of continuous probabilities. We can obtain a discrete outcome by simply applying the **Argmax** operator over the vector to determine the class with the maximum probability. The continuous probabilities are themselves useful because they help to determine the certainty of the network prediction.

1.6.5 Putting them together

So far, we discussed different types of layers in the neural network, their relative merits and how to backpropagate the errors so that parameter adjustments can place. Let us take stock of them now:

- Fully connected layer (FCN).
- Convolutional Neural Network layer (CNN).
- Pooling layer.
- SoftMax layer.

The initial input or image consists of vast amounts of data, but much of it is redundant in relation to the task and the job of the ANN is to aggregate each image down to a handful of conditional probabilities describing whether a CXR should belong to a particular class. The layers mentioned above allow this reduction to be performed in an organised and computationally feasible manner without removing pertinent data or throwing the baby out with the bath water as we progress through each stage.

As we have seen above, feeding the raw pixels of an image directly to a FCN is not a good idea as there will be an unfeasibly large number of parameters to contend with. Therefore, CNNs are usually used as a first point of contact with the raw image data. CNNs exploit the fact that different locations of the image share similar properties so they can make use of the same filters and by extension, they can share the same filter parameters or weights. In subsequent stages of the ANN, the image is gradually aggregated into feature maps of decreasing resolution by means of different techniques such as Pooling. This eventually results in a resolution or feature map size that makes FCNs feasible. Unlike CNNs, FCNs allow each point in the feature map to directly influence the output of the network by creating a unique weighted connection between each point in the feature map to each node in the ANN downstream of the current layer, rather than having shared weights between pixels. Consequently, modern ANNs combine different types of networks in order to take advantage of their merits while maintaining a reasonable level of computational complexity. One well-known example is the Alexnet [5], shown in Figure 1.16. Note how the CNN precedes the FCN as well as how pooling operations are interleaved between CNN operations.

We can easily adapt the structure of the AlexNet to our problem using three modifications.

- The input image resolution will have to be reduced to $512 \times 512 \times 3$ pixels. This will have a knock-on effect on the resolution of the feature maps in subsequent stages.
- The number of SoftMax functions will have to be reduced from 1,000 to 4 to produce four conditional probabilities, corresponding to the four classes described earlier.
- The selection of a class i.e. one class in four is a discrete problem. An **Argmax** operator should be applied SoftMax outputs to generate a discrete outcome.

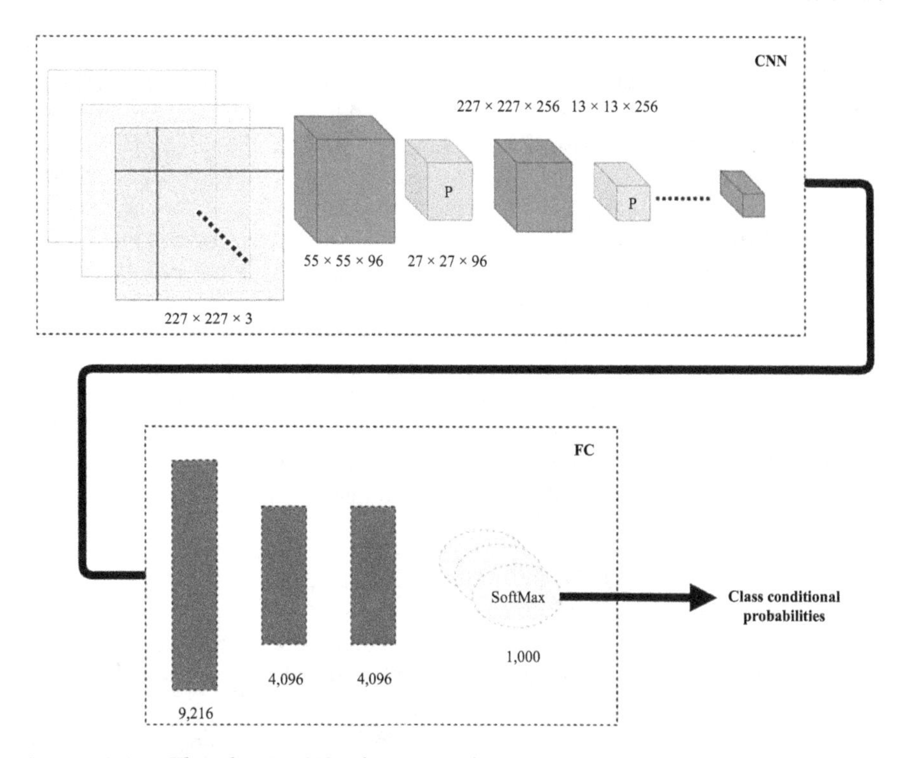

Figure 1.16 *The AlexNet [5] takes in a colour image with the Red-Green-Blue channels, and the resolution of the feature maps is reduced via convolution and pooling operations (feature maps that are pooled are coloured in yellow). The output of the CNN module is flattened into a one-dimensional vector and fed to a Fully Connected network, and finally to 1,000 SoftMax functions to compute the conditional probabilities that belong to each particular class.*

Of course, the network will still need to be trained with enough data before it can be considered fit for purpose. The training method will be discussed in the next section.

1.7 Training and evaluating a neural network

1.7.1 Data organisation

Data is arguably the most important ingredient in ML so much so that it is called the data driven methodology. What sort of data do we need and how should we organise them in order to create a viable ANN? To answer those questions, we will first have to consider the type of ML frameworks available. Broadly, these include Supervised ML, Non-Supervised ML, and Semi-Supervised ML.

Supervised ML makes use of datasets with known results or labels. These labels are also known as the ground truth as they are highly reliable results which might have been obtained from an expert panel. For example, in the case of our CXR classifier, each CXR is accompanied with the diagnosis i.e. whether the CXR belongs to a healthy patient, or a symptomatic COVID infected patient, or other types of disorders. Therefore, the dataset will consist of CXRs and the labels. In the case of unsupervised ML, the dataset consists primarily of raw data and no labels are provided in this case. Finally, the datasets involved in Semi-supervised techniques are partially labelled.

In this chapter, we shall focus on supervised learning as this is perhaps the most common type of ML. By iteratively applying new data (input data and its corresponding label) to the ANN, and subsequently carrying out the forward and backward passes, we will be able to gradually improve the predictive performance of the ANN. This process is known as Training and the primary objective is to minimise ANN prediction error with respect to the reference (desired or expected) values. For this purpose, it is customary to split the data up into three separate subsets, which are defined as follows.

- *Training set*: Data used to iteratively adjust the parameters of the network in order to improve model prediction effectiveness.
- *Development (Dev)/cross-validation set*: This dataset is typically used for model selection (hyperparameter tuning) and validation of the performance of different configurations between training iterations.
- *Testing set*: Used for the final evaluation of the ANN after training has completed for the network.

When data is scarce, 70% of the data is typically used for training and development whereas the remaining 30% is used as part of the Test set as shown in the left pie chart of Figure 1.17. The proportion of training data tend to be much higher for modern ML principally because on the one hand, deep networks with a far larger number of design parameters are usually used, thus requiring more training data. On the other hand, if data is abundantly available, a relatively small Test set is likely to have a representative distribution even if its proportionately less than 30% in size. An example data distribution is shown in the right pie chart of Figure 1.17.

1.7.2 Types of errors and useful evaluation metrics

In this section, we shall talk about how we can evaluate the effectiveness of a predictor or classifier. The plan is to make use of the COVID case study by simplifying it to the original two-class problem. We will then talk about the evaluation metrics in the context of this problem and then generalise it to a multi-class problem. To re-frame the problem, a patient can either be infected with the COVID virus or not – this is defined as the ground truth. Separately, we can build an ANN-based binary predictor that predicts whether a patient is infected with the COVID virus or not. This will then lead to four different scenarios that are depicted in the confusion matrix of Figure 1.18. Specifically, there are two types of errors – False Positive (*FP*) refers to cases where the patient is actually not infected but the predictor wrongly predicts that the patient

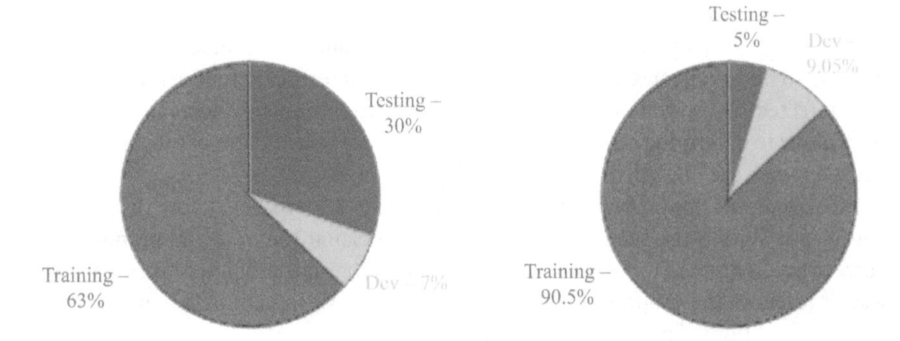

Figure 1.17 Traditional data proportions are shown on the left whereas modern proportions are shown on the right. K-fold validation is used with $K = 10$ *in both cases.*

Actual / Predicted	**Infected**	**Not infected**
Infected	TP	FP
Not infected	FN	TN

Figure 1.18 Confusion matrix of a two-class problem – Infected or Not infected. Four different scenarios could result which are the number of True Positives (TPs), True Negatives (TNs), False Positives (FPs), and False Negatives (FNs).

is infected. On the other hand, False Negative (FN) refers to cases where the patient is actually infected but the predictor wrongly predicts that he/she is not infected. Conversely, True Positives (TPs) and True Negatives (TNs) reflect scenarios when the predictor is correct in predicting infected and not-infected cases respectively.

A simple and straightforward metric is the prediction accuracy. Accuracy is essentially the number of correct predictions as a proportion of the total number of predictions. The total number of predictions is in turn equivalent to the sum of all TPs, TNs, FPs, and FNs, or equivalently (1.47):

$$\text{Accuracy} = \frac{TPs + TNs}{TPs + TNs + FPs + FNs} \times 100 \tag{1.47}$$

However, consider the case when we have 100 data points and corresponding labels. Five of the labels are 'Infected' and the remaining are 'Not Infected'. Let us suppose that we build a predictor that produces four errors among the infected cases and one error among the non-infected cases resulting in four *FPs*, one *FN*, one *TP*, and 94 *TNs*. Applying these figures to (1.47) will produce an accuracy of 95%, which appears to be excellent. On the other hand, the predictor made four errors out of the five infected cases, which is extremely bad. Clearly, predictor accuracy cannot be used alone to determine how effective it is particularly when there is substantial class skew or when the class distribution is highly uneven.

To overcome this problem, we make use of two metrics that are known as precision and recall. Within the scope of this problem, precision refers to the proportion of correct predictions of infections among all cases that are predicted to be infected. Recall refers to the proportion of correct predictions of infections among all actual cases of infections. These metrics are more precisely defined using (1.48) and (1.49). An effective predictor will have high precision and recall values:

$$\text{Precision} = \frac{TPs}{TPs + FPs} \times 100 \tag{1.48}$$

$$\text{Recall} = \frac{TPs}{TPs + FNs} \times 100 \tag{1.49}$$

Returning to our original problem where we have a total of 100 cases and 5 actual infections among them, the precision and recall metrics are 25% and 50% respectively, which accurately indicate that we have a poor predictor. Why are they so different from accuracy? This is because *TNs* is omitted in (1.48) and (1.49) – this makes a difference because *TNs* is very different from the other three types and solely responsible for the class skew in this problem. Rather than considering two separate metrics, we can combine them in a single metric known as the F1 score (1.50). By maximising this score during the training process, we effectively achieve a good balance between Precision and Recall, which is a desirable outcome:

$$\begin{aligned} \text{F1} &= 2 \times \frac{\text{Precision} \times \text{Recall}}{\text{Precision} + \text{Recall}} \\ &= \frac{TPs}{TPs + \frac{1}{2}(FPs + FNs)} \end{aligned} \tag{1.50}$$

Note that the F1 score is particularly useful for datasets with high class skew, which is an exceedingly common problem in the real world. Now that we have explained how the F1 score can be applied to a two-class problem, let us generalise it to a multi-class problem. To do so, we can look at individual classes and consider each as a separate two-class problem. Therefore, for any given class C, an input data point will either belong to class C or it will not i.e. it belongs to class C’, which applies to any data not belonging to Class C. An example of this is shown in Figure 1.19.

Framing the problem in this manner will allow us to work out the F1 score separately for each of the classes. For a problem with *N* classes, we can then compute the aggregated F1 score as a weighted sum of the individual scores as shown in (1.51). The weights are used to adjust the relative importance of particular classes

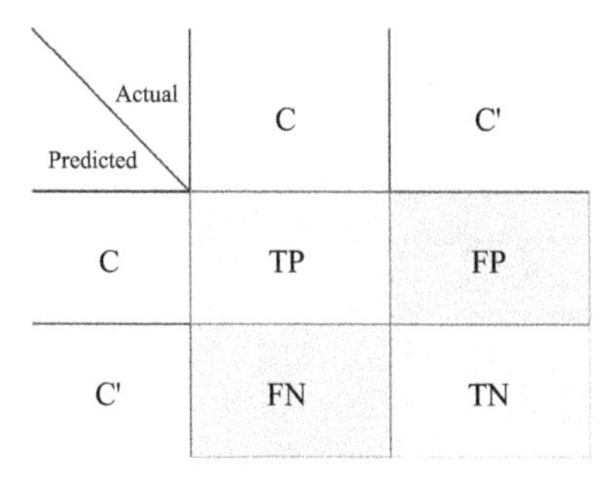

Figure 1.19 Confusion matrix of a two-class problem – Infected or Not infected. Four different scenarios could result which are the number of True Positives (TP), True Negatives (TN), False Positives (FP), and False Negatives (FN).

by increasing its relative weight. In the special case where all classes have equal importance, $w_i = 1/N$:

$$\mathrm{F1}_{\mathrm{agg}} = \sum_{i=0}^{N-1} w_i \mathrm{F1}_i, \sum_{i=0}^{N-1} w_i = 1 \tag{1.51}$$

$$\mathrm{Error} = 1 - \mathrm{F1}_{agg} \tag{1.52}$$

Now that we have described a couple of ways to evaluate an ANN, how can they be incorporated into the training and evaluation process? The most straightforward way would be to divide the randomised dataset into three separate subsets based on the proportions suggested by the right pie chart of Figure 1.17. We can then train the ANN solely using the training set. Depending on depth of the network (which will impact the number of layers and design parameters), the same training set can be applied repeatedly to the ANN to make up for the lack of data, and each training run is described as an epoch. The training score and/or error can then be obtained using (1.52). The trained network can subsequently be applied to the Development and Test sets to obtain their respective F1 scores/errors. Comparisons between the Training, Development, and Test errors are very useful in guiding the ML architectural design and training process, as well as for suggesting adjustments in data composition.

An alternative training method is known as the K-fold cross validation. The pseudo-code of the training procedure is shown in Figure 1.20. In this case, we consider the special case where $K = 10$, where we combine the Training and Development/Cross-validation set into a single set. Subsequently, we partition this dataset into 10 equal parts or folds (Step 3). We then make use of nine folds of the data for training (Step 4.1) and one fold to evaluate the trained network (Step 4.2) – this is equivalent to 90.5% and 9.05% of the dataset, respectively. We then target a different 'fold' in the subsequent iteration and do this exhaustively across all folds. Eventually, we compute the average development score ($F1_T$) and error (e_T) from all iterations. In this way, 95% of the entire dataset (excluding the Test set) is being used for training as well as for validation, so that none of it is being 'wasted'.

K-Fold Validation (K=10)

Inputs: Image Set I, Labels L.

Outputs: Training score $F1_T$. Training error e_T.

1. Concatenate set i and labels L to form Set S.

2. Randomise the order of Set S.

3. Partition Set S into 10 equal parts, each partition being set s_j.

4. For each set s_j in Set S:

    4.1. Apply all partitions $s_j, j \neq i$, to the ANN.

    4.2. Validation: Compute the $F1$ score to obtain $F1_j$.

5. Obtain the mean of the scores $F1_j$ to obtain the Dev score $F1_T$.

6. Obtain the corresponding training error $e_T = 1 - F1_T$.

Figure 1.20 Pseudo-code describing the K-Fold Cross Validation procedure

For each ML problem, there is usually an optimum performance that is just out-of-reach by any ML solution. In the context of the COVID detection problem, this could be the F1 diagnosis score achieved by consensus among a panel of experienced radiologists. The objective of an ML designer is to design an ANN that is able to achieve a score close to the experts. It is certainly possible to design an ML solution that can outperform humans but it becomes difficult to accurately quantify the amount of improvement since our reference labels are derived from humans. In this case, we assume that the expert panel is infallible so the corresponding optimum F1 score is 100%. This is depicted by the dotted line in Figure 1.21.

Assuming that a suitable ML architecture e.g. correct number of layers in the CNN and FCN components has been chosen, the F1 score will gradually increase with an increasing amount of training, as demonstrated by the idealised red Dev score curve in Figure 1.21. Since the parameters are learnt directly via the Training and Dev set, the Dev score will most likely be higher than the Test score since the trained network is exposed to a 'new' dataset that it has never 'seen' before and will therefore perform relatively poorer. As such, the green Test score curve is always below the red curve. The difference between the Optimal and Dev scores is known as the Bias, and the difference between the Dev and Test scores is known as the Variance. We make the following observations about the Bias and Variance.

- An important objective is to design a ML solution with low Bias so that the solution is capable of performing at a level comparable to the experts. A high Bias indicates that the solution is underfitting and therefore ineffective.

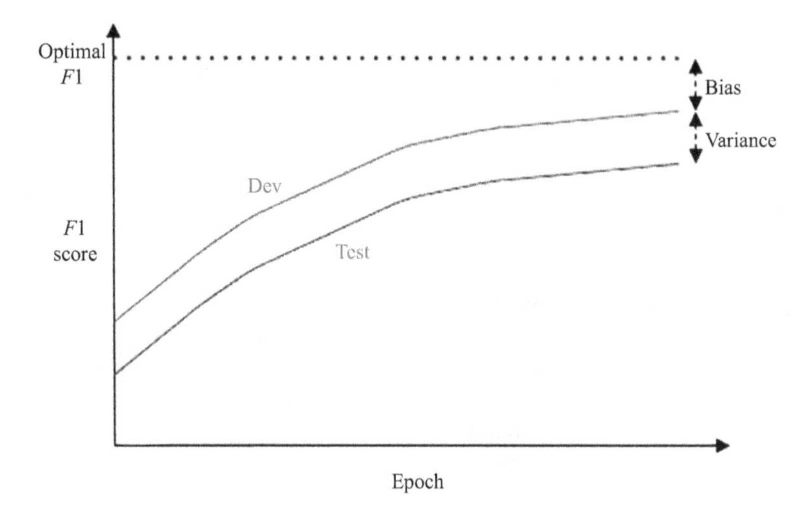

Figure 1.21 Gradual improvement in development and test scores. The bias and variance errors are indicated here as well.

- If the Variance is substantially higher than the Bias, it suggests that the ML solution does not generalise well to new data even though the new data has a similar distribution to the Dev set. Therefore, the solution is overfitting to the Training and Dev datasets.

We introduced two terms here: Underfitting and Overfitting. The best way to explain these terms is by defining a simple two class ML problem with two features x_1 and x_2. Each data point occupies a certain position in the feature space and membership in each class is colour-coded in red and green (Figure 1.22). In this case, the ML solution is a function or hyperplane that discriminates between the red and green points and this hyperplane is represented as a purple line in Figure 1.22(a). Using a simple ANN with a small number of layers, we obtain a straight line that divides the space in two – most of the red and green data points lie to the left and right respectively. However, four of the red points lie to the right and are therefore classified wrongly. We call this situation underfitting, and this leads to a high Bias, which is undesirable.

Zero or Near-Zero error can be achieved by hugely increasing the number of layers in the ANN. In doing so, we create a complex hyperplane e.g. a non-linear function such as a high order polynomial that is shown in Figure 1.22(b). The problem with doing so is that this hyperplane will fit well only to this particular dataset – it will not generalise well when the ANN is exposed to new data such as the Test set where the points might lie in different locations even if they are drawn from the same statistical distribution. This situation is known as overfitting. Overfitting will likely lead to high Variance. A good compromise can be achieved when the hyperplane is neither over nor underfitting (Figure 1.22(c)). We shall talk about a few techniques that can be employed to minimise both the Bias and Variance in the following sections.

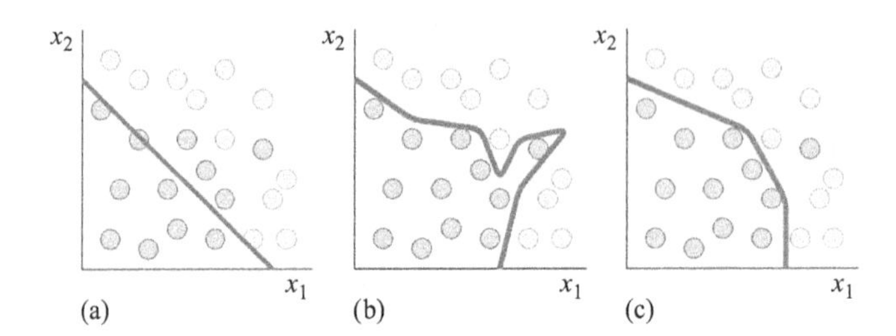

Figure 1.22 A two-class problem illustrated as red and green balls in the feature space described by x_1 and x_2. The ML solution is depicted via a purple line in each case (a) Underfitting leading to high Bias. (b) Overfitting leading to high Variance. (c) Suitable solution that has low bias and generalises well to new data. The red point on the right is an outlier that is located atypically.

1.7.3 ADAM optimisation for bias reduction

At the initial design stages, it is typical to encounter high bias in the Training and Dev sets. Often, the solutions might involve changing the ML framework, increasing the number of layers in the present one, or using different architectures which might lead to more suitable hyperplanes. The problem might also lie with the Stochastic Gradient Descent (SGD) algorithm that is used to find a good combination of parameters or weights that minimises the model prediction error. In general, SGD works well with smoothly varying convex error surfaces (an error surface describes the relationship between the weights and corresponding errors in the error-parameter space) such as the one depicted in Figure 1.2(b), where there is only one global minimum and no local minimas. However, it might fail for real-world error surfaces which contain local minimums and 'ravines' where there is far more variation in one dimension (or weight) relative to the others. We illustrate this notion by plotting the error surface in Figure 1.23. The path taken by SGD is indicated in red – note that not only does it takes several steps to converge, convergence is attained at a local rather than at the global minimum – leading to a high bias.

In order to understand why SGD performs so poorly in Figure 1.23, let us revisit the simple weight update equation that was mentioned earlier – this is duplicated in (1.53) for your convenience. In each SGD step or equivalently, a weight update done by error backpropagation, the current weight is updated using the gradient $\partial e/\partial w$, scaled by the learning rate μ (note that this is done separately for weights x_1 and x_2). Because the order of the data has been randomised, we see the zigzag or oscillatory pattern that is characteristic of the SGD algorithm (red path in Figure 1.23) – only a small component (or projection) of this movement is inclined towards the global minimum. We can reduce the extent of this oscillation by simply reducing μ. However, this will increase the risk of it falling in a local minimum which is clearly the case in Figure 1.23. Consequently, rather than keeping μ constant, it makes sense to use a dynamic scheme that will keep the step size high at the beginning to avoid local

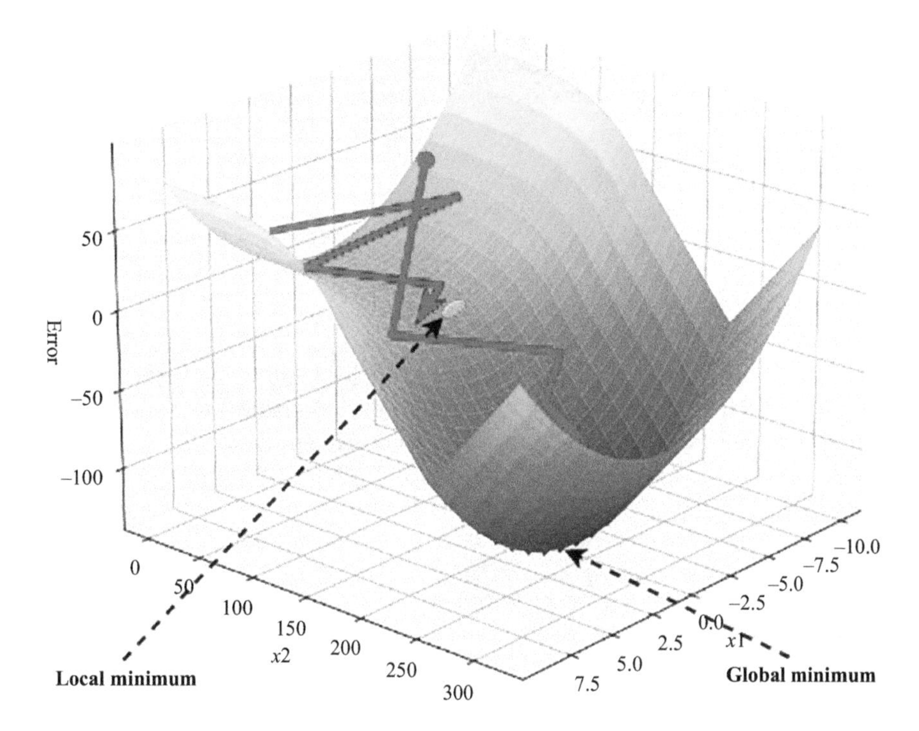

Figure 1.23 Model Prediction error for every combination of weights x_1 and x_2. The red and pink paths are taken by Stochastic Gradient Descent and ADAM Optimised Gradient Descent algorithms.

minimas, and reduce it later as we approach the global minimum so that we reach it quickly and with minimum oscillation:

$$w = w - \mu \frac{\partial e}{\partial w} \tag{1.53}$$

The Adaptive Movement Optimisation Algorithm (ADAM) is an effective dynamic scheme that makes use of two techniques, as follows:

- *Momentum*: It accumulates past gradients by using an averaged exponential weighting scheme, such that consistently increasing/decreasing gradients (or high curvatures) will result in large momentum (akin to a ball rolling down a steep hill) and vice versa.
- *Damping*: Oscillatory movement in gradients is dampened by allowing opposite components of movement to cancel each other and increase step size for non-oscillatory directions of travel and vice versa.

The ADAM method makes modifications to the original weight update equation, resulting in (1.54)–(1.56). Specifically gradients are accumulated with variable v using exponential average weights (1.54), where the typical value of β_1 is 0.9. This has a two-prong effect of accelerating the search at error surfaces with high curvatures (steep slopes or ravines) and creating a damping effect on zigzag or oscillatory paths since the summation of gradients in opposite directions will lead to small values:

$$v = \beta_1 v - (1 - \beta_1)\frac{\partial e}{\partial w} \tag{1.54}$$

$$s = \beta_2 s + (1 - \beta_2)\left(\frac{\partial e}{\partial w}\right)^2 \tag{1.55}$$

$$w = w - \mu\frac{v}{\sqrt{s + \eta}}\frac{\partial e}{\partial w} \tag{1.56}$$

In contrast, the original zigzag vectors will lead to large values in the exponential average weighted sum in (1.55) due to the squared gradient term ($\beta_2 = 0.99$). Conversely, gradient movements along surfaces of steep declines will lead to small values. The combined effect on the modified learning rate μ is a large step size along dimensions of the error surface where there are steep declines (huge change in curvature) and small step sizes where the error surface is essentially flat. In this way, the ADAM method is able to achieve rapid convergence without overshooting the global minimum as shown in the pink path of Figure 1.23. This will facilitate a more effective search for the best parameters and a correspondingly lower bias.

1.7.4 Regularisation for variance reduction

On its own, extended training will create a situation where the model outperforms itself substantially when it is applied to the Training/Dev set, in comparison to the Test set. This is an overfitting situation and occurs when the model weights tend to have excessively large absolute values. Indeed, the impact of the large weights is a hyperplane with excessive flexibility (Figure 1.22(b)). A straightforward variance reduction strategy would be to constrain the size of the weights. We need to do so in such a way that it will not create a 'stiff' hyperplane that will underfit and increase the bias prohibitively (Figure 1.22(a)):

$$e'(w) = e_O(w) + \frac{\lambda}{2}w^2 \tag{1.57}$$

$$\frac{\partial e'}{\partial w} = \frac{\partial e_O}{\partial w} + \lambda w \tag{1.58}$$

$$\begin{aligned} w &= w - \mu\frac{\partial e_O}{\partial w} - \lambda w \\ &= (1 - \lambda)w - \mu\frac{\partial e_O}{\partial w} \end{aligned} \tag{1.59}$$

Therefore, a conceptually simple way would be to regularise the original cost function by adding the weights as a squared penalty term to the original function $e_O(w)$, and scaling it with an adjustable coefficient $\lambda/2$, as we show in (1.57). Subsequently, the modified partial derivative that is required for gradient descent can be worked out using (1.58) – generality is preserved as the new partial derivative or gradient is the sum of the original gradient and the scaled weight. Finally, the complete regularised SGD equation is shown in (1.59). This is known as L2 Regularisation. Essentially, minimising the regularised cost function will reduce the Training/Dev error while keeping the weights small as large weights will result in high cost due to the squared term. The aim of such an approach is to create a model with low variance i.e. one that neither overfits nor underfits the training set so that the model is able to generalise well to new data.

1.8 Conclusion

We gave an overview of ML in the preceding sections of this chapter, explaining its basic composition as well as the fundamental notions for training neural networks. Note that our coverage was not comprehensive as ML is an extremely wide and deep topic and we believe that we will not do it justice if we tried to explain everything in a single chapter. Rather, we made clarity and specificity the priorities at the expense of the breath of coverage. Therefore, we encourage readers to investigate further afield about topics such as ResNets and Recurrent Neural Networks.

Armed with this overview, we hope that readers will better appreciate the application domains that have benefited from ML in the subsequent chapters. Personally, we find this period to be incredibly exciting, inspiring and humbling, when a powerful new technique has been applied to many old and new problems with resounding success – think ML versus the world's top Go player and Depth perception in video using a single camera (which is a traditionally difficult computer vision problem even for stereo camera setups). In this book, we would like to throw the spotlight on the application of ML in the area of Medical Technology, and to do so from many different perspectives, from the frontend sensors used to collect data in order to make possible the application of ML to the crafting of healthcare policy decisions using ML techniques.

References

[1] Hohemberger R, da Rosa CE, Pfeifer FR, *et al.* An approach to mitigate challenges to the Electronic Health Records storage. *Measurement.* 2020;154:107424.

[2] Diaz KM, Krupka DJ, Chang MJ, *et al.* Fitbit®: an accurate and reliable device for wireless physical activity tracking. *International Journal of Cardiology.* 2015;185:138.

[3] An Overview of COVID-19, with emphasis on Radiological Features [paper published on the Internet]. Hong Kong College of Radiologists; 2020 [cited 2021 May 26]. Available from: https://www.hkcr.org/lop.php/COVID19.

[4] Cohen J, Morrison P, Dao L, *et al.* COVID-19 image data collection: prospective predictions are the future. *Journal of Machine Learning for Biomedical Imaging.* 2020;2:1–38.

[5] Krizhevsky A, Sutskever I and Hinton GE. ImageNet classification with deep convolutional neural networks. *Communications of the ACM.* 2017;60(6):84–90.

Chapter 2

Health system planning and optimisation – advancements in the application of machine learning to policy decisions in global health

Oliver Bent[1]

Global health is a domain which presents significant, pressing and timely challenges. Due to the scale of the problems, the potential impact on peoples' lives from even incremental improvement in policy decision making is truly significant. In this chapter, we explore the contributions which may be made through the application of machine learning methods – specifically in the planning and optimisation of population level healthcare interventions. To guide evidence-based policy enhanced through the insights of machine learning models. In order to achieve this, we will focus on existing work in the domain of malaria control policy while also expanding current results on the COVID-19 epidemic. Specifically demonstrating advancements possible through the sharing of data, simulations and compute. While ultimately these factors have enabled engagement in Global Health planning and optimisation, through the application of reinforcement learning methods. This approach has already attracted multiple contributions from machine learning researchers globally, on significant Global Health challenges.

2.1 Model-based decision making

This section will cover a more 'traditional' approach to machine learning (ML) and is based on function approximation [1], statistical learning theory [2] or supervised learning [3] applied to epidemiological modelling. While such approaches are already used for epidemiological modelling to develop phenomenological 'black-box' models which describe disease transmission [4], this will not be the focus. Instead, with access to training data and examples or experience from epidemiological models, we concentrate on how the machinery of data-driven inference may operate in tandem with mechanistic model based insight. We will demonstrate a set of data-efficient ML techniques using training examples from epidemiological model runs to learn functional approximations of computationally expensive epidemiological models. Such ML models learnt from an underlying epidemiological model will be given

[1]University of Oxford, EPSRC Center for Doctoral Training in Autonomous Intelligent Machines and Systems, UK

the term *surrogate* model in this work. The surrogate model is a black-box approximation which will afford us benefits in our handling of tasks of prediction, control and planning.

This approach may be separated into three parts:

1. Surrogate model descriptions for speed, sharing and analysis of computationally expensive models.
2. Function approximation for uncertainty quantification and model calibration for improved predictions.
3. Combination of epidemiological models through committees based on learnt function approximations.

For existing phenomenological prediction approaches, models will aim to generate projections of the future state of the system, based on the available data (or training examples). It may be reasonable for several prediction tasks that we assume no significant change to the underlying epidemiological process, for example due to already imposed intervention plans, evolution of the disease or climate variations. While a challenge remains that if such changes have not already been observed, it is a fruitless task to aim to predict outcomes in the presence of these changes.

The current practice to generate new predictions under significant changes is to rely on the centuries of research, evolving the 'gold-standard' of mechanistic epidemiological models for disease progression. We will therefore place our focus on how ML may be used to infer predictions across multiple policies, where model simulations act as observations under changing intervention plans. Later such ML models learnt from simulations will be used directly in the control and planning process.

As discussed, ML techniques may be used for predictions based on real-world observations alone. Significantly repeated and relevant examples of this have been demonstrated in the context of the COVID-19 pandemic, due to the unprecedented amount of data [5], examples of shared research efforts [6] and global impact [7]. However, all of the observed and collected data allows us only to learn an approximate model based on a single experience or set of observations. Through the addition of model-based observations, these may provide additional training examples or sets of experience to guide planning and control. Such an approach to simulating predictions under varying intervention plans is novel for epidemiological applications but is at the heart of reinforcement learning (RL) for multi-step decision making. Another point which should be made is that if desired to replicate the prediction task for other diseases; the number of real-world observations may be further limited; of far lower quality; and not openly shared, making a repeatable approach to modelling predictions impossible without guidance from mechanistic models. An approach is needed to make such models repeatedly usable, along with the appropriate characterisation of model uncertainty.

Figure 2.1 demonstrates that while prediction models for diseases, such as COVID-19, may be developed in great number, their predictions do not directly agree. This heterogeneity is encouraged with different mechanisms, modelling assumptions and approaches underlying the various models. An ensemble prediction may be chosen to meaningfully combine estimates for an uncertain future, in the case of Figure 2.1

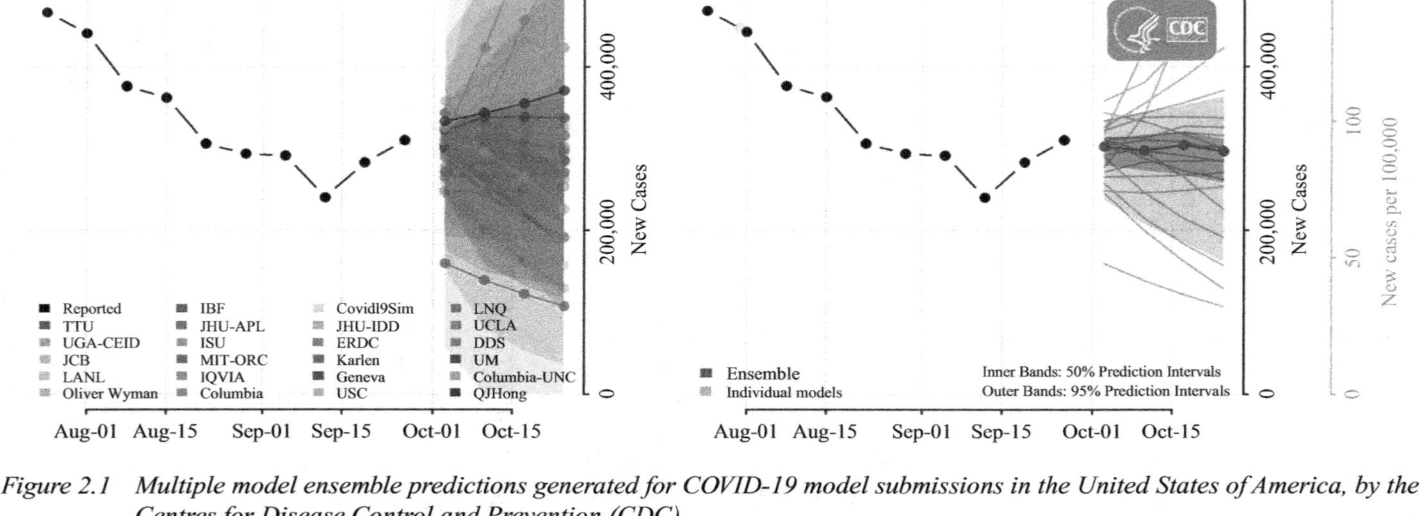

Figure 2.1 Multiple model ensemble predictions generated for COVID-19 model submissions in the United States of America, by the Centres for Disease Control and Prevention (CDC)

this is the median prediction, along with 95% confidence interval. While each prediction is based on the same observational data, some models include additional data with regards to interventions such that a prediction may be made under the current policy. Though these predictions are not exposed as a function of varying policies or intervention plans, should they not be fixed or deterministic, as is the assumption for all presented predictions. The model ensemble is based on the aggregation of these uncertainty measures, with equal contribution from each submitted approach. Ensemble methods [8] have been developed to achieve this without making the assumption that all models are created equal. The often quoted adage of statistician George Box is directly applicable to this scenario [9]

> 'All models are *wrong* but some are useful'

This may be accompanied with the addition that in the case of ensemble forecasts or committee predictions some models may also be more useful than others.

Through the steps outlined we can decompose forecasts as functions of intervention plans, performing ensemble combinations across possible predictions conditioned on intervention, along with providing an unbiased multi-model based estimator for combined predictions under uncertainty. Through this chapter, we hope to motivate that such model structures may undergo semi-automated calibration for predictions, given a base model structure, easing the burden of computation and use of shared models, along with providing insights into the sensitivity of models to changes in input parameters in a manner which is efficient and utilises computations already performed during the calibration process. Mechanistic models may be thought to impose interventions as additional parameterisations to the baseline model parameters and so through a baseline calibration, we may extend calibrated models for the purposes of control and planning.

2.2 ML surrogates for prediction from epidemiological models

Due to heterogeneity, computational cost and lack of a common description of uncertainty for any given epidemiological model, surrogate model approximations provide advantages, which may outweigh the fact that we are introducing errors due to approximation. Though it should be noted these errors are based on our approximation of a model which itself has its own approximation error. Hence it may be argued that a surrogate ML model may accelerate our ability to explore epidemiological control and planning strategies, which nevertheless will always be lower bounded by the epidemiological model's unknown approximation error. ML function approximations provide the following benefits:

- Approximate simulation results may be queried to provide instant feedback for decision-maker or user.
- Any algorithm may query a surrogate model to perform its own simulated planning steps.
- Functional approximation may be used to guide 'true' simulated action selection.

For this work, we will focus on the most data-efficient strategy which naturally incorporates uncertainty quantification in its definition, Gaussian process regression (GPR).

2.2.1 Gaussian process regression

The Gaussian process [10] is a well-known statistical learning model, used extensively for probabilistic non-linear regression. The task of regression aligns with parametric, phenomenological, or black-box epidemiological model design for prediction, specifically how predictions as forecasts, may be learnt of future epidemiological state. Although Gaussian process regression is not a standard modelling approach in epidemiology, there are examples of its use in geospatial modelling [11]. This work will focus on motivating its use in a pipeline for temporal decision-making. Gaussian process regression targets a functional approximation for an underlying stochastic process, a natural fit for capturing uncertainty from epidemiological model training examples.

The outputs used from epidemiological simulations will be considered as stochastic, either through mechanisms of stochasticity within the model or based on stochastic distributions of model input parameters, samples of which generate a stochastic output. First a general overview of Gaussian process regression is necessary before generating the appropriate mapping for prediction tasks.

A Gaussian process is fully specified by its mean function $m(\boldsymbol{x})$ and covariance function $k(\boldsymbol{x},\boldsymbol{x}')$. This is a natural generalisation of the Gaussian distribution, whose mean and covariance are a vector and matrix, respectively. The Gaussian process entertains a probability distribution over some (real) underlying stochastic process $f(\boldsymbol{x})$. Defining:

$$m(\boldsymbol{x}) = \mathbb{E}[f(\boldsymbol{x})] \tag{2.1}$$

$$k(\boldsymbol{x},\boldsymbol{x}') = \mathbb{E}[(f(\boldsymbol{x}) - m(\boldsymbol{x}))(f(\boldsymbol{x}') - m(\boldsymbol{x}'))] \tag{2.2}$$

the Gaussian process is specified as

$$f(\boldsymbol{x}) \sim \mathcal{GP}(m(\boldsymbol{x}), k(\boldsymbol{x},\boldsymbol{x}'))$$

These random variables represent the value of the stochastic process $f(\boldsymbol{x})$ at a location $\boldsymbol{x} \in \mathcal{X}$. The location variable in our application will be over states and actions. We discuss this further in Section 2.5. Judicious choices of kernel function and parameter values need to be made for the successful application of GPR.

2.2.1.1 Choice of covariance kernel function

At the heart of GPR, we expect a level of smoothness in our inference, that is to say neighbouring observations have a reasonable correlation, this correlation weakening from unity with distance between observations. Specifically the value of $f(\boldsymbol{x})$ is strongly correlated to values close to $\boldsymbol{x}$ eg. $f(\boldsymbol{x} + \delta\boldsymbol{x})$ where $\boldsymbol{x}' = \boldsymbol{x} + \delta\boldsymbol{x}$. Intuitively this captures a wide range of well-behaved dynamic systems where we expect similar input conditions to lead to similar outcomes. We consider such responses to be captured by stationary functions such that the expected response of our system is not changing over time, while the remaining choices of the form of the kernel remain somewhat an 'art form', as in the designing of many ML models (e.g. neural network architectures). A common form exists which we'll use as a running example in this work.

This is the Matérn covariance function, specifically the $\nu = 5/2$ covariance function, which is a popular choice in modelling physical systems [12], and included here for the purposes of demonstration of results. Alternative kernel methods and selections or combinations may always be used, but this is not the contribution of the work and so we will stick to describing a single kernel use for the purpose of demonstration. Enumerating all of our hyper-parameters σ_f^2, signal variance is the maximum allowable variance and l the length-scale of the covariance kernel, over which we condition correlations between observations. Finally we specify the value of the kernel function in terms of the input distance $r = |\boldsymbol{x} - \boldsymbol{x}'|$:

$$k_{\nu=5/2}(r) = \sigma_f^2 \left(1 + \frac{\sqrt{5}r}{l} + \frac{5r^2}{3l^2}\right) e^{-\frac{\sqrt{5}r}{l}} \tag{2.3}$$

$k(\boldsymbol{x}, \boldsymbol{x}')$ is a covariance function, used to create the square matrix K, indicating the values of the covariance function at pairs of evaluations.

2.2.1.2 Prediction with stochastic epidemiological observations

The specification of the covariance function implies a distribution over functions. To see this, we can draw samples from the distribution of functions evaluated at any number of points; in detail, we choose a number of input points denoted X_*:

$$f_* \sim \mathcal{N}(\mathbf{0}, \mathbf{K}(\mathbf{X}_*, \mathbf{X}_*))$$

In which X_* is a vector of our test inputs (or locations) of the regression, f_*. Therefore, without observations or training inputs, this fully specifies our Gaussian process prior.

We consider an underlying stochastic or noisy epidemiological function $y = f(\boldsymbol{x}) + \varepsilon$, ε being additive independently identically distributed (i.i.d.) Gaussian noise with variance σ_n^2. Here we have assumed a priori that the underlying stochastic process is unbiased and $m(\boldsymbol{x})$ is $\mathbf{0}$. Without other reasonable prior knowledge, e.g. periodicity, this is a reasonable assumption, along with our beliefs of the underlying process that we have specified through the GP covariance function. Our stochastic epidemiological function (y) is therefore distributed such that:

$$y \sim \mathcal{N}(\mathbf{0}, K(X,X) + \sigma_n^2 I)$$

This now allows us to specify the joint distribution of the observed training values X and the function values at the test locations f_* under the Gaussian process prior:

$$\begin{bmatrix} \boldsymbol{y} \\ \boldsymbol{f}_* \end{bmatrix} \sim \mathcal{N}\left(\mathbf{0}, \begin{bmatrix} K(X,X) + \sigma_n^2 I & K(X,X_*) \\ K(X_*,X) & K(X_*,X_*) \end{bmatrix}\right) \tag{2.4}$$

If there are n training points and n_* test points then $K(X, X_*)$ denotes the matrix of the covariance function evaluations at all pairs of training and test points and X the vector of our training inputs or results from epidemiological runs. The covariance pairings of $K(X,X)$, $K(X_*,X_*)$ and $K(X_*,X)$ follow, respectively, from test points (X_*) and training points (X).

Through constructing the conditional distribution based on the assumption of Gaussian prior observations, this gives us the following:

$$\boldsymbol{f}_*|X,y,X_* \sim \mathcal{GP}(m(\boldsymbol{x}_*), k(\boldsymbol{x}_*,\boldsymbol{x}'_*)) \tag{2.5}$$

In the case of a zero mean function prior this reduces the prediction functions to:

$$m(\boldsymbol{x}_*) = K(X_*,X)\,[K(X,X)+\sigma_n^2 I]^{-1}\boldsymbol{y} \tag{2.6}$$

$$k(\boldsymbol{x}_*,\boldsymbol{x}'_*) = K(X_*,X_*) - K(X_*,X)\,[K(X,X)+\sigma_n^2 I]^{-1}K(X,X_*) \tag{2.7}$$

The above results have direct solutions conditioned on our GP hyper-parameters l, σ_f^2 and σ_n^2, which we shall term $\theta_{\mathcal{GP}}$.

While judicious approximations can be made for these values we will rely on an optimisation scheme to fit them based on observation across tasks. This may be done through maximisation of the log marginal likelihood:

$$\begin{aligned}\log\, p(\boldsymbol{y}|X,\theta_{\mathcal{GP}}) = & -\frac{1}{2}\boldsymbol{y}^T(K(X,X)+\sigma_n^2 I)^{-1}\boldsymbol{y} - \frac{1}{2}\log|K(X,X)+\sigma_n^2 I)| \\ & -\frac{n}{2}\log(2\pi)\end{aligned} \tag{2.8}$$

Performing maximum-likelihood estimation (MLE) based on maximisation through gradient-based optimisation recovers an empirical Bayes estimate to $\theta_{\mathcal{GP}}$. While full Bayesian inference may be performed to generate the posterior parameter distribution, the computations are analytically intractable, approximate methods are therefore deployed, for example Markov Chain Monte Carlo (MCMC) algorithms for sampling the $\mathcal{GP}$ prior [13].

We may see in Figure 2.2 the difference in GP inference based on judicious or expert guided choices (Figure 2.2(a)) of $\theta_{\mathcal{GP}}$ and those learnt through the maximised log likelihood optimisation (Figure 2.2(b)). Our learnt optimal parameter values inflate the epistemic uncertainty of our GPR based on a higher signal variance (σ_f^2).

2.2.2 Action-value function example

This section serves as an extension to the application of Gaussian processes for learning the action-value function $Q(a)$ for the Multi-Armed Bandit problem (Section 2.3.1). Despite stochastic simulation results, the value function of similar actions should be highly correlated and fit the purposes for the use of the method for our application:

$$Q(a) = \mathbb{E}[r|a] = m(a) \tag{2.9}$$

If we consider that any simulated action (a) and set of actions $(\boldsymbol{a})$ return a stochastic scalar reward with mean function $m(\boldsymbol{a})$ and covariance $k(\boldsymbol{a},\boldsymbol{a}') = \mathbb{E}[(R(\boldsymbol{a}) - m(\boldsymbol{a}))\,(R(\boldsymbol{a}') - m(\boldsymbol{a}'))]$, a Gaussian process (GP) can be specified by these mean and covariance functions, where we now consider actions as inputs to the GP:

$$Q(\boldsymbol{a}) \sim \mathcal{GP}(m(\boldsymbol{a}), k(\boldsymbol{a},\boldsymbol{a}'))$$

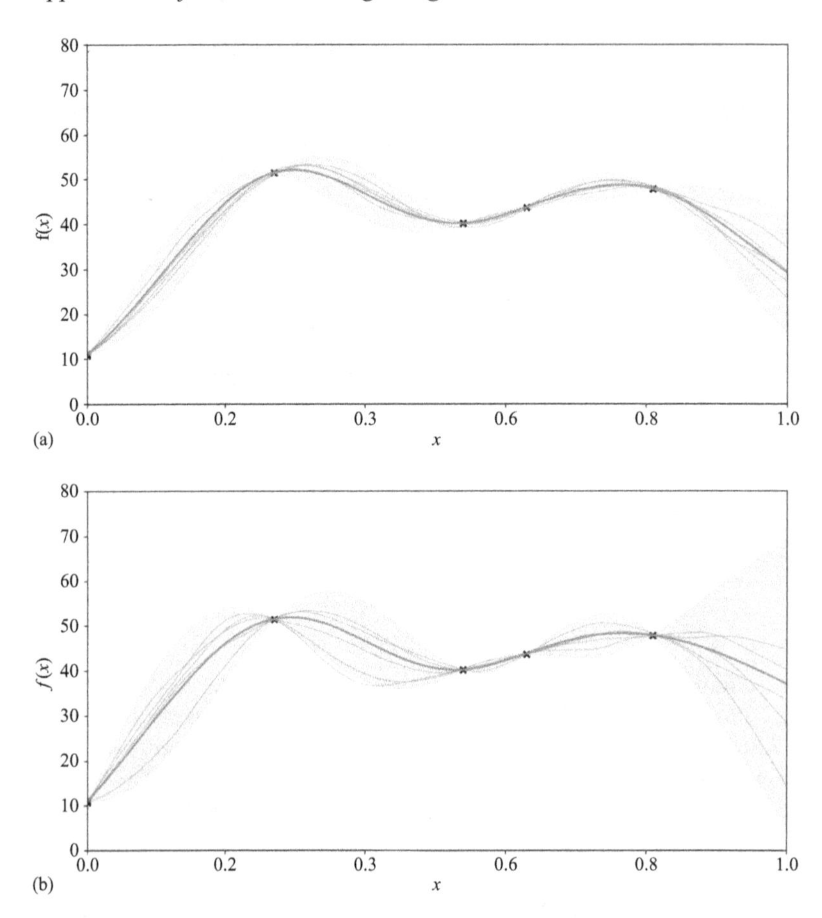

Figure 2.2 Comparison of posterior with selected (a) and learnt (b) MLE parameters, (a) Posterior GP inference based on underlying epidemiological function and judicious choice of $\mathcal{GP}$ parameters $\theta_{\mathcal{GP}}$: $l = 0.25$, $\sigma_f^2 = 100$, $\sigma_n^2 = 0.1$. Including 5 draws from the posterior distribution of functions, (b) based on maximised log likelihood optimisation of GP parameters: $\mathcal{GP}$ parameters $\theta_{\mathcal{GP}}$: $l = 0.38$, $\sigma_f^2 = 1320$, $\sigma_n^2 = 0.02$. Including 5 draws from the posterior distribution.

Gaussian process regression is a supervised learning technique, in which the stochastic scalar rewards $r_t \sim \mathcal{R}^{a_t}$ are used to train a Gaussian process to infer with confidence bounds the action-value function $Q(\boldsymbol{a})$ [10].

The learnt parameters describe the posterior distribution over $Q(\boldsymbol{a})$: or over run t of batch k:

$$m_{t+1}(\boldsymbol{a}) = \boldsymbol{k}_i(\boldsymbol{a})^T(\boldsymbol{K}_i + \sigma^2\boldsymbol{I})^{-1}R_i \tag{2.10}$$

$$k_{i+1}(\boldsymbol{a}) = k(\boldsymbol{a}, \boldsymbol{a}') - \boldsymbol{k}_i(\boldsymbol{a})^T(\boldsymbol{K}_i + \sigma^2\boldsymbol{I})^{-1}\boldsymbol{k}_i(\boldsymbol{a}) \tag{2.11}$$

At each location, $\boldsymbol{a} \in A$, $\boldsymbol{k}_t(\boldsymbol{a}) = [k(\boldsymbol{a}_t^k, \boldsymbol{a})]_{\boldsymbol{a}_t^k \in A_c}$ and $\boldsymbol{K}_t = [k(\boldsymbol{a}, \boldsymbol{a}')]_{\boldsymbol{a},\boldsymbol{a}' \in A_c}$, here σ^2 is the likelihood variance of the GP posterior, where covariance function or kernel $k(\boldsymbol{a}, \boldsymbol{a}')$ may be of Matérn form from Section 2.2.1.1.

As can be demonstrated by a two-dimensional response surface for the action-value function $Q(a)$ in Figure 2.3(a) this is the mean function and the expected performance of a set of two interventions with regards to a performance criterion, in this case the CPDA as outlined in Section 2.4.4.3, along with the upper confidence interval of $Q(a)$ shown in Figure 2.3(b), where regions of lower inferred uncertainty in

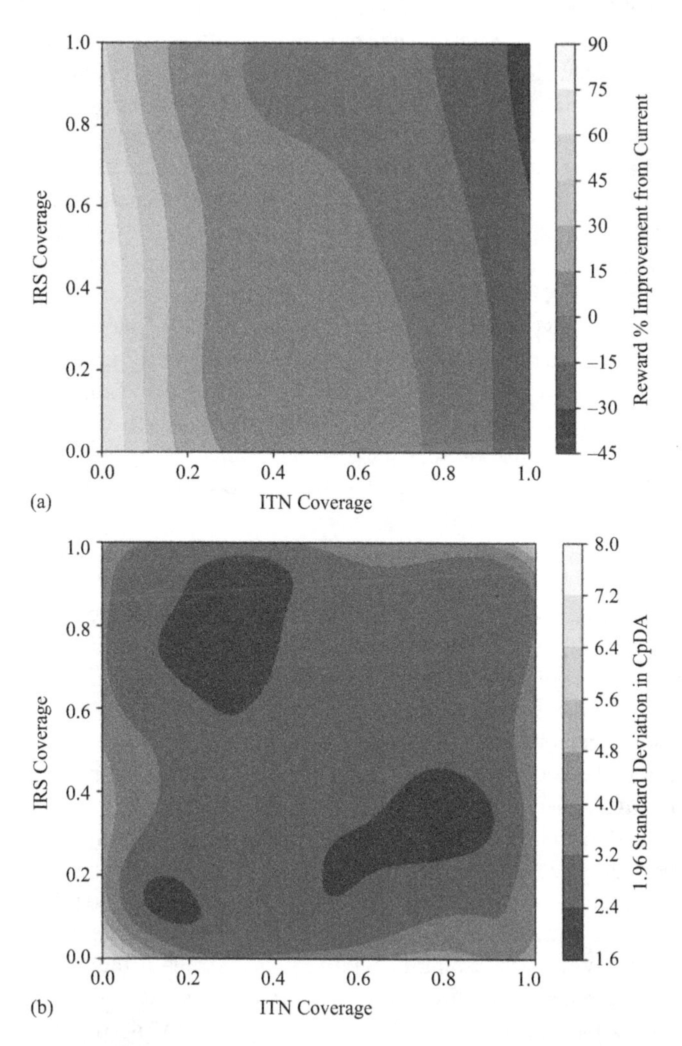

Figure 2.3 *Learnt $\mathcal{GP}$ parameters $\theta_{\mathcal{GP}}$: $l = 0.64$, $\sigma_f^2 = 3,380$, $\sigma_n^2 = 15.6$; (a) $Q(a)$ mean function for OpenMalaria Simulation experiments [14], (b) $Q(a)$ standard deviation bound for OpenMalaria Simulation experiments*

the response surface move down the colour scale to dark blue. This notion of learning a surrogate model or response surface will be extended to sequential decision making in Section 2.5.11.

2.2.3 Epidemiological model calibration

Up to this point, we have assumed a calibrated epidemiological model for the purposes of model-based prediction. In essence a calibrated model implies that the model parameters (θ_m) have been selected, either as single values or distributions. This calibration will likely have been performed by the model developer for the specific disease and location, for which the model has been specified. The process of calibration takes the designed model structure and exposes meaningful parameters (θ_m) to a tuning such that a model achieves higher predictive performance. A model developer may have designed a model with physical parameters in mind and reference the literature for these parameter values, and simply look for the model structure to represent observations. Alternatively a grid-search of parameter values may be performed, which exhaustively evaluates model runs and fixed interval perturbations of the parameter values, specifying a hyper-dimensional grid in the number of parameters. The vertex associated with the parameter set which provides a best-fit (or minimum loss) then being chosen. In this work, we will advocate for an algorithmic approach to achieve best-fit, which may both improve the efficiency of our search for a parameter set, under computational constraints, while also providing additional information with regards to parameter sensitivity and allow us to incorporate additional information around parameter distributions.

For the purposes of this section, we will assume a predefined model structure with exposed model parameters (θ_m), while we may consider the mechanisms of the model as hidden under the black-box abstraction. Simply being given access to a disease model structure will almost always necessitate tuning or *calibration* in order to provide meaningful predictions. Therefore we will look to develop mechanisms which can achieve this in an efficient manner, enabling large-scale use and comparison of epidemiological models.

Any parameter set being calibrated for a mechanistic epidemiological model is often in the order of 10 [15]. Where these parameters will normally have a consistent epidemiological meaning and directly govern the model dynamics, the following list of parameter names consists of examples with more or less obvious physical interpretations; death rate [16], transmission rate [17], incubation period [18], vaccination rate [19], vector mortality rate [20], biting rate [21], Entomological Inoculation Rate (EIR) [22], etc. When performing calibrations for models of the same disease and same location, a mapping of any common parameter sets and to get similar model performance is noted to be a complex task [23]. So we will consider the task of separately calibrating single models for a new geography. To perform such calibrations, reference observational data is required (cases, deaths, vector numbers etc.). We may then interrogate further the impact of parameter values with regards to this observational data and model outputs.

2.2.4 Bayesian optimisation

Bayesian optimisation is a principal global optimisation method driven by the exploration—exploitation paradigm, used in black-box optimisation of computationally expensive functions. Priors on our parameters may build in human expert knowledge from epidemiologists; and an acquisition function [24] which guides the exploration process, reducing unnecessary evaluations of expensive objective functions while maximising information [25]. Importantly Bayesian optimisation leverages approximations in the form of a surrogate model or response surface to remove the computational expense of direct evaluations of the objective function. This surrogate model is almost always a Gaussian process as we developed in Section 2.2.1. While other surrogate modelling methods exist for sequential model-based optimisation, we have developed a consistent set of approaches based on the Gaussian process posterior and will not divert, though methods such as Parzen Estimator approaches exist [26]. As epidemiological models tend not to be written with closed form solutions, we are led to approaches which treat the environment against which a candidate solution is being evaluated as a 'black-box'. Specifically we seek to approximate an optimal solution given that the:

- Objective function ($f(x)$) is *unknown, non-linear, non-convex, non-derivative* and *stochastic.*
- Candidate solutions (x) are a *high dimensional* vector from the set of possible actions. ($\mathcal{A} \subset \mathbb{R}^d$), typically with $d <= 20$ [27].
- Evaluation of $f(x)$ is computationally very expensive.

Therefore framing the process of exploring solutions as an optimisation problem:

$$\max_{x \in \mathcal{A} \subset \mathbb{R}^d} f(x) \tag{2.12}$$

For the purposes of calibration, candidate solutions (x or θ_m) exist across dimensions (d) (often a rectangular hyperspace with uniform ranges $x \in \mathcal{R}^d : a_i \leq x_i \leq b_i$), which are evaluated towards the maximisation of some objective function $f(x)$. The objective function will return a scalar reward which quantifies the performance of the solution.

Algorithm 2.1: Bayesian optimisation Steps

for $t = 1,2,...,n$ **do**
    select new $\boldsymbol{x}_{t+1}$ under acquisition function α [Section 2.2.4.1];
    $\boldsymbol{x}_{t+1} = \underset{x \in \mathcal{A}}{\text{argmax}}\ \alpha(\boldsymbol{x}; \mathcal{D}_t)$;
    run model simulation to provide scalar output y_{t+1};
    new observations $\mathcal{D}_{t+1} = \{\mathcal{D}_t, (\boldsymbol{x}_{t+1}, y_{t+1})\}$;
    Update $\mathcal{GP}$ Posterior: mean and covariance
end

In outline as in Algorithm 2.1, we will use Bayesian optimisation as a sequential optimisation schedule. Through each subsequent model run t, a new set of inputs to the model simulation are chosen $\boldsymbol{x}_{t+1}$ as the result of combining our GP surrogate model with the acquisition function which will be detailed in the following section.

Subsequent evaluations serve to reduce the *epistemic* uncertainty in the GP surrogate model of the loss function and improve our understanding of the captured *aleatoric* uncertainty.

2.2.4.1 Acquisition function

Bayesian optimisation is guided by an acquisition function (α), this may have several functional forms. Popular acquisition functions are Upper Confidence Bound (UCB), Expected Improvement (EI) and Thompson Sampling – they may be grouped in the *optimistic*, *improvement-based* and *information-based* search methods respectively [28]. We will further deal with *optimistic* search in Section 2.5.13 and specifically the action-selection method of for the Multi-Armed Bandit which is analogous to our acquisition function for Bayesian optimisation.

The EI acquisition function is defined by the following equation:

$$\alpha_{EI} = \mathbb{E}\left[\max\left(f(\boldsymbol{x}_*) - f(\boldsymbol{x}^+), 0\right)\right] \tag{2.13}$$

$f(\boldsymbol{x}^+)$ is the best observed loss and $\boldsymbol{x}^+$ the parameters associated with this loss. Based on our GP model:

$$\alpha_{EI} = \begin{cases} (m(\boldsymbol{x}_*) - f(\boldsymbol{x}^+) - \varepsilon)\mathrm{P}(Z) + \sigma(\boldsymbol{x}^*)\rho(Z) & \text{if } \sigma(\boldsymbol{x}_*) > 0 \\ 0 & \text{if } \sigma(\boldsymbol{x}_*) = 0 \end{cases} \tag{2.14}$$

$m(\boldsymbol{x}^*)$ is the prediction mean and $\sigma(\boldsymbol{x}^*)$ the prediction standard deviation at the test location $\boldsymbol{x}^*$ of the Gaussian process. P is the Cumulative Density Function (CDF) and ρ the Probability Density Function (PDF) of the Gaussian distribution and

$$Z = \begin{cases} \frac{m(\boldsymbol{x}_*) - f(\boldsymbol{x}^+) - \varepsilon}{\sigma(\boldsymbol{x}_*)} & \text{if } \sigma(\boldsymbol{x}_*) > 0 \\ 0 & \text{if } \sigma(\boldsymbol{x}_*) = 0 \end{cases} \tag{2.15}$$

Exploration and exploitation may be balanced via the P and ρ terms in the summation of α_{EI} with ε weighting this balance, a greater ε leading to greater exploration. Figure 2.4 plots our EI acquisition function trace as an orange line calculated from our learnt GP introduced previously in Figure 2.2(b), in this case $\varepsilon = 0.01$ and the function is maximised at $x = 1$ which would be selected as the next sample.

2.3 Online learning

Having already discussed the necessity, origin and use of epidemiological models for decision making we will now use the following sections to extend the use of learnt epidemiological models towards learning from simulation. A model is a physical, mathematical, or logical representation of a system entity, phenomenon, or process. A simulation is the implementation of a model over time. A simulation brings a model to life and shows how a particular object or phenomenon will behave. It is useful for

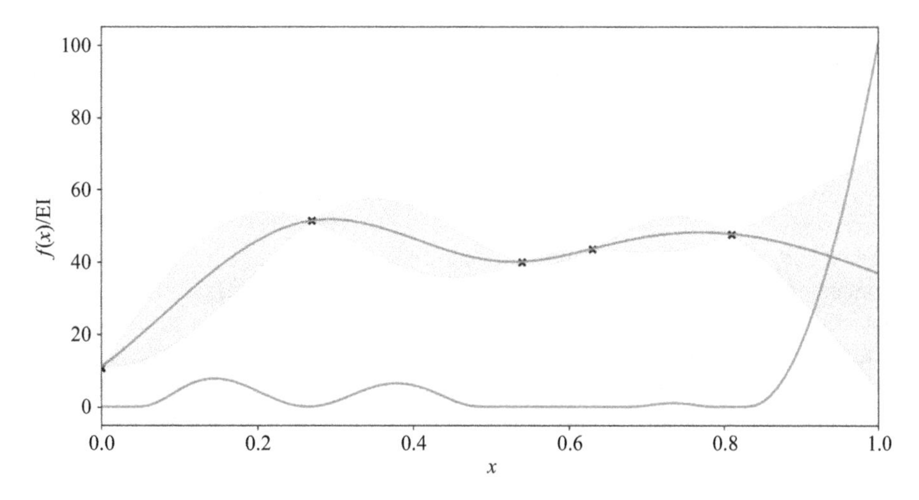

Figure 2.4 Calculation of expected improvement for learnt $\mathcal{GP}$ parameters $\theta_{\mathcal{GP}}$: $l = 0.38$, $\sigma_f^2 = 1{,}320$, $\sigma_n^2 = 0.02$, EI is maximised at $x = 1$

testing, analysis or training where real-world systems or concepts can be represented by a model [29]. With such a description of the real-world, we may now look to train computational systems to learn from simulation.

Extensive scientific research, effort and time may go into creating an exquisitely complex and rich computational model of disease transmission. This will likely be used in the paradigm of *A/B testing* [30] by an epidemiologist (also likely the model creator). The term *A/B testing* hopes to capture the process by which experimentation is performed through a direct comparison, or hypothesis test, between two outcomes *A* or *B*.

Although there is more that can be done, it is proposed that reaching specific policy goals, such as elimination in defined contexts, requires increasingly non-generic advice from modelling. For example research in epidemiological model development looks to motivate the importance of measuring baseline Epidemiology, intervention coverage, vector ecology and program operational constraints in predicting expected outcomes for different combinations of interventions [31]. This does not deviate from the baseline process, that all of these goals may be studied as a series of *A/B tests* without accelerating towards critical solutions, which are needed to solve immediate problems in a timely but principled manner. The rest of this chapter will begin the work of how we may improve our exploration of novel control policies from published or calibrated models.

2.3.1 Stochastic multi-armed Bandit

In formalising an extension to *A/B testing*, we will move towards associative, or one-step Reinforcement learning in the form of the Multi-Armed Bandit (MAB). For direct applicability we will consider a form of the Multi-Armed Bandit under a limited computational budget, specifically a discrete number of epidemiological model runs (n), which may be known or unknown, though each model run is considered to be computationally expensive.

To provide background to the approach of the Multi-Armed Bandit (it is named after the canonical example of the problem), we could interpret this class of problems as analogous to making selections of one-armed Bandits (slot-machines) given a fixed budget of coins or tokens, either known or unknown and the objective being to maximise winnings given unknown stochastic distributions of payouts. Optimal approaches should thus balance exploration and exploitation. Where exploratory actions serve to improve estimates of stochastic rewards, exploitative actions should maximise cumulative reward given the current information and 'model' of the world.

We will consider the MAB problem as a tuple of $\langle \mathcal{A}, \mathcal{R} \rangle$ consisting of K arms or actions a labelled as integers $\{1, 2, ...K\}$. At each round $t = 1, 2, ...n$ an arm is selected:

$$a_t \in \mathcal{A} \in \{1, 2, ...K\}$$

The reward received by an action is a random sample drawn from the unknown reward distributions of the arms $\mathcal{R}^a(r) = \mathbb{P}[r|a]$. Having selected an action a_t at round t the environment generates a reward $r_t \sim \mathcal{R}^{a_t}$ which is assumed to be independent of previous actions or rewards. The goal of any player or algorithm in this setting should be to maximise cumulative stationary rewards across the model runs, where cumulative reward (R_t) is simply the arithmetic sum of rewards across rounds:

$$R_t = \sum_{\tau=1}^{t} r_\tau \tag{2.16}$$

Up until the computational budget of the episode whereby $t = n$. The *action-value* is the mean reward from action a and taken from the reward distribution of the arm, therefore if we drop the dependence on round (t) due to stationarity of rewards, we may state the following:

$$Q(a) = \mathbb{E}[r|a] \tag{2.17}$$

In addition the *optimal value* V^*, for the optimal arm or action a^* is:

$$V^* = Q(a^*) = \max_{a \in \mathcal{A}} Q(a) \tag{2.18}$$

If any player, algorithm or agent had access to these unknown distributions *a priori* they would be able to achieve a maximised expected cumulative reward of:

$$nV^* = nQ(a^*) = \sum_{\tau=1}^{n} \mathbb{E}\left[r_\tau | a^*\right] \tag{2.19}$$

2.4 Running epidemiological simulations as Bandits

Finding an optimal policy from running simulations can be posed as a stochastic Multi-Armed Bandit problem. For example, this approach has been used to develop algorithms aiding the design of clinical trials [32], where action should be made to balance exploitation (positive patient outcomes) and exploration (searching for actions which may lead to a clinical 'breakthrough'). In this chapter, we ground this approach in the determination of the execution of simulations to infer high-performing actions as fixed or deterministic policies for simulated populations.

2.4.1 Time

A clear distinction of simulation time (t_s) and algorithmic time (t) should be made. In the MAB scenario, the notion of time is purely algorithmic and representative of the number of arm pulls or rounds that have been taken. This approach is suitable for taking a one-step, or static decision, from simulated learning, requiring that any change in simulation input is decided before the simulation is performed. For example, a simulation could be run over a 5 year simulated intervention period on a daily time-step resolution. A Bandit approach to the selection of an optimal intervention plan would select from predefined intervention plans, posed as an action a_t to be simulated, receiving a scalar reward based on the epidemiological simulator output r_t, as described in Figure 2.5. The action selection task is not informed by any observation of the simulation state and is reduced to simulating uncountably many intervention plans, which provide stochastic rewards based on the simulators final output. The MAB problem being a single state framing or single epidemiological model online learning problem, with no observed state transitions in this case.

Some epidemiological simulators directly lend themselves to this approach, as they have not been designed to provide observable intermediate simulation values, only a final output result. Through our common generalisation of maintaining the simulator as a black-box, this would follow as the most suitable approach, not requiring any change to the model's internal mechanisms. While simulation time (t_s) may often be on the order of days, the time frame for decision-making could range from years to weeks. Traditionally decisions are made over longer time frames but increasingly

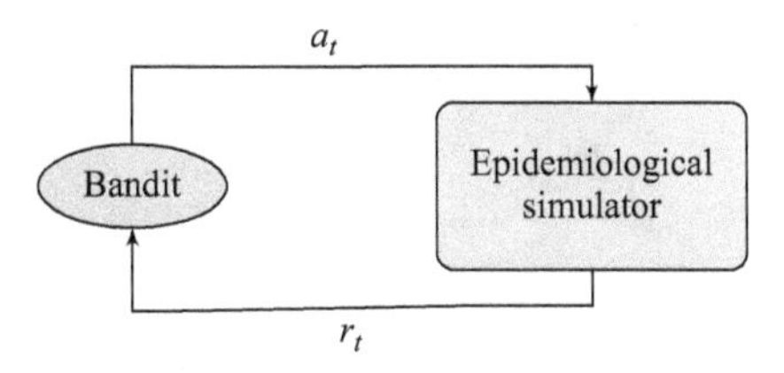

Figure 2.5 Action a_t selected by a Bandit algorithm receives feedback or reinforcement via reward r_t based on the epidemiological simulators outputs

there is pressure for this time horizon to be shortened, specifically in response to global pandemic preparedness.

2.4.2 State

In the MAB framing of the problem, there is no state transition between simulations, this could be reduced to the notion that each simulation upon completion returns to its initial state. We are trying to solve the problem of making one-step control policy recommendations for the simulation intervention period. To consider the single state of the epidemiological model set-up, this may be defined by the model's parametrisation θ_m (which may be uncertain), ultimately defining a model hyper-state, as this is not an observation made directly from the black-box simulator. Each action provides no observation of a physical state of the system, while it does provide information with regards to high performing policies given the model's hyper-state or parametrisation. For example, in Figure 2.6, successive actions (a_{1-5}), may lead to improved inference of a continuous reward distribution, and the transition from $\mathcal{S}_t \rightarrow \mathcal{S}_{t+1}$ defines an improved estimation given a stationary model hyper-state $\mathcal{S}(\theta_m)$.

2.4.3 Action

For the purpose of decision making from epidemiological models, we will define an action (a) as an abstraction based on possible epidemiological model control inputs, the most direct translation of this for the purpose of actionable decision making being a modelled set of interventions. The action-space ($\mathcal{A}$) may be composed of many interventions and therefore be high dimensional, for example efforts

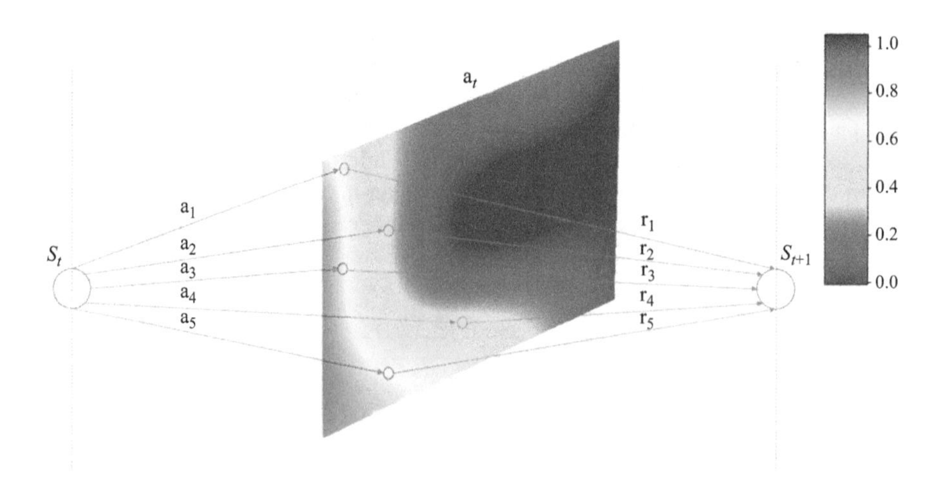

Figure 2.6 A batch of arm pulls for one-step decision making, the state transitions $\mathcal{S}_t \rightarrow \mathcal{S}_{t+1}$ improve estimates of the underlying continuous action value function $Q(a)$, the underlying techniques for inference were introduced in Section 2.2.1

to categorise the number of Nonpharmaceutical Interventions deployed to control the transmission of COVID-19, such as the WNTRAC database, currently contains 6,000 non-pharmaceutical interventions across 261 countries and territories [33]. While in this case, multi-dimensional inputs may be translated through a functional mapping into variations of a single model parameter, such as the infection rate (β) of a compartmental model. Hence often abstractions are required to map a translated intervention to a model change. Here we will not consider specific functional mappings of interventions into model changes, which will be taken as predefined by the epidemiological model developer. Action-based inputs to the epidemiological model will therefore, for generality, be considered as encoded in a purely black-box manner. A component of an action will contain a proposed intervention, its relative strength, coverage or impact on the simulated population ($[0 - 1]$) and timing in the simulation (t_s). Our action-value function based search, achieved through the MAB approach will aim to return the action (a^*) which maximises simulated reward.

2.4.4 Reward

The calculated reward distribution across actions ($\mathcal{R}^{a_t}$) will necessarily have to make use of available model outputs. It is assumed that in order to achieve optimal control of a disease, we aim to reduce the negative outcomes from the disease, where numbers of cases, deaths, complications, etc. may have been modelled, depending on the specific design and original intended use of the model. Any such reward observed from each action $r_t \sim \mathcal{R}^{a_t}$ will be considered as stochastic, based on stochastic epidemiological model outputs, where mechanisms for stochasticity may come from mechanistic model design, utilising probabilistic state transitions or systems of differential equations, where model parameter uncertainty is propagated. We will continue to consider models to have stochastic returns, as we aim to make decisions on uncertain futures and necessitate, that the simulations used provide estimations of uncertainty.

A reward signal could take the simplistic view of these raw stochastic model outputs, for example deaths or infected cases, while to mirror the process of making public health decisions from models we will use a readily available quantitative framework. Such a common metric across diseases and models was developed in the first global burden of disease report (1990) [34], which introduced the term Disability Adjusted Life Year (DALY). For the purposes of learning from simulated outcomes, this provides a meaningful single scalar value for the decision-making process. When making a proposal for mixes of interventions over time, we may now look to maximise the health outcomes with regards to DALYs, while also regulate improved health outcomes with regards to cost.

2.4.4.1 Disability Adjusted Life Years

The Disability Adjusted Life Year (DALY) [35] is a measure defined by the total Years of Life Lost (YLL) due to fatality linked with contraction of a disease, and the number of Years of Life with Disability (YLD) as a result of the disease. Where possible the magnitude of any reward is determined through an economic cost-effectiveness analysis of stochastic epidemiological simulation output, which we will base on the concept of DALYs. Upon completion simulators will return outputs either on an individual or human population level.

2.4.4.2 Costs

Simulated costs are included so that decisions may be made to prioritise the highest impact. As such we should look to promote the most cost-effective solutions on a total simulated population. Where DALYs provide a health outcome for a particular action, different combinations of interventions and coverages may have greatly varying costs. Again, for generality, this will be considered monetary cost. For example administering an Influenza vaccine in the United Kingdom may have a base cost of £5.96. In addition to the base scenario, there may be possible cost savings or further complications, e.g. avoiding a visit to the GP (General Practitioner), costing £31, in the case that a vaccine is administered and a symptomatic case experienced. Or perhaps a £31 cost for treating a side effect of administering the vaccine. Finally, a severe case of hospitalised influenza costs £1,029. All these values have been taken from [36], as an example of enumerated costs to be considered when evaluating cost-effectiveness. If possible, all such events should be priced comprehensively and estimated based on expected outcomes.

Other non-monetary units may be used to place a cost on interventions, but the concept remains the same. This could be taken from an optimal control perspective of a linearised system with quadratic cost, being a function of state and control variables [37]. For example, an optimal and minimum cost solution to the Linear-Quadratic Regulator problem [38] looks to minimise the energy of (2.20) based on control inputs ($u(t)$) and outputs as state observations ($x(t)$) over simulated time (t_s):

$$\frac{1}{2}\int_0^{t_s} \left[x^T(t)Qx(t) + u^T(t)Ru(t)\right] dt \tag{2.20}$$

Where we will take Q and R as stationary matrices, (2.20) may be tuned for the task of decision making from epidemiological models, placing costs on model control inputs or actions ($u(t)$) and model outputs or state ($x(t)$), entries in R act as a direct penalty on inputs ($u(t)$), and entries in Q place costs on negative modelled outcomes such as infections or deaths.

2.4.4.3 Cost-effectiveness

Finally, to complete a general reward calculation, any Bandit algorithm may receive rewards based on the cost effectiveness of an action, balancing both positive and negative outcomes. For this purpose, we define cost effectiveness as the ratio of the relative cost to perform a combination of interventions to the health impact realised from that combination of interventions. The health impact is defined as the DALYs averted (DA), the difference between the DALYs simulated under the action (a_t) and the DALYs simulated under a null action, zero input or intervention. Therefore a calculation of DA requires a minimum of two simulations from a model, in which the action is simulated and compared to the simulated outcomes from no action under the same model parametrisation (θ).

Different decision makers may have different upper limits on considering cost-effectiveness of decisions. For example, the National Institute of Clinical Excellence (NICE) in the United Kingdom will consider interventions to be cost-effective if they cost less than £20,000 per DALY averted. This would add a constraint on which solutions may be feasible for the geography being modelled. Generally, for the application of these techniques, we will make the assumption that the most cost-effective solutions (V^*) are optimal, for decision-makers, as they are the most cost efficient (for resource constrained environments) [39].

2.4.5 Bandit approaches for simulated learning

Each Bandit algorithm outlined in this section performs sequential batch exploration of actions, towards optimisation of an unknown stochastic reward function $\mathcal{R}^a(r)$. At each batch for the round (t), we will choose $j = 1, 2, \ldots, B$ up until the batch size (B) actions $a_t^j \in \mathcal{A}$. Due to the computational expense of calculating $\boldsymbol{r}_t \sim \mathcal{R}^{a_t}$ and the size of the entire action space $\mathcal{A}$, we wish to find solutions of maximal reward in as few batches (n) as possible. The goal being to approximate $a^* = \text{argmax}_{a \in \mathcal{A}_c} Q(a)$ without prohibitively expensive computation for all possible actions, therefore using a subset $A_c \in A$ of the action space. *Note:* we have used n to define both the number of batches and previously the number of rounds for the Bandit problem, this is based on the assumption that evaluation of a batch or a single run requires the same computational time under the ability to perform parallel computations, depending on the availability of the infrastructure, which would necessarily inform the batch size (B), normally of the order number of computational cores. Therefore the descriptions are equivalent while we will use the extension of the Bandit to a batch of actions for these computationally expensive, large action space problems.

2.4.5.1 Upper/lower confidence bound (GP-ULCB)

Based on previous discussion on sub-linear total regret we have extended a Gaussian process upper/lower confidence bound (GP-ULCB) algorithm, inspired by Gaussian process regression (GPR) and work on solutions to the Multi-Armed Bandit problem [40,41]. This is a formulation which combines the natural confidence bounds of Gaussian processes with stochastic MAB problems, and the variants have already been proposed in the form of GP- [42] and GP–PE [43]. The use of Gaussian process regression is detailed in Section 2.2.1, but it is here used to perform inference on a continuous action-value function $Q(a)$, based on successive batches of simulated reward.

The algorithm is initialised with a random sample of a discrete action space ($\mathcal{A}_c$). Subsequent actions are chosen to further explore the action-value function regressed by GPR on all preceding simulation runs. The choice of using both *upper* and *lower* confidence bounds (and LCB respectively) was made due to the large variance often observed in rewards. Specifically, multiple minima and maxima may occur in the action-value function, motivating a search for both optimal and poor actions. In this case, increasing information about poor performing actions may help to avoid taking poor decisions, especially as we have a limited computational budget, increasing the value assigned to exploring poor decisions may be considered necessary

Algorithm 2.2: GP-ULCB

Result: $Q(\boldsymbol{a})$
Input: random discretised actions $\boldsymbol{a}_0 \in \mathcal{A}_c$;
GP priors $m_0 = \mathbf{0}, l, \sigma_f^2, \sigma_n^2$ [Section 2.2.1];
B = batch size, f_m = mixing factor, f_c = masking factor [Section 2.4.5.1];
for $t = 1,2,...n$ **do**
  reset: $\boldsymbol{a}, \boldsymbol{a}_{\text{LCB}}, \mathcal{A}_c$;
  for $j = 1,2,..,B.$ **do**
    if $j < B \times f_m$ **then**
      $a_t^j = \underset{a \in \mathcal{A}_c}{\operatorname{argmax}} \left[m_{t-1}(\boldsymbol{a}) + \beta\sqrt{k_{t-1}(\boldsymbol{a}, \boldsymbol{a}')} \right];$
      mask: $\boldsymbol{a}$ where $|a - a_t^j| < l \times f_c$;
      update: $\boldsymbol{a} \notin \mathcal{A}_c$;
    else
      $a_t^j = \underset{a \in \mathcal{A}_c}{\operatorname{argmin}} \left[m_{t-1}(\boldsymbol{a}) - \beta\sqrt{k_{t-1}(\boldsymbol{a}, \boldsymbol{a}')} \right];$
      mask: $\boldsymbol{a}_{\text{LCB}}$ where $|a - a_t^j| < l \times f_c$;
      update: $\boldsymbol{a}_{\text{LCB}} \notin \mathcal{A}_c$;
    end
  end
  Return: $\boldsymbol{r}_t \sim \mathcal{R}^{a_t}$ [Section 2.3.1];
  Update $\mathcal{GP}$ posterior for $Q(\boldsymbol{a})$: mean $m_t(\boldsymbol{a})$, covariance $k_t(\boldsymbol{a}, \boldsymbol{a}')$ including l [Section 2.2.2] ;
end

for high impact decision making. Should the mixing factor f_m be increased to unity, this approach would be reduced to an only search. Repetition of simulated actions is avoided in a batch via the masking factor f_c, which defines the minimum distance between selected actions within a batch.

In Figure 2.7, we demonstrate the accumulation of batch samples in Algorithm 2.2, with a batch size of 64 applied to the specific problem of learning a deterministic control policy from a malaria model [44]. Qualitatively samples can be seen to shape the expectation ($m_t(\boldsymbol{a})$) in the value function $Q(\boldsymbol{a})$, with a 0.5 mixing factor (f_m), clusters of similar high and low performing solutions emerge. In this case, a fixed masking factor (f_c) of 0.1 has been used. On inspection an adaptive schedule for the masking factor could help to spread out clusters of samples during earlier batches to encourage exploration, with subsequent annealing of this factor placing a focus on exploitation within batches for later rounds.

2.4.6 Extensions to online learning

The techniques presented have been selected and designed with the view to deployment on problems with larger action spaces. More compute time, expansive

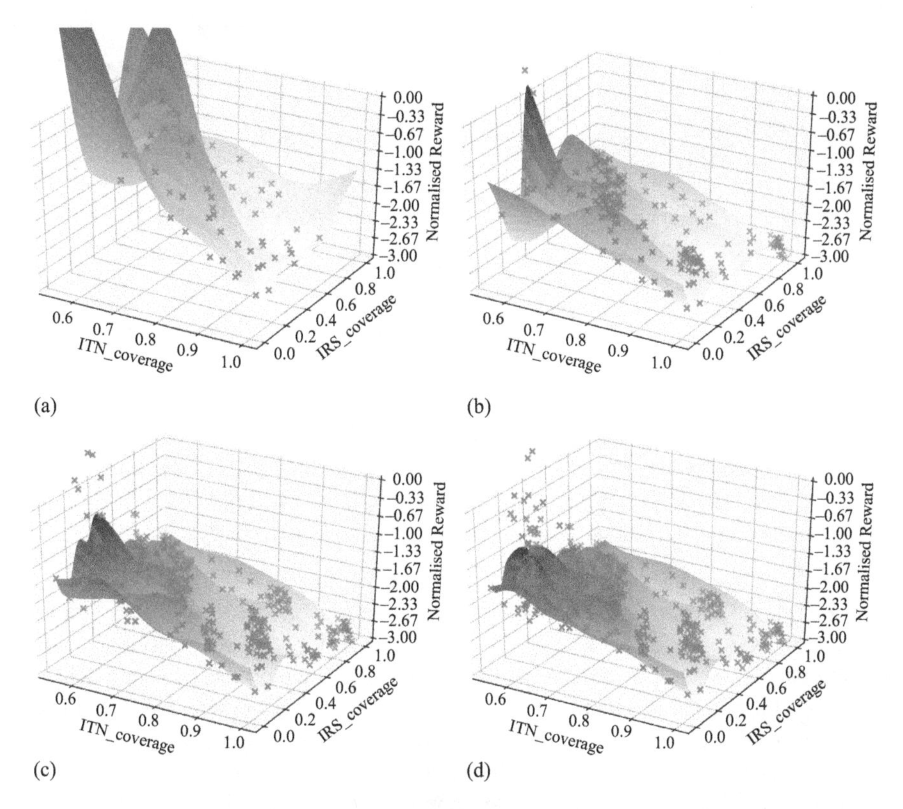

Figure 2.7 Comparison of GP-ULCB samples over subsequent batches from OpenMalaria simulations as outlined in [44]; (a) First (Random) Batch, (b) Third Batch, (c) Fifth Batch, and (d) Seventh Batch

environments and plans are necessary requirements for the real-world human decision maker, while extensive data sources exist across multitudes of possible epidemiological simulation environments. There is also the opportunity to embed online learning approaches deeper into the simulation environment, passing control of simulation resolution parameters in order to allow balancing of computational expense with efficient policy space exploration. This work is viewed as an emerging application for deploying further novel exploration techniques.

Figure 2.8 for example explores the changes in the action-value function generated simply from a perturbation in the timing of the intervention combinations. We'll look to expand on this concept of dynamic policy learning in the rest of this Chapter.

2.5 Reinforcement learning

In Section 2.3.1, the Multi-Armed Bandit framing was presented as a tool in the search for an optimal action-value function $Q(a)$. This use of online learning for prediction and control will now be further developed for the purposes of sequential or multi-step

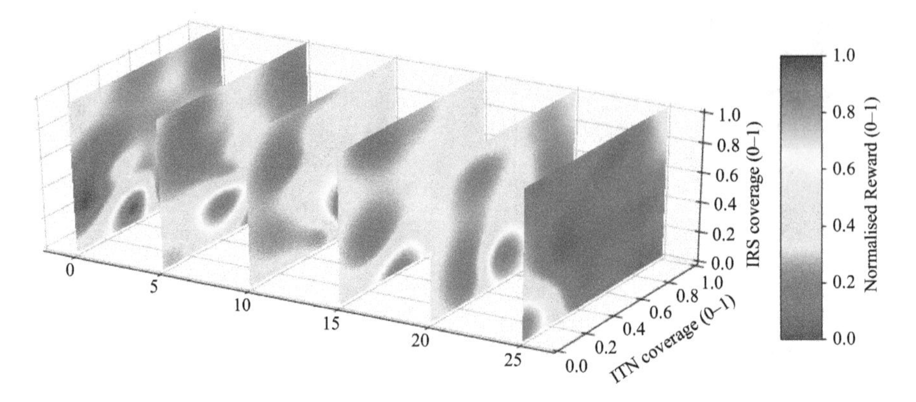

Figure 2.8 Slices of action-value function Q(a) for actions (a) as a combination of Insecticide Treat Net and Indoor Residual Spraying coverage changing through time (days), for OpenMalaria simulations

decision making. Performing predictions and learning control policies for epidemiological models is a decision-making task, which may be constructed as a sequence of actions, generating possible future outcomes and for which there may be delayed rewards. Through the sequential decision-making approach, we will use formalisms to perform predictions across possible futures (from the current observation) and learn a behaviour function based on observations to guide control (via a policy). This development of temporally situated decision making from epidemiological models will be driven through the paradigm of Reinforcement learning.

Ever important in our description of making decisions across time is that this decision-making is problem dependant. To further detail the problem framing, decisions for epidemiology control and planning may be made on varying discrete time intervals. While based on empirical observation and through discussions with domain experts and policy makers, a one-year decision cadence is often used for endemic diseases (HIV, tuberculosis, malaria) supported by organisations such as the Global fund.* This may be guided by monetary constraints, implementation challenges, or simply by the fact that yearly plans, with perhaps a longer horizon overview, are the digested format for the users, although evidently for emerging disease outbreaks the challenge is more immediate and the cadence of decision making reflects this. Ultimately the task is to drive the epidemiological system to asymptotic stability, or eventual eradication, with any delays potentially compounding the number of negative outcomes.

Typically an epidemiological disease model used for decision making will proceed on a simulated time frame of days. As changes in epidemiological state are not required at any finer resolution for the purposes of population level decision making

*https://www.theglobalfund.org/

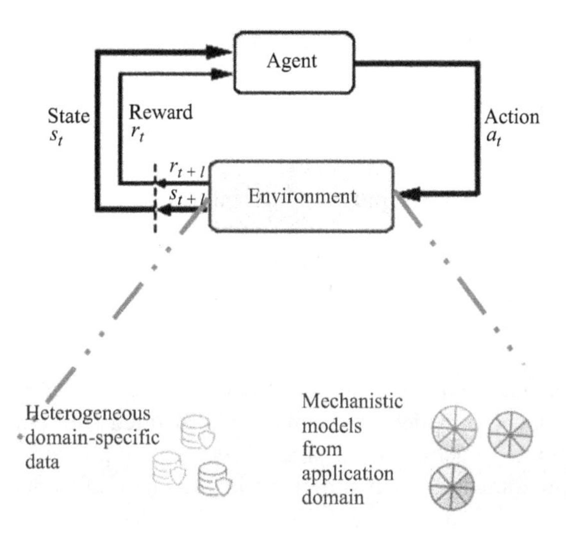

Figure 2.9 Reinforcement learning in environments of epidemiological model simulations. Reward $R_t = r_t$, State $S_t = s_t$, Action $A_t = a_t$.

or assessing health impacts, any particular task will aggregate the data such that it is relevant for the decision-making task.

By positioning epidemiological models as an abstraction of the real world, they may be posed as a simulation environment for an reinforcement learning (RL) agent to interact with. The following sections will attempt to handle this framing of the sequential decision-making task, broken down into the elements of state in Section 2.5.1, action in Section 2.5.2 and reward in Section 2.5.3, whereby a reinforcement learning agent is the algorithmic instantiation of our approach to 'search' for optimal control policies.

As in Figure 2.9, we acknowledge the abstraction that a mechanistic model is driven by disease-specific data. The presence of this data may also act as real-world experience, but is necessary should the model not be calibrated, which we will not discuss extensively here. The presence of real-world experience may also guide an agent's learning in combination with simulated experience, while we acknowledge that real-world experience is significantly more expensive than that which may be generated from a useful epidemiological model. Here we use the term 'expensive' in the direct sense of loss of life and of physical resource. For effective application of RL, we will often be required to simulate much more experience than may be feasible to observe in the real-world. We will now define the observations required by an agent in the RL framing.

2.5.1 State

A state signal is one which provides useful observations to our learning agent. For the purposes of sequential decision making, we require that the state S_t is Markov or has the Markov property [45], which may be summarised in the following quote:

'The future is independent of the past given the present'

Which may also be written as:

$$P[S_{t+1}|S_t] = P[S_{t+1}|S_1, S_2, ..., S_t]$$

The current state (S_t) hence provides all information based on the history ($S_1, S_2, \ldots, S_t$) and is sufficient to perform a prediction of the next state (S_{t+1}). The analogy to epidemiological models would be that this state observation would be sufficient information to restart the model from the point it was stopped, generating the same model prediction as a complete model run for the same time period under the same simulated policy.

The state domain of an epidemiological model may be vast for such purposes, where typically inputs to models may consist of structured files containing, population demographics, epidemiological parameters, climate predictions, etc. This is a problem for our framing, which suffers Bellman's curse of dimensionality [46], meaning that the number of possible states $n_S = |\mathcal{S}|$ grows exponentially with number of state variables. Therefore, useful encodings of state will be required, which do not necessarily retrieve a one to one mapping to the original model but allow us to proceed through the planning process.

A probability of transition $\mathcal{P}^a_{ss'}$ from state s to s' under action a defines our Markov Decision Process (MDP):

$$\mathcal{P}^a_{ss'} = P[S_{t+1} = s'|S_t = s, A_t = a]$$

We revisit this in more detail in Section 2.5.4.

2.5.2 Action

Actions are the mechanisms by which agents may perturb the environment. Given our framing that this environment will be a dynamical system, defined by an epidemiological model, such actions will consist of possible inputs to the model. Specifically for the task of control we aim to perturb the environment based on discrete interventions which may be performed in the epidemiological system. Previously, the definition of an action in Section 2.4.3, of a modelled set of interventions, will be relaxed such that the timing of interventions is not defined a priori before being simulated. Now the agent, as opposed to the Bandit framing of Section 2.3.1 has the chance to perform action selection through interaction with the environment, e.g. during learning. An action may receive delayed rewards in any sequential decision-making task, and we do not expect to reap the benefits of our actions immediately. Instead, any learning method should be able to identify the sequences of actions which lead to positive outcomes. For the purposes of describing a value function (in Section 2.5.7) we will consider actions as a finite set $\mathcal{A}$.

2.5.3 Reward

Much the same as in Section 2.4.4, reward for sequential decision making is as per that for a Multi-Armed Bandit. However, intermediate rewards may be received during interaction, whether delayed or not. The guiding principal in the use of a reward signal for agent learning is captured in Sutton's reward hypothesis [47]:

> All of what we mean by goals and purposes can be well thought of as maximization of the expected value of the cumulative sum of a received scalar signal (reward).

We assume a reward transition function exists, mapping state and action pairs to received stochastic rewards, such that the reward upon taking action a in state s is:

$$\mathcal{R}_s^a = \mathbb{E}[R_{t+1}|S_t = s, A_t = a] \tag{2.21}$$

This reward transition function is necessarily unknown by the agent and must be learnt via interaction, typically via an unknown Markov decision process.

2.5.4 Markov decision processes

Here we combine the concepts of the preceding Sections 2.5.1, 2.5.2 and 2.5.3, extending the Markov property to the complete Markov decision process (MDP) definition, containing States (S), Actions (A) and Rewards (R). For all $S_t, A_t, S_{t+1}, R_{t+1}$, we then have:

$$P(S_{t+1}, R_{t+1}|S_t, A_t) = P(S_{t+1}, R_{t+1}|S_t, A_t, R_t, S_{t-1}, A_{t-1}, ..., R_1, S_0, A_0) \tag{2.22}$$

The result of this summary is that we may restrict our attempts at optimal control to the search for reactive policies (Section 2.5.6). A reactive policy being one that depends only on the current state observation ($S_t = s$) and defines the behaviour function of which action ($A_t = a$) is selected given the observation.

A MDP is a tuple $\langle \mathcal{S}, \mathcal{A}, \mathcal{R}, \mathcal{P}, \gamma \rangle$, where $\mathcal{S}$ is our state space, $\mathcal{A}$ is a finite set of actions, $\mathcal{R}$ a Markov reward process, $\mathcal{P}$ defines state transition probabilities and γ the discount factor $\in [0, 1]$. Therefore, in solving the sequential decision-making task from epidemiological model-based experience, we will require that all environments conserve the Markov property, or may be reduced to a form which is tractable as a MDP.

2.5.5 Cumulated return

The cumulated return G_t is defined as the summed reward after time t is discounted via the factor (γ):

$$G_t = R_{t+1} + \gamma R_{t+2} + \cdots = \sum_{k=0}^{\infty} \gamma^k R_{t+k+1} \tag{2.23}$$

In a finite horizon task, by which we mean that the number of steps k in the learning episode is finite, this becomes a finite summation and it may be argued that all multi-step decision-making tasks in the context of epidemiological prediction, control and planning, are finite horizon tasks. We do not consider systems ad infinitum – perhaps as part of the human condition, we are most interested in the task of decision making over a single human lifetime, though often far shorter horizons, where upon achieving disease elimination, eradication may return our systems to asymptotically stable solutions or no further cases. If we are required to control endemics over a longer horizon this is possibly a far more expensive task, although inexpensive control measures may exist, the length of the planning horizon should influence the decision-making process. For an infinite horizon task that is episodic, instead of continuing, we represent termination to an absorbing state. Such an absorbing state may be mapped to eradication of the disease for the analysis that we perform. The discount factor (γ) may therefore be thought of as placing more weight on rewards which are achieved sooner, while often conditioning against the infinitely delayed splurge [48] phenomenon observed in infinite horizon agent learning tasks. In the absorbing state example of eradication, this particular goal would have to be associated with a zero reward to ensure returns remain finite. For a short horizon task of less than one year, it is likely that the discount factor will be set to unity $\gamma = 1$ and as such not influence the summation in calculating the cumulated return (2.5.5).

2.5.6 Policy

A policy defines an agent's behaviour function. For the purposes of learning within a MDP, policies may be reduced to the reactive type. A reactive policy is one which provides a mapping from the state observed to the action to be taken immediately at that time step. The policy at time step $t, \pi_t(s, a)$ defines a mapping from states to action probabilities such that $\pi(s, a) =$ probability that $A_t = a$ when $S_t = s$, $P(a|s)$. The above specifies a reactive stochastic policy, while for every MDP there exists at least one optimal deterministic reactive policy whereby $\pi(s) = a$. Reinforcement learning methods specify how the agent learns an effective policy as a result of experience. Roughly, the agent's goal is to accumulate as much reward as possible over the long run, therefore reinforcing reactive policies which maximise the reward for the problem. Significantly, it should be noted that policies are stationary, i.e. a function of state observation and independent of time, further meaning that the behaviour function has no knowledge of epidemiological simulation time, only the observations it receives, in principal policies whether deterministic or stochastic are implemented based on observation. Thus our sequence of decisions may be built from a policy which proceeds by evaluating which actions should be chosen based on observations. We will now look to address the mechanisms by which such a policy may be quantified in the search for optimality or control.

2.5.7 Value function

Value functions are the primary tool for reasoning about future reward in a MDP. The base form is the state-value function under a fixed policy π, such that value, the expected cumulated reward, is given as:

$$v^{\pi}(s) = \mathbb{E}_{\pi}\left[G_t | S_t = s\right] = \mathbb{E}_{\pi}\left[\sum_{k=0}^{\infty} \gamma^k r_{t+k+1} | S_t = s\right] \tag{2.24}$$

in which this may be viewed as a prediction based on following a fixed policy regarding future reward conditioned on the current state. For example, in Section 2.2 a single prediction may be based on the current intervention plan and a forward epidemiological model run. The associated cumulated return (Section 2.5.5) would describe the value function of the state (in this case the set of epidemiological model inputs) at the start of model observations. Therefore, the value function for a single prediction will provide expected reward under the fixed policy – the model forecast, and the outcomes of which are quantified by our reward function for the observed states. This is useful for characterising states which lead to high future reward.

Additionally, the action-value of a policy π will be defined as follows:

$$q^{\pi}(s,a) = \mathop{\mathbb{E}}_{\pi}[G_t|S_t = s, A_t = a] = \mathop{\mathbb{E}}_{\pi}\left[\sum_{k=0}^{\infty}\gamma^k r_{t+k+1}|S_t = s, A_t = a\right] \tag{2.25}$$

In this instance, the action-value function may guide the future reward associated with taking action a in state s, which moves us towards learning a reactive policy based on the action-value function $q(s, a)$. This is the foundation for the value-based Reinforcement learning technique of Q-learning [49].

For the purposes of guiding optimal decision making, the properties of optimality can now be defined for our respective value functions:

$$v_*(s) = \max_{\pi} v_{\pi}(s) \tag{2.26}$$

$$q_*(s,a) = \max_{\pi} q_{\pi}(s,a) \tag{2.27}$$

where (2.26) is the condition for the optimal state-value function and (2.27) the optimal action-value function. An optimal policy (π_*) may subsequently be defined by an optimal action-value function as follows:

$$\pi_*(a|s) = \begin{cases} 1 & \text{if } a = \operatorname*{argmax}_{a \in \mathcal{A}} q_*(s,a) \\ 0 & \text{otherwise} \end{cases} \tag{2.28}$$

The Bellman equations [(2.29) and (2.30)] are used to decompose the calculation of value functions into iterative direct solutions [50]:

$$v_{\pi}(s) = \sum_{a \in \mathcal{A}} \pi(a|s)\left(\mathcal{R}_s^a + \gamma \sum_{s' \in S} \mathcal{P}_{ss'}^a v_{\pi}(s')\right) \tag{2.29}$$

$$q_{\pi}(s,a) = \mathcal{R}_s^a + \gamma \sum_{s' \in S} \mathcal{P}_{ss'}^a \sum_{a' \in \mathcal{A}} \pi(a'|s') q_{\pi}(s',a') \tag{2.30}$$

Most iterative solutions for the Bellman equations such as Value Iteration are solved through dynamic programming [46]. We shall continue with a focus on *model-free* approximations to these solutions, which specify agent interactions in an *unknown* environment. Specifically, this means that the transition probabilities $\mathcal{P}_{ss'}^a$ and the Markov reward process $\mathcal{R}_s^a$ for the environment will always be assumed to be unknown, fitting our specifications of interacting with epidemiological models in a black-box manner. This outline will be extended to *model-free* methods, which approximate the environment transitions via samples or observations. The methods used in the search for optimal control policies under unknown environments will be developed further from Section 2.5.9 onwards.

2.5.8 Partially observable MDP (POMDP)

Practically all environments which we interact with, composed from epidemiological models, are abstractions made for the process of decision making. Further we may consider all models as POMDPs and generate state approximations to generically apply learning algorithms to them. Any approximation is required to capture the history of the model run, as is done by the Markov state measure. A MDP has a history-less state measure and at a minimum the POMDP should be able to define the sequential decision-making task in terms of the history [51]. The size of the history will scale in a manner which is worse than the dimensionality of the state measure and may also be referred to as the curse of history: the number of belief-contingent plans increases exponentially with the planning horizon [52]. In this case, an agent or belief state is distinguished from the environment state. If an agent always has access to their own state measure, or beliefs, there are greater opportunities for transfer of approaches, along with providing a general framework for planning under-uncertainty. This will be addressed in Chapter 2, where we propose a plurality of agents to be deployed for epidemiological planning tasks.

2.5.9 Learning sequential surrogate models from episodic simulators

This section will specify learning an agent model or environment from epidemiological model runs. This is necessary as the form of a heterogeneous model will not necessarily be a Markovian environment from which to learn control policies and so it will be necessary to efficiently learn a model using the supervised learning methods from Section 2.2.1. Learning a model approximation may increase our sample efficiency for the application of more general value or policy learning methods. Also, we may generate consistent descriptions of the epidemiological model uncertainty, under which reasoning is being performed, for the purposes of planning or control. Although learning a surrogate model and then performing an agent based approach, which learns a value function or policy as an approximation to the model, introduces two sources of approximation error. Namely the sources of error from the two processes of approximation. We will refer to any surrogate model learnt from epidemiological simulations as $\mathcal{M}$. The surrogate model $\mathcal{M}_{\theta_M}$ is a representation of the MDP tuple $\langle \mathcal{S}, \mathcal{A}, \mathcal{R}, \mathcal{P} \rangle$, based on the parametrisation θ_M, assuming that by defining our own surrogate model structure this means $\mathcal{S}$ and $\mathcal{A}$ are known. Therefore through $\langle \mathcal{R}_{\theta_M}, \mathcal{P}_{\theta_M} \rangle$, we will attempt to approximate the true (unknown) transitions $\mathcal{P}_{\theta_M} \approx \mathcal{P}$ and $\mathcal{R}_{\theta_M} \approx \mathcal{R}$.
Hence,

$$S_{t+1} \sim \mathcal{P}_{\theta_M}(S_{t+1}|S_t, A_t) \tag{2.31}$$

$$R_{t+1} = \mathcal{R}_{\theta_M}(R_{t+1}|S_t, A_t) \tag{2.32}$$

Here we assume (as is common practice) state transitions and rewards to be independent [53], such that:

$$\mathbb{P}[S_{t+1}, R_{t+1}|S_t, A_t] = \mathbb{P}[S_{t+1}|S_t, A_t]\,\mathbb{P}[R_{t+1}|S_t, A_t] \tag{2.33}$$

Estimating $\mathcal{M}_{\theta_M}$ may be reduced to a supervised learning problem based on tuples of experience $\langle S_1, A_1, R_2, ..., S_T\rangle$ [54], decomposing a sequence of decisions as follows:

$$\begin{aligned} S_1, A_1 &\rightarrow R_2, S_2 \\ S_2, A_2 &\rightarrow R_3, S_3 \\ &\cdot \\ &\cdot \\ S_{T-1}, A_{T-1} &\rightarrow R_T, S_T \end{aligned}$$

Learning $\mathcal{R}_{\theta_M}$ as an approximation to $\mathcal{R}$ thence becomes a regression problem and we will continue to use Gaussian process regression for this task. In tandem, learning the approximation of $\mathcal{P}_{\theta_M}$ is a density estimation problem [55], for which we will leverage the Gaussian process structure to generate state transition approximations in the surrogate model.

2.5.10 Prediction – learning a value function

Before learning a model ($\mathcal{M}_{\theta_M}$) of our unknown epidemiological simulator, we need to introduce the mechanism by which we may evaluate policies based on model runs, as this will be equally useful for our purpose of generating the environment abstraction. This mechanism is via learning a value function to define the expected response of the underlying models based on their returns. Now, as we do not have unlimited evaluations to perform over multiple dimensions we must use approximation, therefore the efficiency of samples will have to be improved given a fixed computational budget of B simulations. We have proposed this problem for learning the action-value function in one-step planning using the Multi-Armed Bandit (Section 2.3.1). In continuing under a fixed computational budget, we will now introduce extensions for multi-step planning, in order to provide a common framework to query heterogeneous disease models at scale and approximate a sequential surrogate model, which conserves the Markovian property for sequential decision-making approaches.

In undertaking the prediction task for model-free reinforcement learning via Monte Carlo (MC) methods of sampling, experience may be generated from complete model runs. In the model-free learning sense, we treat the epidemiological model as a black-box, with no knowledge of what we assume to be an underlying MDP, therefore with no knowledge of $\mathcal{R}_s^a$ and $\mathcal{P}_{ss'}^a$, and these will be constructed as approximations in the process of learning model $\mathcal{M}_{\theta_M}$. Monte Carlo searches will learn from complete episodes and therefore a one-shot simulation from a model is not a barrier to our decomposition of the sequential decision-making task.

The value function, based on MC methods, is considered as an empirical mean of the return rather than an expectation and it is employed for episodic MDPs which terminate [56]. This is realistic for our problems as the episode, or decision-making time frame needs to be finite to be meaningful. We will therefore consider non-discounted returns given the decision-making cadence.

Alternatively, temporal difference (TD) learning offers another example of a model-free reinforcement learning method, though not readily applicable to all epidemiological simulations, as temporal difference learning is an online learning technique which performs updates before knowing the final outcome. TD learning may be more appropriate in non-terminating environments, where TD updates may be performed based on the subsequent step reward and state $R_{t+1} + \gamma V(S_{t+1})$.

MC methods do not exploit the Markov property itself, so are effective in non-Markov environments and for generating our approximations from model simulations which may not provide a Markov state. Additionally MC, being based on simulation returns, is a zero bias, high variance estimator which naturally pairs it with the gaussian process in the presence of smoothness between nearby observations. Further MC is a method which generally works well for 'black-box' models [57]. Finally, MC sampling may be used to break Bellman's curse of dimensionality in the size of the state space and the curse of history in the case of a planning from a POMDP [52,58].

In any Monte Carlo method, the law of large numbers dictates that our observations will convergence to our true expected underlying values [59], in our case $V(s) \to v_\pi(s)$ as $N(s) \to \infty$, with updates performed on $V(s)$ after each episode. In this way the number of visitations of each state $N(S_t) \leftarrow N(S_t) + 1$ and the state-value function is updated subject to its current value $(V(S_t))$ and the observed cumulated return (G_t), with the state visitation count $(N(S_t))$:

$$V(S_t) \leftarrow V(S_t) + \frac{1}{N(S_t)}(G_t - V(s_t)) \tag{2.34}$$

In order to generate useful observations of epidemiological state for decision making, we need to further develop an often used framework of the decision tree, where the nodes in a decision tree capture a state approximation for the planning task. Finally, this section on MC predictions or policy evaluations provides the basis for evaluating our *Rollout* policies required in the complete planning approach of Monte Carlo Tree Search (MCTS) laid out in Section 2.5.12.

2.5.11 Simulation-based search – decision trees

In order to learn from black-box epidemiological models, we will impose a tree structure on the decision-making task. Search or decision trees are a common structure used in supervised ML [60], for the purpose of sequential decision making each node of the search tree describes a Markov state S. At this node, we will store the state-action value function $q(s, a)$, along with a visitation count for each action a and the total count for the node $N(s) = \sum_a N(s, a)$, correspondingly each branch of the tree provides a description of our state transitions $\mathcal{P}^a_{ss'}$. Initialisation of the search tree will be considered as setting $Q(s, a) = 0$ and $N(s, a) = 0$.

For this common framing across epidemiological models for decision making we will make abstractions about the task that determines the structure of the solutions developed. In our case, this requires that the cadence at which interventions are to be deployed, along with the number of possible actions, need to be specified and will fully define the structure of the learnt surrogate model. Hence it needs to be guided by judicious choices with respect to the application domain. This choice of decision-making cadence and number of actions directly effects the number of nodes and transitions in the decision tree.

In the following description, it is assumed that for a chosen episode of length k the task is to generate a sequence of k actions. In this case, the episode length defines the depth of the decision tree and is related to what we have informally referred to as the decision-making cadence for the modelled time frame. Subsequently, the history (which is the sequence of actions) may be used to define the state, or each node, in the search tree. This is the simplest encoding of state for the environment being learnt, though at the same time by retaining the history of the sequential decision-making task, especially where each action is of dimension d, it is necessarily to acknowledge an increased complexity to the number of total states described by the history and will look at approximations to break this curse of dimensionality. If we continue to assume a finite set of actions $\mathcal{A}$, the MDP states may be constructed based on the history of actions as follows:

$$S = (a_1, a_2, \ldots, a_i) \mid 1 \leq i \leq k, \ \forall [a_i \in \mathcal{A}] \tag{2.35}$$

In any state, actions may correspond to the implemented interventions and their coverage in the time frame until the next decision. With this definition of $s \in \mathcal{S}$ and $\mathcal{A}$ the transition function $\mathcal{P}^a_{ss'}$ reduces to a deterministic transition function, as actions map to a single state.

For further illustration consider $s = (a_1, a_2, \ldots, a_i)$ with the selection of action a_t governing transition to the next state $s' = (a_1, a_2, \ldots, a_i, a_t)$. Terminal states will necessarily be of length k, the length of the sequence of actions, and the reward for each state will be based on the episodic returns as in Section 2.5.10. The reward of the terminal state will equate to the return for the sequence of actions and to generate such a reward this may be specified on episodic simulation returns, through for example the cost-effectiveness of the policy (the Cost Per Daly Averted (CPDA) calculation is described in Section 2.4.4.3).

We assume a consistent initial state between epidemiological model runs, this is the root node in the search tree and may be thought of as the start of the intervention period of the model run, or even the current state, 'Now' as in Figure 2.10. This should be consistent between different epidemiological models, which aim to describe the same disease and location over the same time frame, for the same decision-making task. This consistency in generating comparisons and learning sequences of interventions is fundamental for an 'apples to apples' comparison of relevant intervention plans. At a minimum, we may wish to assume that the epidemiological models are calibrated with the same data for the same location and may or may not have been validated by an epidemiologist. We will now move on to the algorithmic use of this notion of a search tree in the generation of a model $\mathcal{M}$, which may be used for planning or which may itself be used directly in the task of control in extension to the Multi-Armed Bandit (MAB) for control.

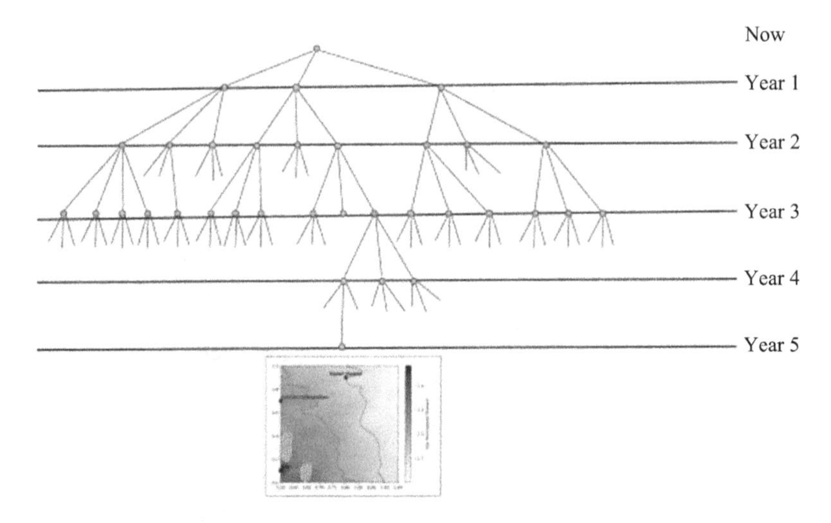

Figure 2.10 Example of cumulated return as a two-dimensional response surface, as a result of a sequence of decisions for 5 years

2.5.12 Monte Carlo tree search (MCTS)

By combining generative predictions via Monte Carlo (MC) methods and the structure of a search tree, MCTS has had success as a planning approach in large state spaces including the ultra-large spaces found in challenging games, most notably the game of Go [61–63]. Any direct optimal solution to such problems is infeasible due to the large size of the state-action space, therefore approximate solvers are used, in the form of MC samples. Further, we wish to restrict samples to relevant regions of the state-action space, which is achieved through the MCTS approach.

MCTS can be broken down into the following four steps:

1. *Selection*: Starting from the root state, select successive actions, appended to the history as states, until a leaf state is reached. The root state is taken as the start of the epidemiological model run and a leaf is any state that has a potential child from which no simulation (*rollout*) has yet been run.
2. *Expansion*: Unless the leaf state is the terminal state, create one (or more) child nodes and choose a next action from one of them. Child nodes are any valid interventions from the leaf state. The process of expansion may be guided by the following notions:
   - in-tree: actions select child states to maximise $Q(s,a)$
   - out-tree: pick actions randomly to explore child states
3. *Simulation*: Complete one random *rollout* from a child node. The *rollout* policy may be as simple as choosing uniform random actions until a terminal state is reached. Any simulation will consist of a sequence of actions of length k.
4. *Backpropagation*: Use the result of the rollout to update information in the nodes on the path using MC updates: $Q(s,a) \leftarrow Q(s,a) + \frac{1}{N(s,a)}(G_t - Q(s,a))$. In this case recording state-action counts $N(s,a)$.

The success of this technique in deterministic games is well documented. To stretch the concept of such a game tree to the process of intervention planning from simulation, we need to be able to efficiently combine the processes of selection and expansion to maximise our exploration of regions of interest. This is detailed in the following section through decomposing the problem into a sequence of Multi-Armed Bandit decisions, aiming to achieve efficiency in the number of simulated runs or experience through learning a model $\mathcal{M}$ as part of the search tree.

2.5.13 Gaussian process regression with selection in MCTS for learning sequential surrogates (GP–MCTS)

Optimism in the face of uncertainty [64] will provide our guiding principle to extending the MCTS approach for the generation of surrogate models in sequential decision making. The generation of a sequential surrogate model using GP–MCTS (Algorithm 2.3) is presented, to produce a surrogate model from simulation results based on inference performed for the state action value function $q(s, a)$ at each node. The value $q(S, A)$ is estimated using regression across a continuous action-space at each node, in order to reduce to a finite set of transitions, and bound the size of the history which may describe the system. $Q(S, A)$ needs to contain both discrete actions and be compressed through density estimation techniques in order to bound the size of the search tree. This search tree will have a depth k, though approximations to the number of states will govern the number of branches in the tree, based on the number of possible transitions. Value function approximation is performed using Gaussian Process regression (GPR), in order to scale the efficiency of model free methods learning from experience. This inference will improve convergence with fewer samples. Rather than a look-up table, we now have a function that is learnt as a $\mathcal{GP}$, which may be constructed in batches of simulations.

We introduce a MCTS approach to generating a sequential surrogate model using the GP–MCTS algorithm, inspired by MCTS and work on the solution to the Multi-Armed Bandit (MAB) problem [40,41] along with the Upper Confidence applied to Trees (UCT) algorithm [65]. This is a formulation which combines the natural confidence bounds of Gaussian processes, for uncertain sequential decision making in stochastic simulation problems. Treating each action selection as a Bandit problem from a state node in the search tree, this structure may be generated from samples, for example, using MCTS with simulators which do not allow for direct manipulation of sequential decisions as they are episodic in formulation. In creating a representative sequential surrogate model the aim is to increase the speed of learning for reinforcement learning-based approaches, which may perform planning from the surrogate $\mathcal{M}$. This has been used to motivate development of agent-based insight into decision making from computationally expensive simulations later in the chapter (Section 2.6), along with external developments based on sharing of efficient sequential surrogate models which is presented in Chapter 2.

Algorithm 2.3 is initialised with a random sample of the action space ($\mathcal{A}$) for each node in the random searches. Subsequent actions are chosen to further explore the search tree nodes which each contain their own learnt GP model based on

---

Algorithm 2.3: GP-UCB-MCTS

Result: $Q(S, A)$
Initialisation;
Search Tree parameters: B = branching or transition number; Tree depth k [Section 2.5.13];
$\mathcal{GP}$ priors $\theta_{\mathcal{GP}}$ [Section 2.2.1.2];
All $Q(s, a) = 0$ [for other initialisation values see Section 2.5.13.1];
for $t = 1, 2, \ldots, end$ **do**
  for $j = 1, 2, \ldots k$ **do**
    Selection;
    if $Q(S^j, A) = 0$ **then**
      A_t^j random action $\in \mathcal{A}$;
    else
      $A_t^j = \underset{a \in \mathcal{A}}{\text{argmax}} \left[m_{t-1}(\boldsymbol{a}) + \beta \sqrt{k_{t-1}(\boldsymbol{a}, \boldsymbol{a}')} \right]$ [Section 2.4.5.1];
      $S^j \rightarrow S^{j+1}$ based on compression for B choices [Section 2.5.13];
    end
    Expansion;
    if $j=k$ **then**
      Simulation;
      $\boldsymbol{a}_t$;
    else
      Repeat *Selection*;
    end
  end
  Return: G_t;
  State-action count: $N(S, A) \leftarrow N(S, A) + 1$;
  Backpropagation;
  $Q(S^j, A^j) \leftarrow Q(S^j, A^j) + \frac{1}{N(S^j, A^j)}(G_t - Q(S^j, A^j))$;
  Update posterior $Q(S_t, A_t)$: mean $m_t(\boldsymbol{a})$, covariance $k_t(\boldsymbol{a}, \boldsymbol{a}')$;
end
$Q(S, A) \rightarrow q_*(S, A)$

MC returns from preceding simulation runs. In order to balance exploration and exploitation in generating the tree we make use of the UCB heuristic introduced in Section 2.5.13.

Further heuristics have been included for practical realisations of this approach, such as a branching factor (B) commonly of value 3 has been used along with a uniform discretisation of the action space ($\mathcal{A}$) for each node, as visualised in Figure 2.11. In this sequential surrogate model described by a tree, each node consists of mean and covariance for the estimated action value function ($Q(S, A)$). This approximation

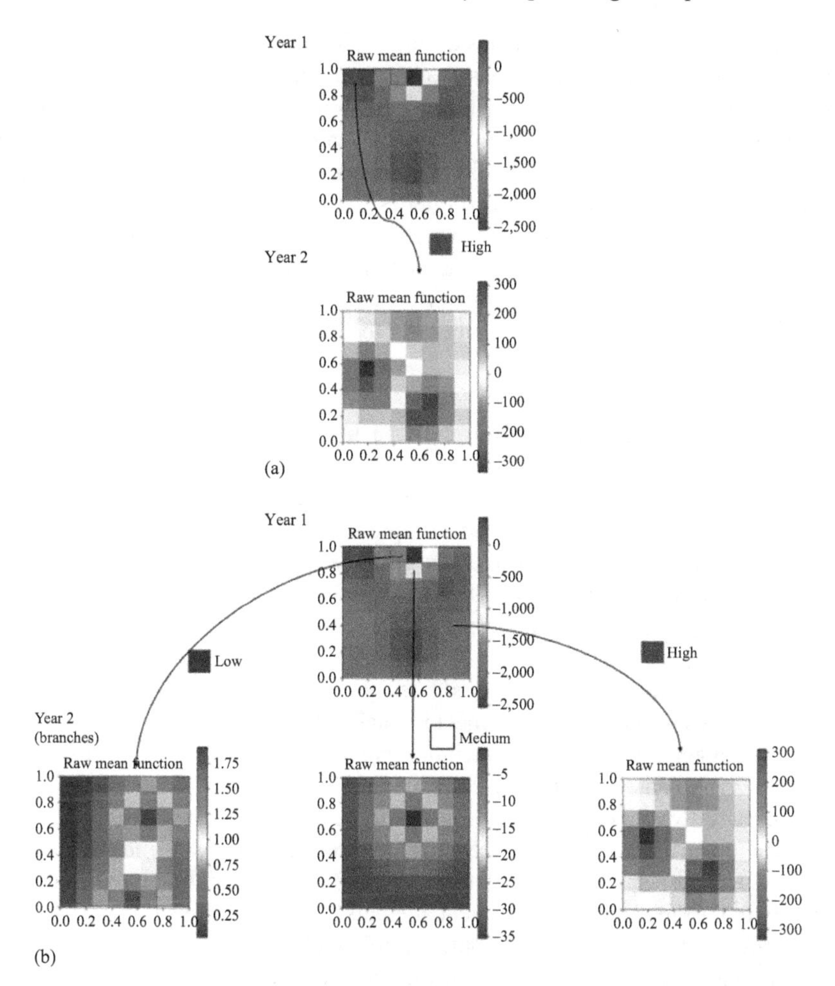

Figure 2.11 Visual description of generation of sequential surrogate $Q(S, A)$ with regards to the mean $m(\boldsymbol{a})$ for each state. (a) Discretised 2D action-value function for $Q(a)$ showing mean $m(a)$. (b) Transitions based on branching (B) of order three for High, Medium and Low value actions.

of $q(s, a)$ for the epidemiological model may be sampled at query time by any agent code. Additionally an ensemble may be generated across $Q(S, A)$ for different epidemiological models where the same tree structure is learnt, though the application of these techniques will not be described here.

2.5.13.1 Incorporating domain knowledge into MCTS

This section is included to highlight that generating new search trees based on available epidemiological models may be sped up in the multi-model scenario. The most immediate mechanism to achieve this during the learning process, including domain knowledge, is to initialise the tree based on the learnt nodes for other models, assuming the same structure persists. Therefore, the initialisation of $Q(S,A)$ for each of the nodes, $N(S,A)=0$ will help to focus the search to policies we believe, from other models, to be better, without altering the convergence properties of the Monte Carlo approach. In this case, the learnt nodes from another model may act as a GP prior for the action selection task in further Monte Carlo simulations. While it is also conceived that trees may be later combined at each node based on the committee predictions, boosting our ability to quantify uncertainty in the problem and therefore have an effective multi-model description for the decision-making task.

2.6 Control – optimal policies and planning

Now we return to the task of learning a policy in extension to the developments of Section 2.5.9 in learning a surrogate representation of the sequential decision-making task from epidemiological models. We have already stated that Reinforcement learning is complicated by the fact that the reinforcement, in the form of rewards (Section 2.5.3), is often intermittent and delayed [66]. The agent may perform a long sequence of actions (Section 2.5.2) before receiving any reward. This leads to the difficulty in solving the temporal credit assignment problem to attribute credit or blame to actions when the reinforcement is finally received [67]. The policy (Section 2.5.6) that an agent learns via interaction represents a particular plan that indicates the best action to take in every possible state that it observes.

There are two fundamental approaches to designing an agent for problems in which no information is given with regards to the transition probabilities and rewards ($\mathcal{P}^a_{ss'}$ and $\mathcal{R}^a_s$, respectively), this defines the domain of model-free reinforcement learning. In the first approach, the agent attempts to learn the transition probabilities and rewards and then construct an optimal policy off-line, using a method such as Bellman's value iteration in (2.26). For the second approach, the agent attempts to learn an optimal policy by constructing an evaluation function for use in selecting the best action to take in a given state [68]. The agent through learning this evaluation function cannot predict what the state transition resulting from a given action will be, its evaluations based on determining whether this new state leads to greater future rewards or return compared to the selection of another action.

2.6.1 Optimal policy learning

Learning an optimal policy requires the specification of an action-value function in addition to the value function of the prediction task (Section 2.5.10). We have shown in the previous section how we may generate a sequential surrogate model from epidemiological simulations and this may be used for planning as outlined previously in Chapter 2. The purpose of this section is to specify common domains of learning methods, specifically model-free methods, in the hope that generalisable approaches may be developed. Results will be shown based on learning from computationally less expensive sequential surrogate models, learnt from sets of experience and also in extension to less computationally expensive, population level, deterministic compartmental models.

2.7 Comparing predictions from multi-step and one-step methods with direct experience

As we explore more nuanced decision-making, varying the timing of interventions has been previously ignored due to complexity. Specifically incorporating timing adds to the combinatorial complexity of the problem, further stressing the need for data efficiency and also the volume of data needed to make useful inference. When making the best policy decision we can see in Figure 2.8, that for decisions of cost-effectiveness, optimal decisions may vary slightly even across the course of the month. Such responses have not been previously studied and are now being supported by the presented infrastructure as well as extending into the space of geospatial decision making. We may observe empirically in Figure 2.12 that the search for multi-step as opposed to single-step control policies provide high programmatic relevance when we evaluate the cost of the policy based on it reaching a desired performance. In this case, for an underlying malaria model, we are able to reduce the cost by 25% to reach the same final target prevalence, or in the presence of the same budget reach a final prevalence of 50% of the target. Given the highlighted evidence of the benefits afforded by multi-step control and casting the generation of intervention plans as a sequential decision-making task, we will now progress the search for

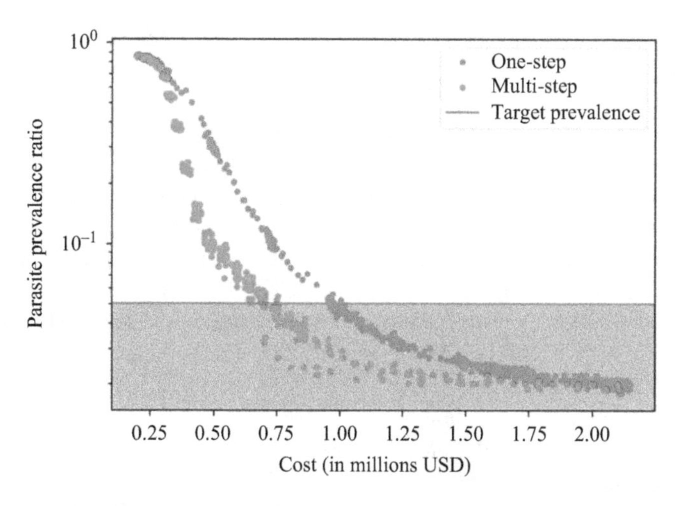

Figure 2.12 Ability of 'Best' multi-step policies (GP-UCB-MCTS Section 2.5.13) to consistently outperform static policies (GP-ULCB Section 2.4.5.1) for the same simulation environment

refined approaches and systems to accelerate the development of optimal policies for epidemiological control.

References

[1] Friedman JH. An overview of predictive learning and function approximation. In: *From Statistics to Neural Networks*. New York, NY: Springer; 1994. p. 1–61.

[2] Vapnik V. *The Nature of Statistical Learning Theory*. New York, NY: Springer Science & Business Media; 2013.

[3] Kotsiantis SB, Zaharakis I and Pintelas P. Supervised machine learning: a review of classification techniques. *Emerging Artificial Intelligence Applications in Computer Engineering*. 2007;160(1):3–24.

[4] Lawson AB. *Statistical Methods in Spatial Epidemiology*. New York, NY: John Wiley & Sons; 2013.

[5] Roser M, Ritchie H, Ortiz-Ospina E, *et al*. Coronavirus pandemic (Covid-19). In: *Our World in Data*; 2020.

[6] Ray EL, Wattanachit N, Niemi J, *et al*. Ensemble forecasts of coronavirus disease 2019 (Covid-19) in the US. MedRXiv. 2020.

[7] Nicola M, Alsafi Z, Sohrabi C, *et al*. The socio-economic implications of the coronavirus pandemic (COVID-19): a review. *International Journal of Surgery (London, England)*. 2020;78:185.

[8] Dietterich TG. Ensemble methods in machine learning. In: *International workshop on Multiple Classifier Systems*. New York, NY: Springer; 2000. p. 1–15.

[9] Box GEP. Robustness in the strategy of scientific model building. In: Launer RL, Wilkinson GN, editors. *Robustness in Statistics*. London: Academic Press; 1979. p. 201–236. Available from: http://www.sciencedirect.com/science/article/pii/B9780124381506500182.

[10] Williams CK and Rasmussen CE. Gaussian processes for regression. In: *Advances in Neural Information Processing systems*; 1996. p. 514–520.

[11] Vanhatalo J and Vehtari A. Sparse log Gaussian processes via MCMC for spatial epidemiology. In: *Gaussian Processes in Practice*; 2007. p. 73–89.

[12] Cornford D, Nabney IT and Williams CK. Modelling frontal discontinuities in wind fields. *Journal of Nonparametric Statistics*. 2002;14(1–2):43–58.

[13] Titsias MK, Lawrence N and Rattray M. Markov chain Monte Carlo algorithms for Gaussian processes. In: *Inference and Estimation in Probabilistic Time-Series Models*; 2008. p. 9.

[14] Bore NK, Raman RK, Markus IM, *et al*. Promoting distributed trust in machine learning and computational simulation. In: *IEEE International Conference on Blockchain and Cryptocurrency, ICBC 2019*, Seoul, Korea (South), May 14–17, 2019; 2019. p. 311–319. Available from: https://doi.org/10.1109/BLOC.2019.8751423.

[15] McCarthy KA, Wenger EA, Huynh GH, *et al.* Calibration of an intrahost malaria model and parameter ensemble evaluation of a pre-erythrocytic vaccine. *Malaria Journal*. 2015;14(1):1–10.

[16] Keeling MJ and Rohani P. Estimating spatial coupling in epidemiological systems: a mechanistic approach. *Ecology Letters*. 2002;5(1):20–29.

[17] McLean AR and Anderson RM. Measles in developing countries Part I. Epidemiological parameters and patterns. *Epidemiology & Infection*. 1988;100(1):111–133.

[18] Anderson RM. The epidemiology of HIV infection: variable incubation plus infectious periods and heterogeneity in sexual activity. *Journal of the Royal Statistical Society: Series A (Statistics in Society)*. 1988;151(1):66–93.

[19] Alexander ME, Bowman C, Moghadas SM, *et al.* A vaccination model for transmission dynamics of influenza. *SIAM Journal on Applied Dynamical Systems*. 2004;3(4):503–524.

[20] Moore S, Shrestha S, Tomlinson KW, *et al.* Predicting the effect of climate change on African trypanosomiasis: integrating epidemiology with parasite and vector biology. *Journal of the Royal Society Interface*. 2012;9(70):817–830.

[21] Chitnis N, Hyman JM and Cushing JM. Determining important parameters in the spread of malaria through the sensitivity analysis of a mathematical model. *Bulletin of Mathematical Biology*. 2008;70(5):1272.

[22] Yamba EI, Tompkins AM, Fink AH, *et al.* Monthly entomological inoculation rate data for studying the seasonality of malaria transmission in Africa. *Data*. 2020;5(2):31. Available from: https://www.mdpi.com/2306-5729/5/2/31.

[23] Ferris C, Raybaud B and Madey G. OpenMalaria and EMOD: a case study on model alignment. In: *Proceedings of the Conference on Summer Computer Simulation*; 2015. p. 1–9.

[24] Wang Z, Zoghi M, Hutter F, *et al.* Bayesian optimization in high dimensions via random embeddings. In: *Proceedings of the International Joint Conference on Artificial Intelligence*; 2013; p. 1778–1784.

[25] Osborne MA, Garnett R and Roberts SJ. Gaussian processes for global optimization. In: *3rd International Conference on Learning and Intelligent Optimization LION3*, vol. x; 2009; p. 1–15. Available from: http://www.intelligent-optimization.org/LION3/online_proceedings/94.pdf http://www.robots.ox.ac.uk/~mosb/OsborneGarnettRobertsGPGO.pdf.

[26] Bergstra J, Bardenet R, Bengio Y, *et al.* Algorithms for hyper-parameter optimization. *Advances in Neural Information Processing Systems*. 2011;24:2546–2554.

[27] Frazier PI. A Tutorial on Bayesian Optimization. arXiv preprint arXiv:180702811; 2018.

[28] Shahriari B, Swersky K, Wang Z, *et al.* Taking the human out of the loop: a review of Bayesian optimization. *Proceedings of the IEEE*. 2015;104(1): 148–175.

[29] Press DAUU. *Test and Evaluation Management Guide*. Defense Acquisition University Press; 2001.

[30] Fisher RA. Design of experiments. *BMJ*. 1936;1(3923):554–554.

[31] Brady OJ, Godfray HCJ, Tatem AJ, *et al.* Vectorial capacity and vector control: reconsidering sensitivity to parameters for malaria elimination. *Transactions of the Royal Society of Tropical Medicine and Hygiene*. 2016;110(2):107–117. Available from: https://academic.oup.com/trstmh/article-abstract/110/2/107/2578714.

[32] Villar SS, Bowden J and Wason J. Multi-Armed Bandit models for the optimal design of clinical trials: benefits and challenges. *Statistical Science: A Review Journal of the Institute of Mathematical Statistics*. 2015;30(2):199.

[33] Suryanarayanan P, Tsou CH, Poddar A, *et al.* WNTRAC: artificial intelligence assisted tracking of non-pharmaceutical interventions implemented worldwide for COVID-19. arXiv preprint arXiv:200907057; 2020.

[34] Lopez AD and Murray CC. The global burden of disease, 1990–2020. *Nature Medicine*. 1998;4(11):1241–1243.

[35] Murray CJL and Lopez AD. *The Global Burden of Disease: A Comprehensive Assessment of Mortality and Disability from deceases, Injuries and Risk Factors in 1990 and Projected to 2010*, vol. 1. Cambridge, MA: Harvard University Press; 1996. p. 1–35.

[36] Turner D, Wailoo A, Nicholson K, *et al.* Systematic review and economic decision modelling for the prevention and treatment of influenza A and B. In: *NIHR Health Technology Assessment Programme: Executive Summaries. NIHR Journals Library*; 2003.

[37] Kalman RE. Contributions to the theory of optimal control. *Boletin de la Sociedad Matemática Mexicana*. 1960;5(2):102–119.

[38] Bemporad A, Morari M, Dua V, *et al.* The explicit linear quadratic regulator for constrained systems. *Automatica*. 2002;38(1):3–20.

[39] Gunda R and Chimbari MJ. *Cost-Effectiveness Analysis of Malaria Interventions Using Disability Adjusted Life Years: A Systematic Review*. BioMed Central Ltd.; 2017.

[40] Auer P and Ortner R. UCB revisited: improved regret bounds for the stochastic Multi-Armed Bandit problem. *Periodica Mathematica Hungarica*. 2010;61(1 and 2):55–65.

[41] Auer P, Cesa-Bianchi N and Fischer P. Finite-time analysis of the Multiarmed Bandit problem. *Machine Learning*. 2002;47(2–3):235–256.

[42] Srinivas N, Krause A, Kakade SM, *et al.* Gaussian Process Optimization in the Bandit Setting: No Regret and Experimental Design. arXiv preprint arXiv:09123995; 2009.

[43] Contal E, Buffoni D, Robicquet A, *et al.* Parallel Gaussian process optimization with and pure exploration. *Lecture Notes in Computer Science (including subseries Lecture Notes in Artificial Intelligence and Lecture Notes in Bioinformatics)*. 2013;8188 LNAI(PART 1):225–240.

[44] Bent O, Remy SL, Roberts S, *et al.* Novel Exploration Techniques (NETs) for malaria policy interventions. In: *Proceedings of the Thirty-Second AAAI Conference on Artificial Intelligence (AAAI-18)*; 2018. p. 7735–7740. Available from: https://www.aaai.org/ocs/index.php/AAAI/AAAI18/paper/view/16148.

[45] Markov AA. The theory of algorithms. *Trudy Matematicheskogo Instituta Imeni VA Steklova*. 1954;42:3–375.

[46] Bellman R. *Dynamic Programming*. Santa Monica, CA: RAND Corp; 1956.

[47] Sutton RS, Barto AG. *Reinforcement Learning: An Introduction*. Cambridge, MA: MIT Press; 1998.

[48] Platzman LK. *Finite Memory Estimation and Control of Finite Probabilistic Systems*. Cambridge, MA: Massachusetts Institute of Technology; 1977.

[49] Watkins CJCH. *Learning from Delayed Rewards*. PhD Thesis, University of Cambridge, England; 1989.

[50] Bellman RE and Dreyfus SE. *Applied Dynamic Programming*. Princeton, NJ: Princeton University Press; 2015.

[51] Littman ML. *Algorithms for Sequential Decision Making*. Rhode Island, RI: Brown University Providence; 1996.

[52] Pineau J, Gordon G and Thrun S. Anytime point-based approximations for large POMDPs. *Journal of Artificial Intelligence Research*. 2006;27: 335–380.

[53] Strens M. A Bayesian framework for reinforcement learning. In: *ICML*. vol. 2000; 2000. p. 943–950.

[54] Atkeson CG and Santamaria JC. A comparison of direct and model-based reinforcement learning. In: *Proceedings of International Conference on Robotics and Automation*. vol. 4. New York, NY: IEEE; 1997. p. 3557–3564.

[55] Dayan P and Hinton GE. Using expectation-maximization for reinforcement learning. *Neural Computation*. 1997;9(2):271–278.

[56] Szepesvári C. Algorithms for reinforcement learning. *Synthesis Lectures on Artificial Intelligence and Machine Learning*. 2010;4(1):1–103.

[57] Silver D and Veness J. Monte-Carlo planning in large POMDPs. In: Lafferty J, Williams C, Shawe-Taylor J, *et al.*, editors. *Advances in Neural Information Processing Systems*. vol. 23. Curran Associates, Inc.; 2010. p. 2164–2172. Available from: https://proceedings.neurips.cc/paper/2010/file/edfbe1afcf9246bb0d40eb4d8027d90f-Paper.pdf.

[58] Kaelbling LP, Littman ML and Cassandra AR. Planning and acting in partially observable stochastic domains. *Artificial Intelligence*. 1998;101(1–2): 99–134.

[59] Yao K and Gao J. Law of large numbers for uncertain random variables. *IEEE Transactions on Fuzzy Systems*. 2016;24(3):615–621.

[60] Breiman L, Friedman J, Stone CJ, *et al. Classification and Regression Trees*. London: CRC Press; 1984.

[61] Brügmann B. Monte Carlo Go. Citeseer; 1993.

[62] Gelly S and Silver D. Combining online and offline knowledge in UCT. In: *Proceedings of the 24th International Conference on Machine Learning. ICML '07*. New York, NY: Association for Computing Machinery; 2007. p. 273–280. Available from: https://doi.org/10.1145/1273496.1273531.

[63] Coulom R. Efficient selectivity and backup operators in Monte Carlo tree search. In: *International Conference on Computers and Games*. New York, NY: Springer; 2006. p. 72–83.

[64] Bubeck S and Cesa-Bianchi N. Regret analysis of stochastic and nonstochastic Multi-Armed Bandit problems. arXiv preprint arXiv:12045721; 2012.

[65] Gelly S and Wang Y. Exploration Exploitation in Go: UCT for Monte-Carlo Go. NIPS: Neural Information Processing Systems Conference On-line trading of Exploration and Exploitation Workshop, Dec 2006, Canada; 2006.

[66] Dean T, Basye K and Shewchuk J. Reinforcement Learning for Planning and Control. In: Minton S, editor. *Machine Learning Methods for Planning*. Burlington, MA: Morgan Kaufmann; 1993. p. 67–92 (CHAPTER 3). Available from: http://www.sciencedirect.com/science/article/pii/B9781483207742500081.

[67] Sutton RS. Temporal Credit Assignment in Reinforcement Learning. Ph.D. Dissertation, Dept. of Computer and Information Science, University of Massachusetts, Amherst, MA. 1985.

[68] Precup D. Eligibility traces for off-policy policy evaluation. In: *Computer Science Department Faculty Publication Series*; 2000; p. 80.

Chapter 3

Health system preparedness – coordination and sharing of computation, models and data

Oliver Bent[1]

Following on the previous chapter of machine learning (ML) advancements for optimisation and planning in global health policy decisions, here we discuss the mechanisms that enable such an approach through the global sharing of data, simulations and compute. This has been done through both institutional and crowd-sourced participation in global health challenges. Motivating a platform which enables trusted contributions and coordination of shared resources. Specifically demonstrating results applied to malaria endemic countries and the COVID-19 pandemic, and how such approaches may be extended towards future epidemic preparedness. Through this work, we aim to envision new approaches and systems to answer future global health challenges.

3.1 Computation

Computational Epidemiology [1] brings the afforded by computational methods in simulating challenging physical systems [2] specifically to those which dictate disease spread, intervention strategies and epidemic preparedness. At the core of this work, we have assumed a pre-developed epidemiological model, the practice of computational Epidemiology may marry together the processes of both developing an epidemiological model and designing a computational infrastructure to support the computational requirements of running model simulations. A canonical example of this for the reader is the work of the Institute of Disease Modelling (IDM), who have their own computational infrastructure (COMPS)* to support their own epidemiological model (EMOD) [3]. Instead we will focus on an infrastructure which may agnostically achieve the proposed tasks given any abstract epidemiological model format and the minimum required information for executing a model simulation.

To motivate the significance of the discussion of computation in congress with combating disease transmission, this year (2020) the Covid-19 pandemic received the

[1]University of Oxford, EPSRC Center for Doctoral Training in Autonomous Intelligent Machines and Systems, UK
*https://comps.idmod.org/

singular support of the largest volunteer supercomputer on earth [4] as part of the Folding@home project [5], a close relative of the Search for Extra Terrestrial Intelligence (SETI) @home project [6], which itself stopped distributing tasks after close to 20 years of using distributed computing for this further sighted research effort.[†]

More generically High-Performance Compute (HPC) clusters, which aggregate their compute nodes using high bandwidth networking and support a high-degree of inter-process communication, are ubiquitous across scientific research [7]. This work presents results from single instances to multiple nodes but there is no reason that this could not be permanently housed in a HPC setting, simply by the application justifying the compute being continuously available if necessary. Therefore these methods may be developed and used on individual laptops, shared servers and all the way up the scale to shared HPC systems.

In Chapter 2, we have extensively discussed the design of algorithms to perform inference to limit the impacts of computational expense. We will proceed to outline a design for a system to combine the insights provided by computational simulation with ML approaches to move towards programmatic relevance of the work.

3.1.1 A proposed infrastructure

This section provides an overview on the updated system architecture and deployment, as displayed in full in Figure 3.1.

3.1.1.1 Terms

To be precise in the language used we have laid out common system terms to be used in the proceeding sections:

- *Job*: A single model simulation.
- *Experiment*: A group of *jobs* with common properties, for example with same geographic location or user.
- *Intervention plan*: A list of intervention actions with their percentage coverages and deployment time.
- *Decision variables*: Disease control outcomes which we often related to rewards (Section 2.5.3) e.g. prevalence, CPDA, number of deaths averted, etc.

Along with the current implementation supporting five different types of user, including:

- *Policy maker*: A person who is required to make decisions based on epidemiological model-based evidence.
- *Expert modeller*: A user who makes and maintains the epidemiological models utilised for decision making.
- *Novice modeller*: A person who is not familiar with epidemiological model development, and is able to learn from interacting with deployed models.

[†]https://setiathome.berkeley.edu/

- *Machine learning practitioner*: A person who creates agent-based learning algorithms, for example as a competitor in Section 3.2 or the most likely group reading this work.
- *Admin*: A person who manages the whole technology stack presented.

3.1.1.2 Design principles

To design what we will term the platform for disease prediction, control and planning, and address the challenges outlined, the following seven paradigms were leveraged.

3.1.1.2.1 Microservices

Microservices is an architectural style inspired by service-oriented computing that has recently started gaining popularity [8]. For our specification, this provides a general design principle to breakdown the system into its component parts, then deployed as microservices, enabling flexible deployment and improved support for testing of interdependent components. The use of microservices also has the added benefit of permitting teams of developers to work more collaboratively, as they promote distinct segments of the codebase, exposing possibilities for readily sharing components of the work.

3.1.1.2.2 Frameworks

This involved the transfer of the original (Python) codebase to Java to take advantage of the Spring Boot framework.[‡] The Spring Boot framework enables the reuse of existing, well-vetted design patterns for microservice deployment [9].

3.1.1.2.3 Containers

Containers are a lightweight virtualization method for running multiple isolated Linux systems under a common host operating system and a new ecosystem has emerged around the Docker[§] platform to enable container based computing [10]. Docker is not extensively used in the HPC community and we will harness containers to deploy the developed microservices. Using tools like Docker makes it easy to deploy microservices at scale in heterogenous compute environments, achieving an approach which may be accessible, at whatever scale, for others to also replicate the work.

3.1.1.2.4 Cloud computing

Due to its wide accessibility as a commodity we can harness cloud computing as Infrastructure as a Service (IaaS) or Software as a Service (SaaS) to deploy system components at scale, in the location or specific data center best suited for the target users.

3.1.1.2.5 Application programming interface (API)

An API provides users with a library/module which exposes the microservices for their agent algorithms to interact directly with. At present, we have implemented a Python API which is deployed via github, and intended to be installed via the pip utility. This is extensively demonstrated with regards to the competition setting in Section 3.2.

[‡] https://spring.io/projects/spring-boot
[§] https://docker.io

3.1.1.2.6 Distributed ledger technology (DLT)

DLT deployed as a blockchain provides a decentralized and distributed platform that is useful in storing transactions aimed at improving trust, transparency and integrity among the participants in a fragmented ecosystem (e.g. disease modelling community). We can demonstrate the ability to store, share and maintain auditable logs and records of each step in the simulation process, showing how to validate results generated by computational workers [11]. This provides modellers with the ability to track who and how their models are accessed and used. If the compute is distributed, trust issues could arise on the results of a given job. The Hyperledger Fabric (HLF) [12] allows for a platform to perform validation of these results.

3.1.1.2.7 User interface

Consideration has been given for a front-end platform that allows sharing, calibration and running of models at scale, providing an access point for users who may not consider all of the lower level technical details of the microservices which make up this platform for disease prediction, control and planning.

3.1.2 Platform components

Using the introduced design approaches, the platform was framed into the following ten component classes contained within Figure 3.1. Each of these platform components could be deployed onto distributed computers running on a network.

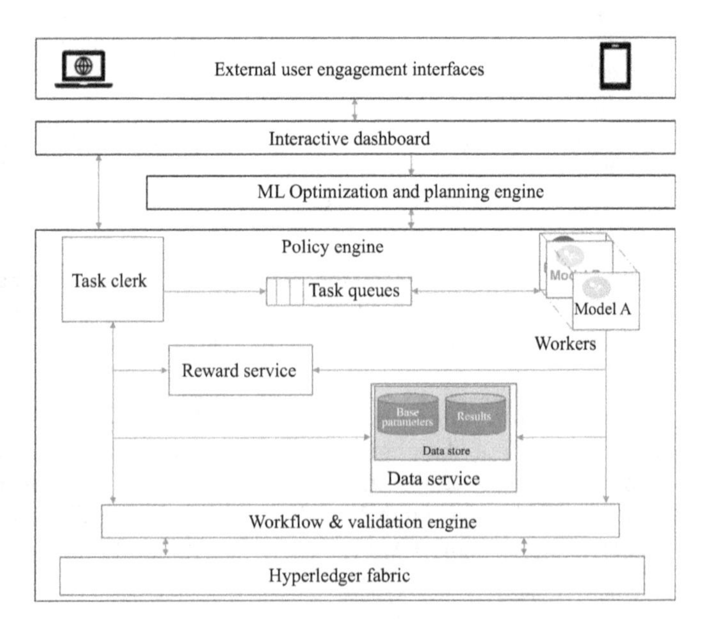

Figure 3.1 Conceptual architecture of a platform for ML-based prediction, control and planning from epidemiological models [13]

3.1.3 Performance results

This platform permits workers to be added (or subtracted) on demand, thereby scaling the compute available to evaluate intervention plans. In the original work, scaling was supported by running additional experiments independently on additional computers. In this case, all resources are partitioned across the machines which precluded past results from being shared across experiments at run time, and introduced duplicity in experiments which were run. Finally, idle compute resources on one machine were not available for use on another.

In our approach, the same compute could be pooled to support multiple experiments. Further, computational resources running on different networks could also be pooled together. This was demonstrated with workers deployed in a Kubernetes cluster in the United Kingdom, HPC Virtual Machines in a data centre in Canada, and on a bare metal machine in the United States of America. A total of 200 workers spread across these networks were used to re-run the experiments developed in [14] as shown in Figure 3.2. This result of using Bandit approaches for one-step control, later deemed agents in Chapter 2, questions what is the most cost-efficient combination of two interventions (ITN and IRS). The presentation is of a response surface where better policies based on simulation results are dark red. The principle takeaway being that as shown in [14], there are distinct recommendations provided by the independently implemented ML approaches in comparison with both the current and recommended policies.

3.1.4 Example: technical approach for competitions

This section will now outline the use of the design approach of the API to harness the results of epidemiological simulations at scale for the competitive outline of

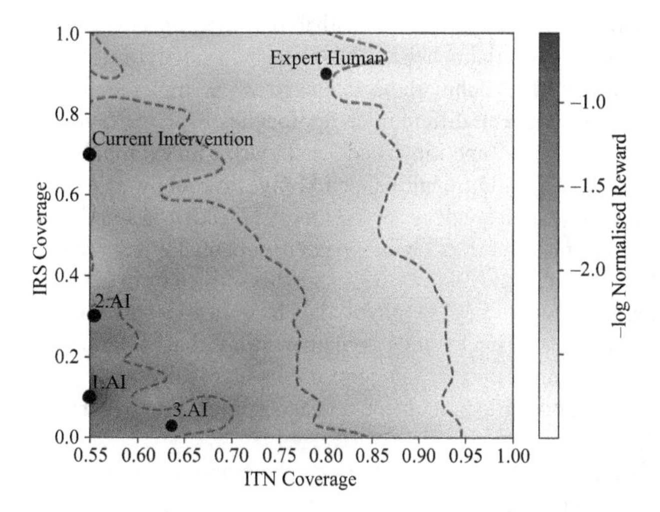

Figure 3.2 Combined results from [14] showing AI (optimal), current, and human expert recommended solutions

Section 3.2. Through pioneering projects like the OpenAI Gym [15], there is an established paradigm of how the reinforcement learning community expects to interface their agent algorithms with simulation environments. In this work, we implement the gym paradigm to provide a common interface for development and testing of agent algorithms to be submitted as solutions to the competition simulation environments.

Our formulation for these environments is novel in that it features a client–server architecture which supports sharing of resource intensive elements, as well as providing a mechanism to support transparency and trust [16], the elements of which have been discussed in Section 3.1.1.

3.1.5 Environment web service

To provide access to the environment in a scalable manner, a new client–server architecture was developed which hosts the environments as another model type which is exposed as a web service. The web service is written as a modular codebase, developed with the Python-based microservice *flask* [17], and leverages the *blueprint* [18] module's functionality to provide an extensive platform to concurrently host multiple environments. Each subdirectory contains the code and/or data required to implement the environment, with only a one-line stub in the base web service code to import that environment object into the common namespace. The surrogate model is used as the description of the epidemiological model for the purposes of planning and easily shared with multiple participants, providing a single competitive reference environment, which is archived and may be returned to, while not overly abstracting the problem.

3.1.6 Competition API

The interface for competitor interaction with the competition environments took inspiration from OpenAI Gym, launched in 2016 [19], a tool-kit designed to create a unified API in which many RL benchmark problems could be implemented, so allowing for direct comparisons between different RL approaches. OpenAI Gym has itself drawn inspiration from earlier competitions and benchmarks, for example the Arcade Learning Environment [20]. Additionally, OpenAI Gym currently provides extensions to gym environments in areas such as control theory [21,22] and simulated robotics [23]. Users may also implement their own gym environments for any arbitrary task via the unified Python-based gym API [24]. Although this service from OpenAI Gym was not used directly, the competition API mirrored its form to allow cross compatibility with existing reinforcement learning code implementations that participants may have had access to. This version of the environments used are not fully 'gym-compliant' as they do not currently import the gym object, and accordingly do not harness gym-defined *spaces* [16].

The API assumes that an agent class has been implemented by the end-user to allow their chosen RL algorithm to interface with the competition environments. The primary tasks of the environment are to produce observations given the current state and to ingest actions from the agents to produce reward signals, mapping the required functions for an environment class to our underlying surrogate models in the following manner:

Listing 3.1 OpenAI gym-like environment

```
nextstate, reward, done, _ = env.evaluateAction(env_action)
```

- env – Instantiates an environment object with a limited number of evaluations (runs)
- env.reset() – Starts a new sequence or policy evaluation of the environment object
- observation (nextstate) – This returns the users next state based on the evaluated action
- reward – The reward given by the environment associated with the performed action
- done – A Boolean value denoting if the end of the episode has been reached
- info (_) – This remains unassigned in the API implementation

With the existence of significant work to create open RL benchmarks, this enabled competitors to leverage much published code in exploring solutions to the Malaria environments. The novel environment for epidemiological models was intended to allow data scientists and RL practitioners to expand the type of ML approaches for policy learning and bring reinforcement learning to other real-world applications, outside of games and simulated dynamics. The environment is coupled with a language-specific API which implements a client library for participants to access methods which evaluate an intervention or a sequence of interventions (which we have referred to as a deterministic policy). These methods are analogous to the step method common in gym environments. A reward (a scalar value) is provided as a result for either the intervention or the policy. This reward value contains the signal, which enables a participant's algorithm to learn the utility in performing certain interventions as they engage in the challenge activity.

The API is designed to integrate with an agent class implemented by the end-user allowing their chosen RL algorithm to interface with the competition environments. The principle functionality of an environment is to produce observations and representative transitions based on actions as input from external agents, along with providing a scalar reward signal to these agents and mapping the required functions for an environment class to our underlying surrogate models in the following manner.

The first API was implemented in Python, however leveraging the *reticulate* package, it has been used to support R language competition activities as well. In the future it is envisioned to have implementations in multiple languages natively [16]. The web service architecture will make such libraries easy to implement and to support as HTTP functionality is pervasive in modern programming languages.

3.1.7 Example code

Code for random policy generation and submission using the competition API was shared for competitors to accelerate their developments and test submission of a solution. This gives an example of using the API to evaluate an entire policy in Listing 3.2.

Following the code example, a submission file may be generated using the competition library, by passing the CustomAgent class implemented by the competitor (eg. Listing 3.2) and generating a CSV file (example.csv) which is uploaded onto the site to generate a public leader-board score, through the EvaluateChallengeSubmission class of Listing 3.3.

These examples therefore serve to demonstrate that along with other examples on the competition website [25] and documentation, a few lines of code written by the competitor within a *generate* method were sufficient to participate.

Listing 3.2 Example random agent submission

```
from sys import exit, exc_info, argv
import numpy as np
import pandas as pd

from netsapi.challenge import *

class CustomAgent:
    def __init__(self, environment):
        self.environment = environment

    def generate(self):
        best_policy = None
        best_reward = -float('Inf')
        candidates = []
        try:
            # Agents should make use of 20 episodes in each training
                run, if making sequential decisions
            for i in range(20):
                self.environment.reset()
                policy = {}
                for j in range(5): #episode length
                    policy[str(j+1)]=[random.random(),random.random()]
                candidates.append(policy)

            rewards = self.environment.evaluatePolicy(candidates)
            best_policy = candidates[np.argmax(rewards)]
            best_reward = rewards[np.argmax(rewards)]

        except (KeyboardInterrupt, SystemExit):
            print(exc_info())

        return best_policy, best_reward
```

Listing 3.3 Submission file generation using the competition library

```
EvaluateChallengeSubmission(ChallengeSeqDecEnvironment, CustomAgent,
    "example.csv") #How scoring files generated by participants
```

3.1.8 Related work

There are a variety of methods and platforms that provide scalable compute resources and infrastructures which can be applied to support decision making from epidemiological models.

Deep Learning as a Service (DLaaS) [26] is a unified platform that provides developers, data scientists/engineers, with a mechanism to use popular deep learning libraries (e.g., Caffe, Torch and TensorFlow) in one-go, with minimal effort, helping to minimise unnecessary effort to configure the infrastructures needed for deep learning based computations. To do so, DLaaS leverages the benefits of cloud service and API economies, enabling the training of deep learning models using one or more deep learning libraries with user specified amounts of resources or budget, as well as proving additional utilities for the user like control of the training progress.

Like DLaaS, MLaaS is another architecture for a ML as a service type of model. It provides compute infrastructure and ML libraries to create flexible and scalable models [27]. MLaaS can support multiple data sources and create a variety of models using various algorithms, parameters and training sets via web-based APIs. However, MLaaS's architecture is currently designed for specific analytical libraries (limited to predictive modelling), making it difficult to extend it for our use in our infrastructure.

A similar concept has been discussed in other numerous and varied approaches in the expanded literature e.g. [28–30]. PredictionIO [28], for example, is the first open standard architecture that combines multiple Machine learning processes into a distributed and scalable system where each ML process can be accessed via APIs and a dedicated GUI as well, including integration with a variety of big data platforms (e.g. in Hadoop and Spark). PredictionIO also provides a template gallery, allowing users to reuse templates for various ML applications. Another open-source platform that compiles popular statistical analysis libraries of R is OpenCPU [30]. However, OpenCPU lacks key features such as flexibility, ease-of-use, and economy of scale and hence it is difficult (or requires expertise) to configure (including acquisition of the requisite data), execute, and interpret data, models and results.

Following this detailed overview of an infrastructure and systems approach for prediction and control from epidemiological models, the next chapter will consider the results of exposing the development of ML approaches in optimisation and planning to open competition.

3.2 ML competitions for health system preparedness

Competition is a theme tightly woven into the advancement and demonstration of AI. Significant efforts have been taken by organisations to demonstrate publicly above-human performance machine intelligence in games. Notable examples include Deep Blue (Chess) [31], Alpha GO (Go) [32] and Alpha Zero (Chess, Shogi, Go) [33]. In addition, grand challenges have been posed as open competitions (often with significant funding or investment) such as the Neflix prize ($1M) [34] and the DARPA

Grand Challenges ($Multi M) [35]. These efforts all raise awareness and stimulate immediate and further research about their respective topics.

There is also an emerging class of problems promoted through dedicated ML platforms such as Kaggle‖ and CodaLab,¶ inviting engagement from a global community of Data Scientists to engage in problems that matter [36]. ML challenges are not always focussed on algorithmic advancements or theoretical contributions. They tend to generate solutions of a more practical nature, specific to a proposed problem, rather than a generalisable result, with the aim of the challenge to insight the imagination of others to pose novel solutions to the problem.

Of greatest relevance to our objectives in epidemiological control, we look to the subset of competitions which are based on reinforcement learning (RL). The first RL challenges were launched in 2006 [37] out of a necessity to create benchmarks and common environments to support results, sharing and publication. This use of the competitive framework has harboured a competitive spirit in demonstration and publication of research in RL, especially in dominant topics such as Deep Reinforcement Learning, where the Arcade Environment [20] for Atari 2600 games is a commonbench mark in Deep RL, in much the same way that the benchmark data set Imagenet [38] shaped Deep learning research in Computer vision.

A set of problems may have arisen in the RL domain, from a focus on reporting benchmark performance, but no standardised evaluation or reporting framework [39] has accompanied such benchmarks and this is potentially damaging for research, leading to concerns regarding the reproducibility, reusablility, and robustness of Reinforcement learning algorithms or approaches [40].

Direct comparisons between different RL algorithms are inherently difficult as the authors of algorithms may spend a lot of time tuning them to work well in specific domains and overfit to environments [41], even when they are (by design) supposed to learn in an active manner from zero experience. What is required is a way to perform unbiased comparisons across various learning tasks, as was proposed early on in RL competitions. Unbiased comparisons for static data sets are a well-established practice in other areas of data science but the interactive nature of RL means that no fixed data sets or experience are possible. However, more standard notions of Machine learning evaluation (for static data sets), based on a train-test split [42], may still be applied to entire learning environments, whereby a competitor develops an algorithm or learning agent on a limited set of representative training environments, and then reports learning performance on another set of similar, but non identical testing environments. This principle is adopted for many RL competitions, often over multiple generalised domains. Each domain corresponds to a different type of problem. Considering, for example the well-used example of helicopter control [43], training and testing are performed on specific domain instances of the problem, such as helicopters with different physical characteristics. The competition format can then provide an unbiased comparison between algorithms, that truly tests their ability to learn [44].

‖https://www.kaggle.com/
¶https://competitions.codalab.org/

Our motivation is hence not just to tackle a new class of problem through RL, but to frame the entire problem in a competition. This provides the benefit of unbiased evaluation of new approaches to the epidemiological domain, which may provide novel answers.

The rest of this chapter will report on the mechanics of executing a global series of RL competitions, motivated by the sequential decision-making task for optimal malaria control, abstracted into a set of reusable environments through the techniques for learning a model developed in Chapter 2. The competition environment code-base presented here is open source, along with accompanying educational material [16] and the infrastructure required to run the competitions at scale is discussed in Section 3.1.

3.3 Planning from learnt models

Planning is defined formally as policy improvement based on Agent interaction with a model of the environment [45]. This is especially important as we may not be learning by interacting in real-time with the 'real' environment for challenges in epidemiological decision making. The process of planning necessitating that plans should be developed in silico, based on model evidence, before implementation. We propose that a competitive framework may be used to perform the task of planning in a distributed manner, each competitor developing an agent to interact with a model of the 'real' environment for the epidemiological decision making task. A learnt surrogate model of underlying epidemiological models is what is shared with competitors. The reasons that a learnt surrogate model is developed for sharing are due to the following:

- Lowering the computational cost associated with directly running epidemiological models.
- Abstraction allows for ease of sharing without requiring expert knowledge to run a model.
- Quantification of uncertainty may be consistently applied.
- Combination of insights from multiple epidemiological models may be generated.

There are several examples of competitions and leading AI research achievements where it is thought access to the latest computational hardware [46], along with highly motivated specialised teams [47], led to the most successful contributions to challenging problems [48]. Although all global challenges will necessarily not have a substantial monetary budget or team of experienced researchers tackling the problem. Through sharing an abstracted model with limited numbers of evaluations and requiring little computational resource may lower the barriers to contribution and increase the numbers of researchers able to work on challenging problems [49,50]. Our goal is to generate examples of distributed policy learning from epidemiological environments, not generating one single strong learner or team, but achieving strength through heterogeneity and lowering the barriers to access or contribution.

As in Figure 3.3, an integration of planning into the life-cycle of learning allows us to generate a union with the objective of control, in addition to the control policies learned through direct RL methods. This life-cycle of learning was initially posed in

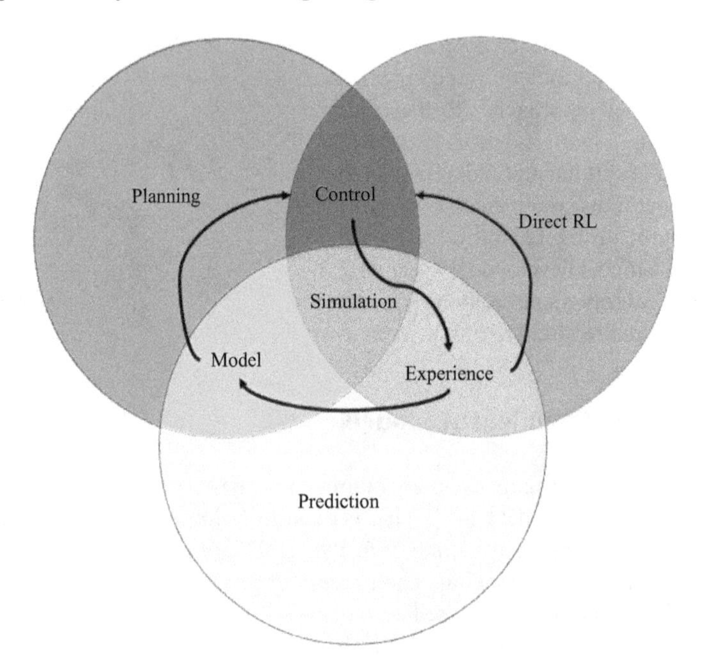

Figure 3.3 Inspired by the work Planning, Learning and Acting [52]. Blue*: Learning a value function and policy based on simulated experience from an epidemiological model. Yellow: Prediction and necessarily learning a model based on epidemiological simulations and experience.* Red*: Planning from a model to learn a policy which in turn may be used for further simulations to generate more experience.*

the Dyna architecture [51], an integrated approach to learning, planning and reacting. High performing algorithms or agents to the planning task may also learn from further experience or direct interaction with epidemiological models, substituting planning based learning to performing direct RL and consequently generate further experience. In this way, it can be seen that the process of learning may be entirely deployed in an iterative manner for further improvements, based on the principle of sharing epidemiological models as environments, along with the straightforward deployment of learning agents which algorithmically look to solve these environments, either directly or via model-based approximations. We will now proceed to further detail the instantiations of model-based planning based on the set-up of a global series of ML competitions.

3.4 KDD Cup 2019 and other competitions

The KDD Cup is a renowned ML competition which began in 1997 [53] and has produced several notable examples of benchmark data sets and ML challenges

[34,54–56]. This work presented here was accepted as the KDD Cup—Humanity RL track competition 2019 [25]. Over a 4-month period the competition attracted 248 teams, 296 competitors and a total 735 submissions, along with sponsorship from the Association for Computing Machinery (ACM) for prize money totalling $25k. The competition was advertised to competitors as follows:

"*We are looking for participants to apply ML tools to determine novel solutions which could impact malaria policy in Sub Saharan Africa. Specifically, how should combinations of interventions which control the transmission, prevalence and health outcomes of malaria infection, be distributed in a simulated human population....*"

This competition was uniquely framed as a Reinforcement learning problem, the first in the history of the KDD Cup since its inception in 1997 [57]. Participants were expected to submit high performing solutions to the sequential decision making task shown in Figure 3.4. For this competition, actions (defined as coverages of Insecticide Treated Net (ITN) and Indoor Residual Spraying (IRS) interventions receive stochastic and delayed rewards, which are resource constrained by both the monetary and computational costs associated with implementing and observing the impact of an action. The environments used by the competitors were developed as sequential surrogate models as outlined in Chapter 2. Due to the open nature of the competition, submissions were encouraged from participants who may not have had previous experience in RL problems, with supplementary material, tutorials and Q&A sessions provided and administered through the competition platform.**

This activity was executed as a three-phase, 4-month long activity, featuring multiple distinct environments as well as a single-blind submission. During the first two phases, the environments were used for both training and evaluation on a shared leader-board, while in the final proving phase, submissions were evaluated on an environment unseen by the participants. The reward utilised for the activity was the measure of cost effectiveness of the considered intervention policies in terms of Cost Per DALY Averted (CPDA).

We now proceed to provide further detail into the competitive process and its evaluation with regards to submission of algorithmic approaches or agent towards our proposition of distributed planning for the task of epidemiological decision making.

3.4.1 Evaluation framework

In distinction to previous RL competitions [37,44] which served as guidance for the work, a new evaluation framework was proposed for the KDD Cup competition [58], which required that the final evaluation was performed on a 'held-out' environment to which competitors did not have access. For this purpose, it was necessary to engineer the infrastructure and framework to accept code submissions on unseen environments requiring:

1. That competitors adhere to specific code submission requirements.
2. There was an automated and secure infrastructure to execute these submissions and report on results.

** https://compete.hexagon-ml.com/practice/rl_competition/37/

Action

$A_S = [a_{ITN}, a_{IRS}]$ where $a_{ITN} \in [0, 1]$ and $a_{\mathrm{IRS}} \in [0, 1]$

Reward

$R_\pi \in (-\infty, \infty)$

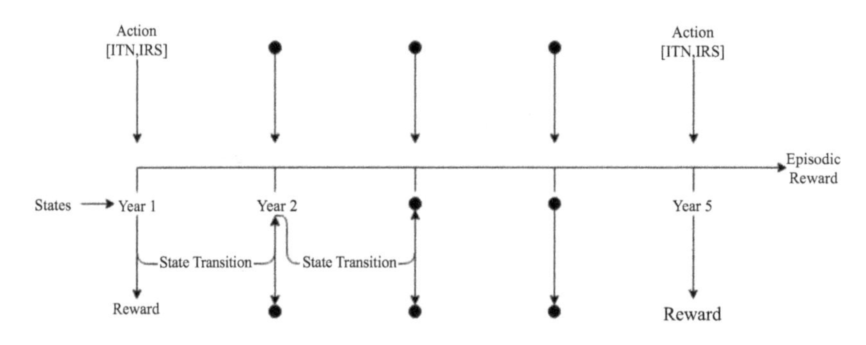

Figure 3.4 Sequential decision making task – competition framing

The evaluation framework was applied to three phases of competition which will be detailed before discussing further submission and scoring.

3.4.1.1 Feedback phase – test environment

The feedback phase was used to provide the chance for competitors to learn, read the 'Getting started code' examples, follow tutorials, blogs and engage in the discussion forum, principally consisting of running the example code given previously in Section 3.1 specifically Listing 3.1.7 and generating a submission file to post a public leader-board score. After a period of a month over which approaches were iterated, improved and refined, a selection was then made such that the top scoring 100 submissions on the public leader-board were able to proceed to the next phase (for details on scoring, see Section 3.4.2). Those competitors with a top 100 score were required to make a valid code submission to proceed to the following check phase. The upper bound of 100 submissions also provided a reasonable limit to the instantiations of code which would have to be run, along with providing a check that the formatting of the submissions was clear to competitors before their final evaluation submission. The test environment used was in fact an analytic function, based on the Bird function [59], taken from the optimisation literature. This multi-modal function can mimic some of the dynamics observed in epidemiological simulation models but be run efficiently at scale on a new web-service for all of the competitors. The additional practical reason for this was that further testing and validations were being run on the sequential surrogate models used in latter stages. In order to generate a sequence from the Bird function and conserve a notion of resource efficiency in the reward signal, this took the functional form of Listing 3.4 (where $x \rightarrow a_{ITN}$, $y \rightarrow a_{IRS}$ and $ff(x, y) \rightarrow R$).

Listing 3.4 Sequential bird function for test environment

```
def ff(x,y):
  x = np.clip(x,0,1)
  y = np.clip(y,0,1)
  x *= -10
  y *= -10
  return - random.uniform(.9,1.1)*(sin(y)*exp((1-cos(x))**2) +
       cos(x)*exp((1-sin(y))**2) + (x-y)**2)

 def ffprime(x,y,oldx,oldy):
  return ff(x*(1-oldx), y*(1-oldy))
```

3.4.1.2 Check phase – prove environment

The top 100 participants ranking on the public leader-board from the feedback phase were taken forward to have access to the proving environment. With no public leader-board in the check phase, a private dashboard logs a competitor's progress, so that approaches may be iterated upon, while the results of this are not shared with other competitors. Unlike the feedback phase, competitors lose all visibility of what other competitors are doing. To progress to the final verification phase competitors had to submit their final code and dependencies to be run in the verification phase. The prove environment was generated based on samples from a low-resolution mechanistic malaria model. This produced the appropriate sequential simulation surrogate model structure from thousands of simulations selected, based on Monte Carlo Tree Search (MCTS) runs from the epidemiological model, specifically using the method of GP-UCB-MCTS (outlined in Section 2.5.13). To achieve generality in their approach each competitor was expected to develop approaches with good performance in both the test and prove environments, before having their submissions ranked on the final unseen evaluation environment.

3.4.1.3 Verification phase – evaluation environment

Final submissions for the verification phase were run, validated and verified by the organisers on an evaluation environment. The evaluation environment was modelled based on high-resolution observations from the same mechanistic malaria model scenario as used for the check phase, additionally a hidden environment was used for further blind validation of the submissions. The hidden environment was a combination based on combining all samples from both the prove and evaluation environments and relearning a new sequential surrogate model. The top ten participants of the competition were honoured with prizes, and emphasis was placed by the organisers on fair evaluation, including provision for detection of any anomalous behaviour in submissions, which would fall outside the rules of the competition. A final ranking and leader-board of all the submitted solutions was openly posted on the competition platform. Figure 3.5(b) gives the cumulative normalised scores of the top ten submissions over all the three stated environments (including the fourth hidden environment).

3.4.2 Submission and scoring

We now proceed to provide more details on the submission and scoring of each phase. Each submission was evaluated based on ten runs, with each run consisting of ten episodes, and the learnt greedy policy over a run was submitted along with its episodic reward. This is indicative of the limited experience for any algorithm to learn a high performing policy from direct query of the underlying simulation models, although participants had access to a computationally less expensive surrogate model description of the underlying epidemiological models. This framing of the problem is novel compared to existing RL benchmarks, which do not typically penalise the number of evaluations in reporting the performance of an approach. There is no sharing of information allowed between runs, so every submission learns from zero initial experience during each run. In the live challenge format (Section 3.4.3) only a single environment was used, and participants' scores were determined from Comma Separated Value (CSV) files generated by their algorithms. In the multi-phase format, participants submitted the same formatted CSV file except for in the last phase.

In the last phase of the multi-phase format, participants submitted a compressed file containing their algorithm code for an independent assessment on remote machines. Code submissions were limited to a list of standard Python libraries which could be pre-installed on the infrastructure, limiting not only the requirement of the machines to be connected to the Internet for package installs but also any malicious behaviour which may be hidden in the submissions. If submitted code could not be run, the team was contacted. If minor remediation or sufficient information was not provided in order to run the code, the submission was ultimately removed, though in the 2019 KDD Cup competition this was only the case for a single submission. The reported score was taken as the median of the episodic rewards from the ten greedy policies learnt over the ten runs.

Figure 3.5(a) shows the distribution of CSV file submissions over the competition duration, while Figure 3.5(b) gives insights into the relative performance of the top ten teams over the three stages, ultimately it is not clear that single solutions were the best performing for all environments, though in Section 3.4.3 we interrogate this further with regards the effects of tuning an agent submission to a single environment.

3.4.3 Other competitions

In 2019, culminating with the NeurIPS conference [60], a series of similarly themed events were performed at conferences with different audiences. In all cases, competitions have been framed in the same manner, where the participants have been asked to write an agent algorithm which interacts with the challenge environment in fixed number of samples, learns and recommends a policy (the sequence of interventions) identified by their solution as best. Each intervention has its own behaviour and parametrisation in the model as they act on individual simulated humans in the environment.

In the framing, interventions are defined as the coverage of Insecticide Treated Nets (ITN) and distribution of IRS Indoor Residual Spraying (IRS) at given times in a 5-year period. The state of the environment changes as a result of the interventions which have been performed, as well as the dynamics which have been implemented.

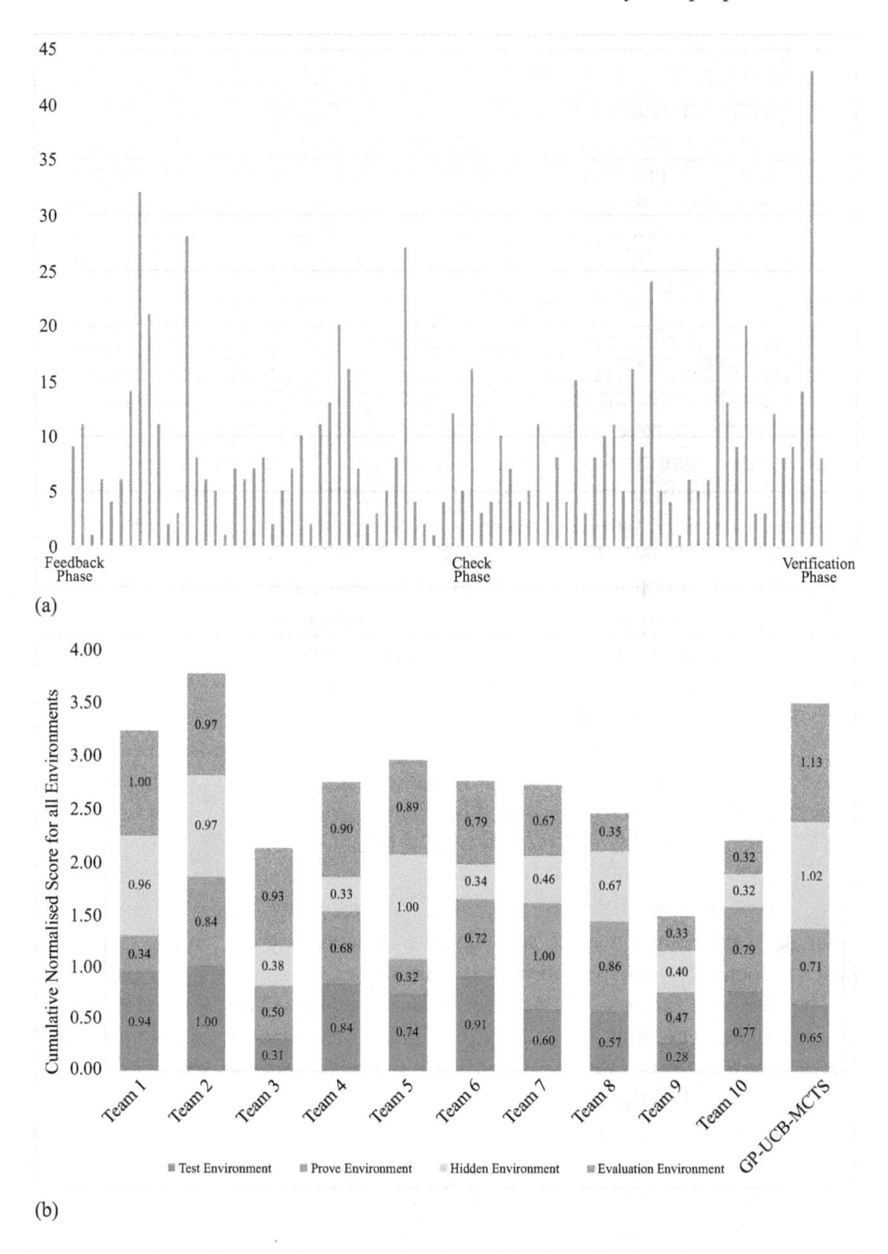

Figure 3.5 KDD Cup analysis. (a) Number of daily CSV file submissions up until the submission of code for verification and final scoring. (b) Cumulative normalised scores of top ten teams on all competition environments (additionally included for comparison the relative performance of algorithm GP-UCB-MCTS detailed in Section 2.5.13).

A state signal from the underlying environment is presented as a variable from which the algorithm may learn a policy. Accordingly, the rewards derived from interventions were learned as a function of the interventions selected, and the time in which they were selected (see Figure 3.4). The participant's code was given a fixed compute budget of 2,000 samples from the environment which could be used for learning.

The problem was framed at a level where assert is accessible to the community. With several possible approaches to the problem, its framing will remain deliberately non-prescriptive. Also, due to our API abstraction for tackling the problem in a ML guise, little domain knowledge is required of malaria transmission and control, or its simulation (which was left to the competitor's interests).

Because of the selection of the client–server approach to access the environment and subsequently accessing the remotely hosted environments, it is not considered a distinct advantage to have considerable local compute to run the models in a competitor's considered algorithm. Such model complexity would be discouraged by design as it is unlikely a competitor can generate enough data in the competition time frame to train such a model, with each submission taking hours to run. As all models are learning from surrogates of physics-based simulators developed by epidemiological modellers, the results found should be interpretable by the malaria modelling community. The complexity of a particular strategy should also be confined to real-world feasibility, which is ensured through constraints on timing, cost and number of actions which may be performed. All of which for simplicity were contained in the evaluation metrics, to generate a reward signal.

To date, the malaria scientific community has no clear answer for an optimal strategy in a particular location, which we believe may efficiently be found through computationally intelligent solutions, and challenges like ours. As such, submitted algorithms may act as a black-box in terms of application to epidemiological decision-making problems, with regards to this framing, algorithmic results which outperform other strategies in simulation may be ranked and presented. In continuation from the challenge, these results may be subsequently validated by epidemiologists, with a long-term focus on implementation of the results.

The data used for these competitions is stationary based on the state observation with stochastic returns. Also, due to the high dimensionality of combinations of actions consisting of an intervention strategy, the possible set of solutions cannot be exhaustively searched through the platform, so there is little concern for over-fitting the possible data in the environment. Specifically, competitors will have to grapple with the paradigm of exploration and exploitation of high performing strategies under limited resources in terms of both compute and time. We hope that given the stochastic multi-dimensional nature of the problem, this will discourage over-tuning of any particular model for a simulator due to the noisy results returned.

3.4.3.1 Indaba Hackathon

The Deep Learning Indaba Conference [61] was used as a chance to engage directly with 40 teams over the course of a conference in Nairobi, Kenya. Competitors used solely the evaluation environment (Section 3.4.1.3) to post submissions to a public leader-board published around the conference venue. This live competition setting had different incentives as the reward for participation was principally knowledge based, the problem framing being used as an instructional environment for reinforcement learning. The top three submissions at the end of the conference were honoured in the closing ceremony receiving certificates and small prizes. It is clear from Table 3.1 that submissions obtained a far higher performance through direct tuning on the previously hidden evaluation environment. Daily lab sessions in excess of 2 h interacting with the teams suggested that techniques tended to focus on tuned improvements on Q-learning and Policy Gradient-based approaches as an Introduction to RL [61].

3.4.3.2 NeurIPS

This particular competition did not gain the same number of submissions, organised as a live competition it used the same concept and is included for completeness. The event differed from the other two competitions in a lack of funding and incentives for competitors such as prizes and it serves as an example that for substantial competitive submissions, additional considerations should be made to ensure attention in a busy conference schedule over a one week period. Essentially, this was not an event which could compete for attendees time at the venue and it was reportedly difficult to attract attention over the course of the event. The organiser was unable to travel to the venue, running operations remotely and this did not elevate some of the challenges in gaining submissions. The work could still be presented at the conference competitions workshop but with four RL competitions competing for participants' time, all had a lack of engagement compared to what was expected (typically each attracting less than 10 submissions), this work for comparison registering 8 teams, 6 competitors and 2 submissions [60]. Framed as a live, multi-day activity, this challenge format was executed as a single phase event with data from a mechanistic malaria model used to provide the input loaded by the web service. Synthetic data was used to bootstrap

Table 3.1 Relative scores of top ten submissions for the same environment when agent code is submitted for evaluation on an unseen environment (KDD Cup) and generated CSV files are uploaded by competitors for a live environment (Indaba Hackathon and NeurIPS)

	Top 10 Scores 2019 Competitions		
Position	KDD Cup	Indaba Hackathon	NeurIPS
1	27.513	44.997	23.754
2	26.572	43.575	–
3	25.581	42.436	–
4	24.791	38.344	–
5	24.552	36.564	–
6	21.686	33.579	–
7	18.369	33.232	–
8	9.756	31.760	–
9	8.987	30.411	–
10	8.808	30.223	–
Last (position)	−2.548(30)	10.486(17)	23.754(1)

the malaria model for this activity and the model output was an approximation of the problem domain based on epidemiological model simulations. The reward signals in this activity were the parasite prevalence, and a measure of cost effectiveness of the considered intervention policies (Cost Per DALY Averted (CPDA)).

3.5 Collaboration from competition

At the start of this chapter, we acknowledged that competition inherently serves to increase the pace of research and innovation, given the ground work we have laid in a demonstration of how competition may be harnessed towards achieving this for planning from epidemiological models. We will now include an extension to the family of approaches within ML which maximise algorithmic performance via collaboration, sharing or more commonly the ensemble. Ensemble methods improve the predictive performance of a single model by training multiple models and combining their predictions [62], this combination will be considered as a weighted vote learnt by the ensemble method [63]. Through a principled combination of competitive submissions, we now also classify shared epidemiological models as a competitive submission, in the spirit of the received competitive agent algorithms which we have so far focussed on in this chapter.

3.6 Example: analysis of successful competition approaches

In this section, we provide an overview of the top approaches taken by competitors, these were influenced by material shared to competitors, which focussed on the techniques of Q-learning [64], and Policy gradients [65] which we are yet to discuss and also a Genetic algorithm [66] method, all as examples using the competition environments to generate submissions for the stages described in Section 3.4.1.

Algorithmic innovations looked to increase the sample efficiency of the approaches and improve performance within the limited number of episodes for the competitions. The main way this was achieved was by mirroring the work used to generate the environments (unknown to the competitors) through using model estimations centred on Gaussian process surrogates and Bayesian optimisation to maximise episode reward based on policies learnt during the training runs from the provided environment. While such approaches would introduce additional approximation error, based on approximating an approximation, under the problem framing these approaches proved to be the most successful model-based learning approximations.

The repositories are included for Figures 3.6(a),[††] 3.6(b),[‡‡] 3.6(c),[§§] and 3.6(d),[‖‖] each contains the agent code from which their approaches were evaluated and a workshop poster for the KDD cup 2019 workshop [58].

[††] https://github.com/bach1292/KDD_Cup_2019_LOLS_Team
[‡‡] https://github.com/quanjunc/KDD-CUP-RL-8th-solution
[§§] https://github.com/federerjiang/KDD2019-Humanity-RL-Track-
[‖‖] https://github.com/luosuiqian/submission

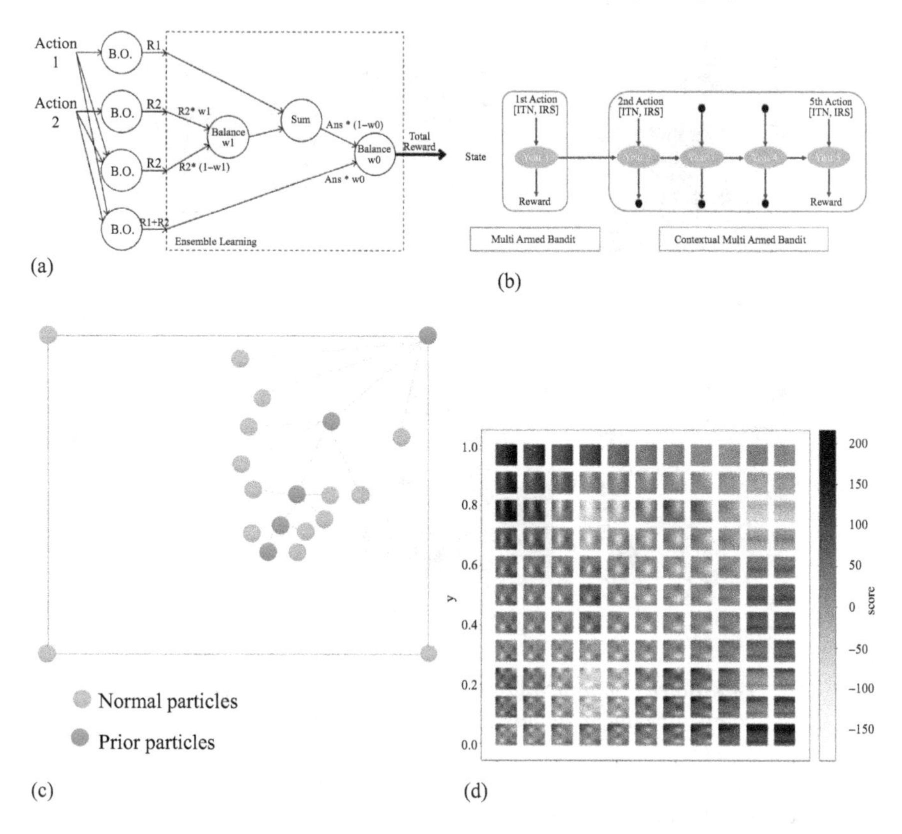

Figure 3.6 *Sample of successful approaches for the malaria competition sequential decision making task; (a) Bayesian optimisation with a forward boosting network [47], (b) Thompson sampling bandit method, (c) fast search with prior particles (FSPP), and (d) gridded random search.*

We now consider the results of submitted approaches for the verification phase. In Figure 3.7, we see consensus in actions chosen for each year based on the agent samples. High-frequency actions tended to balance the components of ITN and IRS interventions with reasonable frequencies of selections of 'bang-bang' policies based on one of the components of the actions being either on-or-off e.g. $[0, 1]$, $[1, 0]$ or $[0, 0]$ in [ITN, IRS], curiously this is an optimal control strategy for infectious disease transmission with Nonpharmaceutical Interventions under linear costs (Section 2.4.4.2) [62]. This visual description combines information to generate a unique set of insights that policies for malaria control based on the used epidemiological models have a clear pattern in algorithmically searched solutions, qualitatively suggesting that cost-efficient strategies should inversely weight ITN deployments to IRS coverages, with greater initial IRS coverages in this particular model environment.

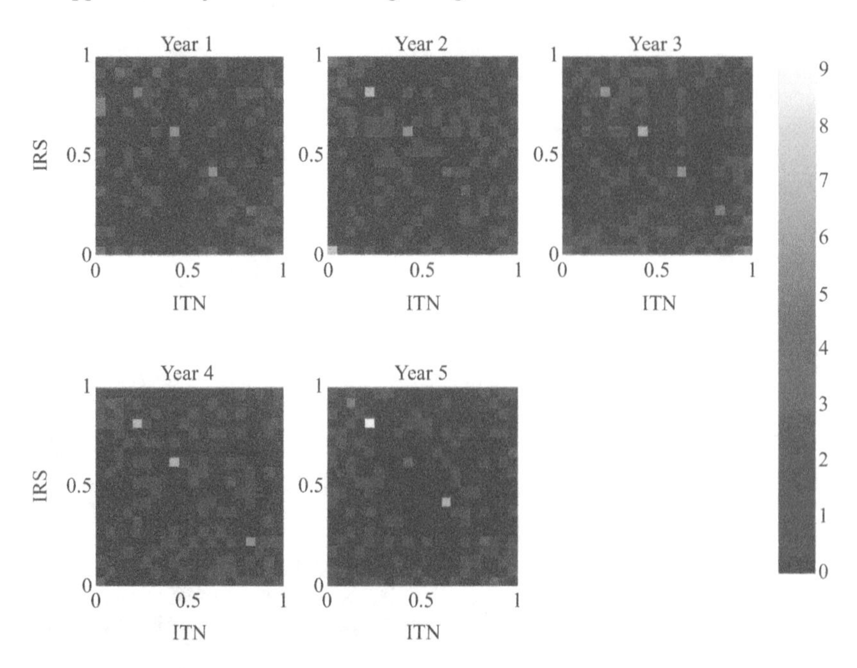

Figure 3.7 Two-dimensional histograms for the yearly actions generated by competitor agent submissions during the verification phase (Section 2.4.1.3)

3.6.1 Conclusions on competitions for health system planning

In the health domain, competitions such as Learning to Run [68] and AI for Prosthetics [69] competitions created by the Stanford Neuromuscular Biomechanics Lab promote the creation of a controller that enables a physiologically based human model to navigate complex obstacles within a short duration of time and simulate changes in gait that occur after receiving a prosthesis in order to develop approaches that may be generalized to health conditions impacting movement. Additionally, the Personalized Medicine: Redefining Cancer Treatment competition [70] set the task of classifying clinically actionable genetic mutations. More specifically this task may be stated as: given the genetic mutations of a cancer tumour, classify the mutations which accelerate tumour growth as separate from the neutral mutations.

In contrast, we developed a novel competition applying ML to the global challenge of malaria elimination, using agent-based models to learn from simulation and demonstrate novel strategies for the elimination of malaria in an endemic location. We believe that this challenge builds upon interest across the ML community for interpretable, scalable, and repeatable models which solve real-world challenges, with higher level motivation provided by the promise of ML assisting in the elimination of a disease, which for centuries has been missing the medical breakthrough to achieve this goal. Such a challenge applying ML to epidemiological models, specifically malaria, has never been run to the knowledge of the organisers. This applied form of ML treating such models as a simulator is also novel.

3.6.2 Human-in-the-loop

The environment abstractions of learnt model approximations presented have been adequate for promoting engagement, evaluation and demonstrating potential for new benchmarks and approaches and there is a critical need to support human-in-the-loop evaluation for any final application or decision, which could be made from such a process. To this end, the evaluation of a domain specialist or policy maker is required, effectively pairing an intelligent software agent not only with the environment and compute to explore new policies but also with direct feedback from a domain specialist. By providing a scalable platform that supports such evaluations, this could encourage the community to benchmark performance on tasks outside of the traditional RL playground and lead to a better understanding of AI guided decision making processes [71].

References

[1] Gorder PF. Computational epidemiology. *Computing in Science & Engineering*. 2009;12(1):4–6.
[2] Feynman RP. Simulating physics with computers. *International Journal of Theoretical Physics*. 1982;21(6/7):467–488.
[3] Bershteyn A, Gerardin J, Bridenbecker D, *et al.* Implementation and applications of EMOD, an individual-based multi-disease modeling platform. *Pathogens and Disease*. 2018;76(5):fty059.
[4] Kratzke N. Volunteer down: How COVID-19 created the largest idling supercomputer on earth. *Future Internet*. 2020;12(6):98. Available from: https://www.mdpi.com/1999-5903/12/6/98.
[5] Beberg AL, Ensign DL, Jayachandran G, *et al.* Folding@Home: lessons from eight years of volunteer distributed computing. In: *2009 IEEE International Symposium on Parallel & Distributed Processing*, IEEE, 2009. p. 1–8.
[6] Anderson DP, Cobb J, Korpela E, *et al.* SETI@ home: an experiment in public-resource computing. *Communications of the ACM*. 2002;45(11):56–61.
[7] Alnasir JJ. Distributed computing in a pandemic: a review of technologies available for tackling Covid-19. *ADCAIJ: Advances in Distributed Computing and Artificial Intelligence Journal*. 2020:11:19–43.
[8] Dragoni N, Giallorenzo S, Lafuente AL, *et al.* Microservices: yesterday, today, and tomorrow. In: *Present and Ulterior Software Engineering*. New York, NY: Springer, 2017, p. 195–216.
[9] Walls C. *Spring Boot in Action*. Shelter Island, NY: Manning Publications, 2016.
[10] Jacobsen DM and Canon RS. Contain this, unleashing docker for HPC. In: *Proceedings of the Cray User Group*, 2015, p. 33–49.
[11] Bore NK, Raman RK, Markus IM, *et al.* Promoting distributed trust in machine learning and computational simulation. In: *IEEE International Conference on Blockchain and Cryptocurrency, ICBC 2019*, Seoul,

Korea (South), May 14–17, 2019, 2019. p. 311–319. Available from: https://doi.org/10.1109/BLOC.2019.8751423.

[12] Androulaki E, Barger A, Bortnikov V, *et al. Hyperledger Fabric: A Distributed Operating System for Permissioned Blockchains*. CoRR. 2018. abs/1801.10228. Available from: http://arxiv.org/abs/1801.10228.

[13] Wachira C, Remy SL, Bent O, *et al.* A platform for disease intervention planning. In: *2020 IEEE International Conference on Healthcare Informatics (ICHI)* – to appear. IEEE, 2020.

[14] Bent O, Remy SL, Roberts S, *et al.* Novel Exploration Techniques (NETs) for malaria policy interventions. In: *Proceedings of the Thirty-Second AAAI Conference on Artificial Intelligence, (AAAI-18)*, 2018. p. 7735–7740. Available from: https://www.aaai.org/ocs/index.php/AAAI/AAAI18/paper/view/16148.

[15] Brockman G, Cheung V, Pettersson L, *et al.* Openai Gym, 2016. arXiv preprint arXiv:160601540.

[16] Remy SL and Bent O. A global health gym environment for RL applications. In: *NeurIPS 2019 Competition and Demonstration Track*. PMLR, 2020. p. 253–261.

[17] Grinberg M. *Flask Web Development: Developing Web Applications with Python*. Sebastopol, CA: O'Reilly Media, Inc., 2018.

[18] Perras J. *Flask Blueprints*. Birmingham: Packt Publishing Ltd, 2015.

[19] OpenAI. OpenAI Gym Environments. Accessed 2019. https://gym.openai.com/envs/.

[20] Bellemare MG, Naddaf Y, Veness J, *et al.* The arcade learning environment: an evaluation platform for general agents. *Journal of Artificial Intelligence Research*. 2013;47:253–279.

[21] Moore AW. *Efficient Memory-Based Learning for Robot Control*, Ph.D. thesis, Computer Laboratory, Cambridge University, 1990.

[22] Barto AG, Sutton RS, and Anderson CW. Neuron-like adaptive elements that can solve difficult learning control problems. *IEEE Transactions on Systems, Man, and Cybernetics*. 1983;5:834–846.

[23] Erez T, Tassa Y, and Todorov E. Infinite horizon model predictive control for nonlinear periodic tasks. Manuscript under review. 2011;4.

[24] Rohde D, Bonner S, Dunlop T, *et al. RecoGym: A Reinforcement Learning Environment for the Problem of Product Recommendation in Online Advertising*; 2018. Available from: https://github.com/criteo-research/reco-gym.

[25] Bent O and Remy S. *KDD Cup 2019*, 2019. https://www.kdd.org/kdd2019/kdd-cup.

[26] Bhattacharjee B, Boag S, Doshi C, *et al.* IBM deep learning service. *IBM Journal of Research and Development*. 2017;61(4):10. Available from: http://ieeexplore.ieee.org/document/8030274/.

[27] Ribeiro M, Grolinger K, and Capretz MAM. MLaaS: machine learning as a service. In: Li T, Kurgan LA, Palade V, *et al.*, editors. *14th IEEE International Conference on Machine Learning and Applications, ICMLA* 2015, Miami, FL, December 9–11, 2015. IEEE, 2015. p. 896–902. Available from: https://doi.org/10.1109/ICMLA.2015.152.

[28] Chan S, Stone T, Szeto KP, *et al.* PredictionIO: a distributed machine learning server for practical software development. In: He Q, Iyengar A, Nejdl W, *et al.*, editors. |it 22nd ACM International Conference on Information and Knowledge Management, CIKM'13, San Francisco, CA, October 27–November 1, 2013. ACM, 2013. p. 2493–2496. Available from: http://doi.acm.org/10.1145/2505515.2508198.
[29] Baldominos A, Albacete E, Saez Y, *et al.* A scalable machine learning online service for big data real-time analysis. In: *2014 IEEE Symposium on Computational Intelligence in Big Data (CIBD)*, 2014. p. 1–8.
[30] Ooms J. The OpenCPU System: Towards a Universal Interface for Scientific Computing through Separation of Concerns. CoRR. 2014;abs/1406.4806. Available from: http://arxiv.org/abs/1406.4806.
[31] Campbell M, Hoane Jr AJ, and Hsu Fh. Deep blue. *Artificial Intelligence.* 2002;134(1–2):57–83.
[32] Silver D, Huang A, Maddison CJ, *et al.* Mastering the game of Go with deep neural networks and tree search. *Nature.* 2016;529(7587):484–489.
[33] Silver D, Hubert T, Schrittwieser J, *et al.* A general reinforcement learning algorithm that masters chess, shogi, and go through self-play. *Science.* 2018;362(6419):1140–1144.
[34] Bennett J and Lanning S. The Netflix prize. In: *Proceedings of KDD cup and workshop*, vol. 2007. New York, NY, 2007. p. 35.
[35] Thrun S, Montemerlo M, Dahlkamp H, *et al.* Stanley: the robot that won the DARPA Grand Challenge. *Journal of Field Robotics.* 2006;23(9):661–692.
[36] Wagstaff K. Machine Learning that Matters, 2012. arXiv preprint arXiv:12064656.
[37] Whiteson S, Tanner B, and White A. The reinforcement learning competitions. *AI Magazine.* 2010;31(2):81–94.
[38] Deng J, Dong W, Socher R, *et al.* Imagenet: a large-scale hierarchical image database. In: *2009 IEEE Conference on Computer Vision and Pattern Recognition.* New York, NY: Ieee; 2009. p. 248–255.
[39] Khetarpal K, Ahmed Z, Cianflone A, *et al.* Re-evaluate: Reproducibility in evaluating Reinforcement learning algorithms. In: *[ICML Workshop 2018], Reproducibility in ML Workshop*, 2018.
[40] Henderson P, Islam R, Bachman P, *et al.* Deep Reinforcement Learning that Matters, 2017. arXiv preprint arXiv:170906560.
[41] Whiteson S, Tanner B, Taylor ME, *et al.* Protecting against evaluation overfitting in empirical reinforcement learning. In: *2011 IEEE Symposium on Adaptive Dynamic Programming and Reinforcement Learning (ADPRL).* IEEE, 2011. p. 120–127.
[42] Hall MA. *Correlation-Based Feature Selection of Discrete and Numeric Class Machine Learning.* Hamilton, New Zealand: University of Waikato, Working paper 00/08. 2000.
[43] Abbeel P, Coates A, Quigley M, *et al.* An application of reinforcement learning to aerobatic helicopter flight. In: *Advances in Neural Information Processing Systems*, 2007. p. 1–8.

[44] Dimitrakakis C, Li G, and Tziortziotis N. The reinforcement learning competition 2014. *AI Magazine*. 2014;35(3):61–65.

[45] Sutton RS. Integrated architectures for learning, planning, and reacting based on approximating dynamic programming. In: *Machine Learning Proceedings 1990*. New York, NY: Elsevier, 1990. p. 216–224.

[46] Ahmed N and Wahed M. The De-democratization of AI: Deep learning and the Compute Divide in Artificial Intelligence Research. Working Paper. 2020.

[47] San Martín-Rodríguez L, Beaulieu MD, D'Amour D, *et al.* The determinants of successful collaboration: a review of theoretical and empirical studies. *Journal of Interprofessional Care*. 2005;19(Supp 1):132–147.

[48] Al Quraishi M. AlphaFold at CASP13. *Bioinformatics*. 2019;35(22): 4862–4865.

[49] Nguyen VB, Karim BM, Vu BL, *et al.* Policy Learning for Malaria Control, 2019. arXiv preprint arXiv:191008926.

[50] Alam Khan Z, Feng Z, Uddin MI, *et al.* Optimal policy learning for disease prevention using Reinforcement learning. *Scientific Programming*. 2020;2020:1–13.

[51] Sutton RS. Dyna, an integrated architecture for learning, planning, and reacting. *ACM Sigart Bulletin*. 1991;2(4):160–163.

[52] Sutton RS and Barto AG. *Reinforcement Learning: An Introduction*. Cambridge, MA: MIT Press, 1998.

[53] Rosset S and Inger A. KDD-cup 99: knowledge discovery in a charitable organization's donor database. *ACM SIGKDD Explorations Newsletter*. 2000;1(2):85–90.

[54] Tavallaee M, Bagheri E, Lu W, *et al.* A detailed analysis of the KDD cup 99 data set. In: *2009 IEEE Symposium on Computational Intelligence for Security and Defense Applications*. IEEE, 2009. p. 1–6.

[55] Dror G, Koenigstein N, Koren Y, *et al.* The Yahoo! music dataset and KDD-cup 2011. In: *Proceedings of KDD Cup 2011*. PMLR, 2012. p. 3–18.

[56] Kohavi R, Brodley CE, Frasca B, *et al.* KDD-Cup 2000 organizers' report: peeling the onion. *Acm Sigkdd Explorations Newsletter*. 2000;2(2):86–93.

[57] KDD. KDD Cup Archives, 1997. https://www.kdd.org/kdd-cup.

[58] Zhou W, Roy TD, and Skrypnyk I. The KDD Cup 2019 Report. *SIGKDD Explor Newsletter*. 2020 May;22(1):8–17. Available from: https://doi.org/10.1145/3400051.3400056.

[59] Mishra SK. Global optimization by differential evolution and particle swarm methods: evaluation on some benchmark functions. In: *MPRA*, 2006. Available at SSRN 933827.

[60] Bent O and Remy S. NeurIPS Live Malaria Challenge 2019, 2019. https://neurips.cc/Conferences/2019/CallForCompetitions.

[61] Bent O and Remy S. Indaba Malaria Hackathon 2019, 2019. http://www.deeplearningindaba.com/hackathon-2019.html.

[62] Sagi O and Rokach L. Ensemble learning: a survey. *Wiley Interdisciplinary Reviews: Data Mining and Knowledge Discovery*. 2018;8(4):e1249.

[63] Dietterich TG. Ensemble methods in machine learning. In: *International Workshop on Multiple Classifier Systems*. New York, NY: Springer, 2000. p. 1–15.

[64] Watkins CJCH. *Learning from Delayed Rewards*. PhD Thesis, University of Cambridge, 1989.

[65] Sutton RS, Mcallester D, Singh S, *et al.* Policy gradient methods for reinforcement learning with function approximation. In: *Advances in Neural Information Processing Systems 12*, 1999. p. 1057–1063.

[66] Algorithms G. Computer programs that "evolve" in ways that resemble natural selection can solve complex problems even their creators do not fully understand. In: *Holland in Scientific American*. 1992. p. 66–72.

[67] Lin F, Muthuraman K, and Lawley M. An optimal control theory approach to non-pharmaceutical interventions. *BMC Infectious Diseases*. 2010;10(1): 1–13.

[68] Kidziński Ł, Mohanty SP, Ong CF, *et al.* Learning to run challenge solutions: Adapting Reinforcement learning methods for neuromusculoskeletal environments. In: *The NIPS'17 Competition: Building Intelligent Systems*. New York, NY: Springer, 2018. p. 121–153.

[69] Kidziński Ł, Ong C, Mohanty SP, *et al.* Artificial intelligence for prosthetics: challenge solutions. In: *The NeurIPS'18 Competition: From Machine Learning to Intelligent Conversations*, 2019, p. 69.

[70] Kukovacec M, Kukurin T, and Vernier M. Personalized medicine: redefining cancer treatment. In: *Text Analysis and Retrieval 2018 Course Project Reports*, p. 69.

[71] Yadav D, Jain R, Agrawal H, *et al.* Evalai: Towards Better Evaluation Systems for AI Agents, 2019. arXiv preprint arXiv:190203570.

Chapter 4

Applications of machine learning for image-guided microsurgery

Prathamesh Chati[1], Andrew Whitaker[1] and Jonathan Silva[1]

Microsurgery is a subfield of general surgery that requires visualization of the patient anatomy via microscopes. Common microsurgical procedures include complex wound reconstruction, nerve and blood vessel repair, and microvascular transplantation [1]. Microsurgical procedures focus on microstructures like blood vessels and nerve endings, necessitating the use of highly precise instruments. Beyond the visual perspective that advanced microscopes provide surgeons, navigating and manipulating these structures is extremely challenging. Fortunately, over the past few decades, advancements in image-guidance systems have provided surgeons with enhanced navigational support while operating. This enhanced perspective gained by surgeons has led to safer and more minimally invasive procedures [2].

In this chapter, we will review the stages of developing image-guided systems for microsurgical procedures. We will emphasize the use of supervised and unsupervised machine learning models when designing image-guided systems, and how machine learning methods can enhance the development process. At each stage of the development process, we will provide a surface-level description of several mathematical and statistical methods within the scope of these machine learning models being used. We will also discuss unresolved challenges that are barriers in the field and how researchers are developing advanced techniques to enhance the precision and accuracy of existing image-guidance systems. Finally, we will conclude our discussion by reviewing existing and emerging research for the development of image-guided microsurgical systems and some limitations within the field. The stages needed to realize image-guided microsurgery are listed below.

1. Preoperative collection of imaging data
2. Image preprocessing to improve quality
3. Image segmentation to delineate key structures
4. Registering structures to patient anatomy
5. Display and visualizing the co-registered images

[1]Department of Biomedical Engineering, Washington University, Missouri

4.1 Preoperative data collection

Preoperative tomographic images are essential for diagnosis and "big picture" planning of a minimally invasive surgical procedure. Doctors can rarely directly view the surgical site. High-resolution images provide the ability to construct a mental or visible interior display of a patient's anatomy. The most common types of tomographic images include computer tomography (CT), magnetic resonance imaging (MRI), X-ray imaging, ultrasound, and nuclear imaging. Despite their extensive application, these imaging modalities still require doctors to navigate different challenges. However, new types of imaging system are currently being developed to manage these complications. In this section, the challenges associated with each type of imaging modalities will be discussed as well as recent technological advancements.

Computer tomography uses a computer process to combine a series of X-ray measurements from different angles in order to generate cross-sectional slices of the bones, blood vessels, and soft tissue within the body [3]. MRI generates the mapping images of organs by utilizing strong magnetic fields to align the protons and analyze their response [4]. The 2D images being 2D are not ideal for educating patients on their anatomy, planning surgical procedures and evaluating surgical success. Through post-processing, these cross-sectional slices can generate 3D volumetric representation of their anatomical structures. The algorithms needed to automate this processing are still being improved. Any error in creation of the 3D model could harm the quality of care provided [5]. In 1990, Hildebolt *et al.* created a 3D reconstruction of a skull from CT scans that could generally be accurate within 1 mm when compared to 2D scans However, the study only compared surface anthropologic measurements [6].

Another challenge with CT and MRI scans is the time gap between the initial scans and the patient's procedure. Scans are usually collected days or even weeks before the surgery. Thus, the innate difference between the preoperative and intraoperative environment limits the validity of these scans. The significance of discordant scans is vital in the neurosurgical field. High precision is required because of the brain's complex structure and the organ's progressive deformation. There exists the common phenomenon of brain shift caused by tissue manipulation, gravity, tumor size, loss of cerebrospinal fluid and the application of medication [7]. Brain shift greatly affects the accuracy of image guided systems needed for surgery. Today, the most common solution for the differences in anatomical structures between preoperative scans and the surgical procedures is intraoperative scanning. However, intraoperative scans can introduce new sources of error, including resolution [7].

Preoperative images normally have higher resolution for anatomical structures but require longer scanning duration. Predictably, the intraoperative images contain lower resolution and can be attained more quickly. Ultrasound and X-rays are lower-cost alternatives that are often implemented for real-time anatomical vision. The noninvasive, portable, and economic nature of ultrasound allows it to be a common choice for intraoperative monitoring. The imaging type is equally as significant of the timing of the image. For the procedure, the preoperative images can be merged with intraoperative images from other modalities. For example, the combination of CT and Ultrasound, CT and MRI or PET and CT. However, the challenge that comes with fusing imaging is the natural shift of structures. These shifts are caused by respiratory movement, patient position, the surgery and natural body movement [8].

In microsurgeries, CT scans are often preferred. These scans provide benefits including accessibility and precise bony detail resolution. Though CT scans lack contrast resolution for soft-tissue structures. Contrast resolution is defined as the ability to distinguish between different amplitudes of adjacent structures [9]. However, CNS-related microsurgeries favor MRI for their higher soft-tissue contrast resolution, more precise vascular visualization, and fewer interfering artifacts [10].

The demand for higher resolution imaging modalities has increased for microsurgeries. Most CT scans offer spatial resolution from 0.3 mm to 1 mm, a higher spatial resolution than the 1.5 mm resolution of most MRI scanners. Furthermore, the temporal resolution for CT scans is under a second [11,12]. As a reminder, temporal resolution is defined as the duration of time for acquisition of a single frame of a dynamic process. Some hospitals possess equipment that achieve higher quality resolution derivatives of MRI, CT such as ultra-high field MRI and micro-CT. Specifically, micro-CT scans have spatial resolution of 70 μm compared to ultra-high resolution MRI [13]. Micro-CT is similarly applied in cases for ultrasound imaging. Ultrasound provides lower resolution of 0.5 mm [14]. However, one of the downfalls of micro-CT scans are the extensive scan duration. It can take 2 min to 14 h depending on the density of the micro-CT scan which is much longer than ultrasound [14]. Furthermore, the challenge of CT and micro-CT scans is an accumulation of radiation for the patient.

Another imaging process for microsurgeries is optical coherence tomography (OCT) that detects backscattered infrared light. The anatomical visualization delivers a high sensitivity with precise spatial resolution of 10 μm and 2–3 mm depth. Compared to the other imaging modalities, OCT has superior resolution and more instrument compatibility in intraoperative surgeries. Since OCT is built upon optical communication technology, it has no risk of ionizing radiation. Additionally, the real-time modality can be used to construct 3D anatomical structures [15]. The fast-paced technology (300,000 A-scans per second) combined with different angles has seen to produce high-resolution volumetric structures for otolaryngology microsurgeries [16]. A study conducted by Weiser *et al.* studied high-quality live video of 3D OCT [17]. They tested a syringe above the skin of a finger to simulate surgical procedure. In Figure 4.1, the needle position can clearly be seen on top of the elastic skin of the fingertip.

Figure 4.2 displays the captured outputs for several common imaging techniques. Once these images have been captured, they must be processed and segmented appropriately; a process which is described next [18–21].

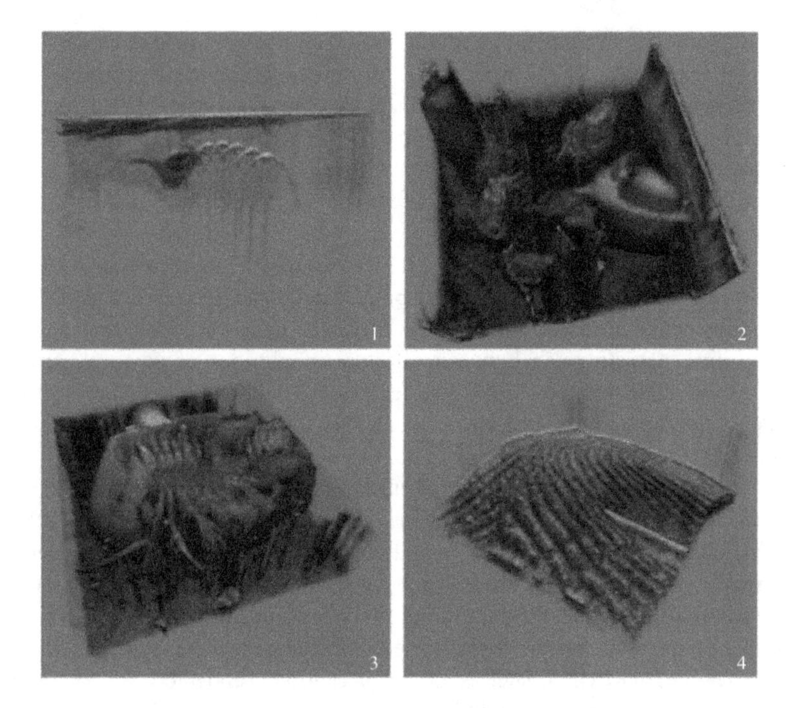

Figure 4.1 *3D OCT image of (1) spiralling wire, (2) Daphniae, (3) filter legs dynamics, and (4) syringe over finger. Reference: https://www.ncbi.nlm.nih.gov/pmc/articles/PMC4230855/.*

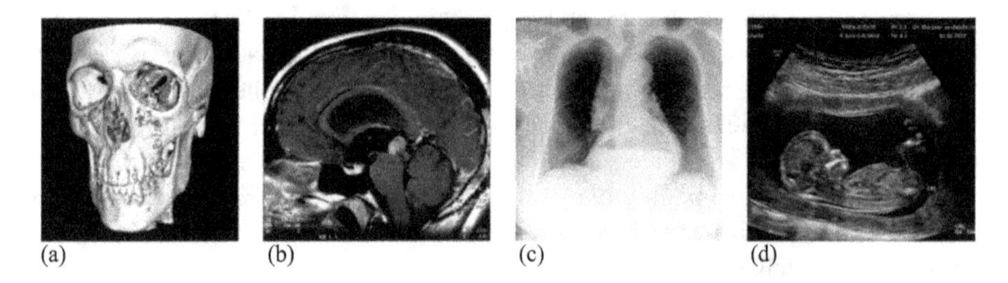

Figure 4.2 *Preoperative image collection methods displaying (a) a 3D CT scan of the brain and vertebrae, (b) an MRI scan of the brain, (c) an X-ray fluoroscopy scan of the lungs, and (d) a fetal ultrasound scan*

4.2 Preprocessing

Image preprocessing focuses on reducing the noise, or interference, in the image and strengthening the contrast between the foreground and background. This is a vital step for enhancing the image quality and preparing the image for the next stage, segmentation. Image segmentation broadly relies on using quantitative measures to distinguish certain target features from the background, such as a set of nerve endings or a group of cells embedded from a tissue biopsy. Thus, reducing the noise within an image ensures a more efficient segmentation process. The two image preprocessing methods we will be reviewing are listed below.

1. Noise reduction
2. Contrast adjustment

Prior to reviewing these preprocessing methods, however, we will briefly review the concept of an intensity histogram, which graphically displays the intensity distribution of an image.

4.2.1 Intensity histograms

Image preprocessing and a majority of the future stages for developing image-guided systems rely on algorithms and functions that take numerical inputs (vectors, arrays, etc.). For these processes to dissect and alter an image, the image must first be quantified. One popular method is by constructing a pixel-level histogram that displays the distribution of intensities. A pixel can be considered as the fundamental component of the image. A group of pixels that vary in characteristics form a larger image. An image can be represented by a 2D matrix with values corresponding to the intensity of each pixel. When an algorithm or function alters the image, the new intensity matrix is calculated and displayed as a histogram. This provides a numerical characteristic that functions can optimize and serves as a visible marker of change. Figure 4.3 displays a cranial CT scan and its respective intensity histogram [22].

In the histogram displayed above, the x-axis corresponds to the intensity values while the y-axis corresponds to the frequency of each intensity value. We will be focusing most of our analysis on grayscale images, which contain pixels with intensity values between 0 (black) and 255 (white). Now that we have briefly reviewed intensity histograms, we can continue our discussion of image preprocessing methods. We begin by reviewing the first step, noise reduction.

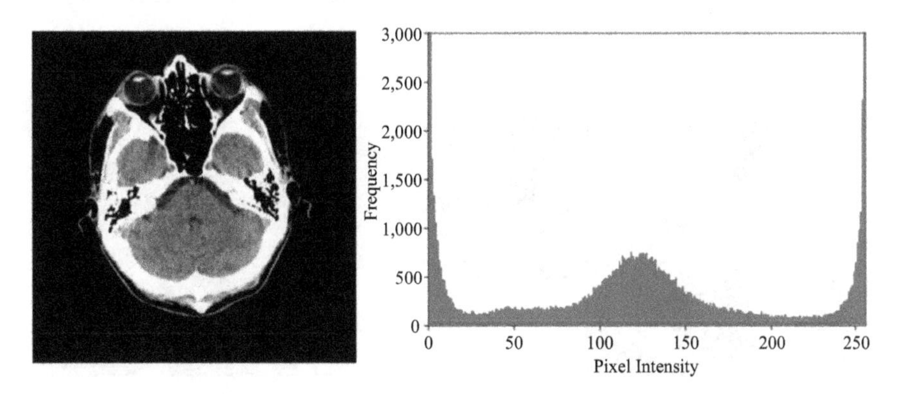

Figure 4.3 A 2D grayscale cranial CT scan (left) and its corresponding intensity histogram (right)

4.2.2 Noise reduction

The first component of image preprocessing quantifies and reduces the level of noise to enhance the accuracy of the final visual model. Noise is the statistical interference that obscures an image and can be classified as either global or local. Global noise refers to interference across the entire image – this could range from the brightness of an image to the overall resolution – and can typically be characterized by certain overlaying patterns. Global noise originates from the methods or tools used to acquire an image. Local noise refers to pixel-specific interference – this type of noise is usually sporadically dispersed throughout the image and requires more granular approaches [23].

Global versus local characteristics: Although global and local characteristics are discussed separately throughout this chapter, the distinction between global and local characteristics is fuzzy. Thus, many global methods and characteristics can pertain to the local level as well, and vice versa.

4.2.2.1 Global noise reduction

Reducing global noise involves using functions to quantify the noise of an interfering feature at the global level of the image (contrast, intensity, etc.). For example, say an acquired image displays an intensity gradient such that there is a contrast imbalance across the entire image – intensity, in this case, is the global level feature causing interference. A polynomial function can be fitted to quantify the global intensity gradient, which can then be subtracted from the total intensity matrix of the entire image. This gradient-based noise reduction method is reviewed by Iskakov *et al.* and shown in Figure 4.4 [23].

Figure 4.4 involves fitting a polynomial function that approximates, or quantifies, the global intensity gradient. This process first involves constructing an intensity matrix $I(i,j)$ for an m by n 2D grayscale image (measured in pixels). For our intensity matrix $I(i,j)$, $i = 1, \ldots, m$ and $j = 1, \ldots, n$, and each component (i,j) has an integer value between 0 and 255 (representing the intensity range of a grayscale image). Next, we can construct a general polynomial function $f(i,j;B)$, where B represents

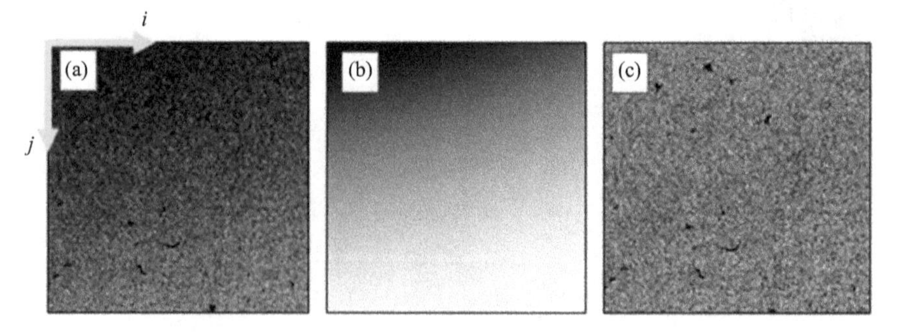

Figure 4.4 The original image (a) is denoised based on the intensity gradient (b), and resulting in image (c)

the coefficients or parameters to be fitted (i.e. $B = \{b_0, \ldots, b_k\}$ for a polynomial with k components). Iskakov *et al.* display the following second-order polynomial as an example of a second-order general function:

$$f(i,j;B) = b_0 + b_1 i + b_2 j + b_3 i^2 + b_4 j^2 + b_5 ij \tag{4.1}$$

To find the coefficients for a generalized polynomial function, we can employ the standard least-squares regression (least-squares) method. The least-squares method is popularly used to determine the line of best fit for a given set of points. In this case, we are identifying the optimal parameters for an applied intensity gradient across the entire image. Once the global intensity gradient has been quantified, it can be subtracted from the original intensity matrix to yield a noise-reduced image. In the expression below, I_R represents the reduced image, I represents the original intensity matrix, and f represents the approximated intensity gradient polynomial:

$$I_R = I - f \tag{4.2}$$

Notice how a general polynomial function can characterize the intensity gradient for the entire image, hence, the additional intensity at each pixel is considered global noise. Since global noise applies to the totality of the image, the strategy discussed above focuses on, first, identifying and quantifying a spatial pattern corresponding to global noise – whether that be an intensity gradient, a change in contrast, or an increase in brightness – then second, subtracting the quantified representation of the global noise from the quantified value of the whole image. Although global approximation methods are useful, noise can often be sporadic, following no quantifiable global pattern. In these situations, local noise reduction techniques are beneficial for removing pixel-level interference.

4.2.2.2 Local noise reduction

Local noise occurs at the pixel-level of an image and can either be randomly dispersed throughout an image or systematic, where a function can be used to characterize the additive noise for each pixel. Local noise reduction methods require increased calibration, necessitating a balance between the type and strength of the filter being applied. More robust methods extend beyond pixel-level patterns and also evaluate a defined area surrounding a given pixel (i.e. there is a spatial component to local noise reduction). Each individual local area can then be manipulated accordingly to achieve area-specific noise reduction. Prior to reviewing local denoising techniques, let there are several main types of local noise [24].

1. Gaussian noise is derived from the initial acquisition of the image. As the name describes, each pixel has additive statistical noise with a probability density function (PDF) that can be mapped to a normal (Gaussian) distribution.
2. Speckle noise is also derived from the initial acquisition of the image, however, it is often multiplicative rather than additive. Speckle noise does not follow a gaussian curve, and therefore, requires different denoising approaches than those used to clean gaussian noise [25].
3. Salt-and-pepper noise consists of randomly dispersed pixels with intensities of either 255 (white or salt) or 0 (black or pepper).

4. Poisson noise (shot noise) originates from the electromagnetic waves used for image acquisition. The presence of Poisson noise is spatially and temporally random as a result of the fluctuations of photons.

Note, methods to resolve each type of noise are dictated by the degree and distribution of interference throughout the image. We will outline four main local noise reduction techniques, which are listed below. The four listed techniques are considered spatial domain filtering methods since they account for neighboring pixels.

1. Mean filter
2. Median filter
3. Gaussian filter
4. Bilateral filter

Prior to reviewing the filtering methods listed above, however, we will discuss the concept of image convolution, which uses weighted kernels to generate local changes to a pixel, and filtering strength, which determines the types of features one is aiming to preserve upon denoising.

4.2.2.3 Image convolution

Image convolution is the process of adjusting a pixel value by the weighted linear sum of its neighboring pixels. The weights, or coefficients, of the linear combination are provided by a kernel, which is a matrix of values with the same size as our pixel window under inspection [26]. The values within a kernel differ based on the filtering method being used and do not have to be constant – some filtering techniques alter kernel values based on the intensity values of neighboring pixels. Figure 4.5 presents a simple visualization of image convolution using a kernel [27].

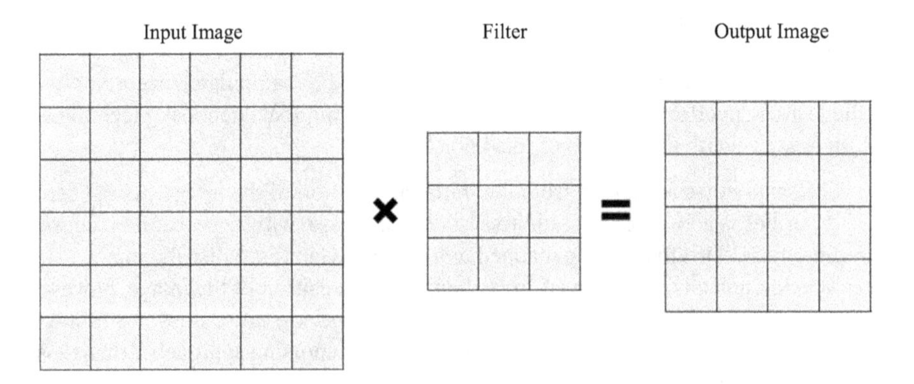

Figure 4.5 Each pixel in the input image is altered by the filter, yielding the resulting image

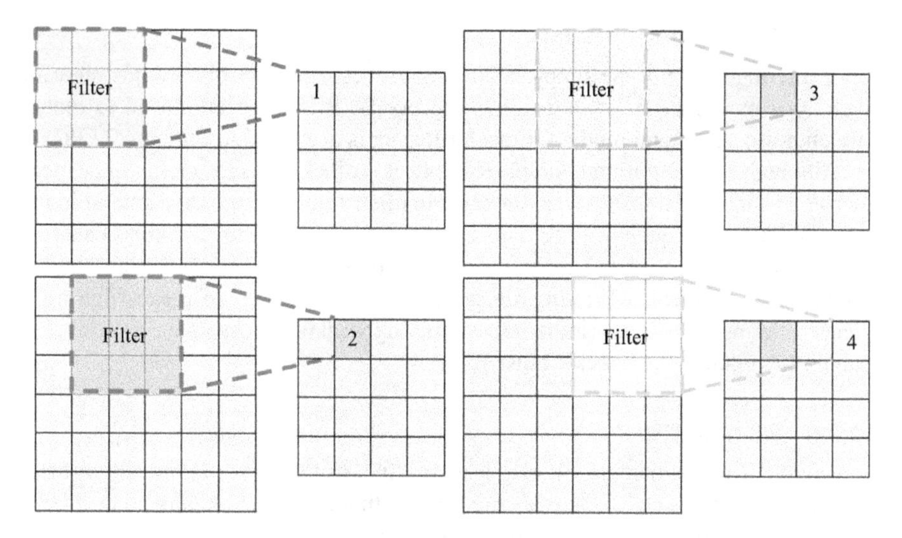

Figure 4.6 An animated graphic demonstrating the process of image convolution. The colored area on the larger matrices indicates a 33 kernel moving across each pixel of the image. The resulting smaller matrix provides a filtered pixel based on the weighted linear combination of the kernel values and neighboring pixels.

Mathematically, image convolution can be represented by the expression below [28]:

$$g(x,y) = \sum_{dx=-a}^{a} \sum_{dy=-b}^{b} \omega(dx,dy) f(x+dx, y+dy) \tag{4.3}$$

In the expression above, $g(x,y)$ corresponds to the filtered image, which is derived by taking the weighted linear combination of neighboring pixels $(x+dx, y+dy)$ from the original image $f(x,y)$ using weights within kernel ω. The overarching image convolution process can be represented in Figure 4.6 [27].

4.2.2.4 Filtering strength

Filtering strength corresponds to the pixel frequency being removed. Filtering can be broadly categorized as either low pass or high pass. Low pass filtering, or smoothing, can be used to remove high-frequency noise, while high pass filtering can be used to remove low frequency noise. Typically, low pass filtering is used during the noise reduction stage as it "blurs" or "smooths" the input image. High pass filtering is often used during segmentation and thresholding, with the goal of sharpening features such as edges, corners, and contours [29].

Now that we have reviewed image convolution and filtering strength, we can discuss the aforementioned local noise reduction techniques beginning with mean filtering.

4.2.2.5 Mean filter

Mean filtering is one of the simplest filtering processes and involves adjusting each pixel value by the mean of its neighboring pixels. It is often employed to reduce Poisson noise. A general model for mean filtering is displayed in Figure 4.7 [30].

Although mean filtering is simple to apply, it is often not as efficient. Since mean filtering is a linear denoising technique, it can often result in excessive smoothing of images with high frequency noise. This causes a reduction in the visibility of sharper features such as edges, contours, and texture. In these scenarios, non-linear filtering methods such as median filtering are preferred. An extension of mean filtering is Wiener filtering, which has the same overarching weaknesses as mean filtering, but enables the reduction of speckle noise [31].

4.2.2.6 Median filter

As mentioned above, median filtering is a non-linear filtering technique that is more efficient at preserving edges and texture. The methodology behind a median filter is nearly identical to that of mean filtering, with the exception of using the median of neighboring pixels as opposed to the mean. This process is shown in Figure 4.8 [30].

Notice the difference between the final values of the mean and median filtered matrix. Median filtering is helpful for reducing salt-and-pepper noise.

Figure 4.7 Mean filtering applied to the central pixel in the left grid, resulting in the single value on the right grid

Figure 4.8 Median filtering applied to the central pixel in the left grid, resulting in the single value on the right grid

4.2.2.7 Gaussian filter

As the name suggests, Gaussian filtering, or Gaussian smoothing, is a denoising method used to blur an image via a Gaussian function and remove Gaussian noise. It harnesses a gaussian kernel that can be represented by the following expression [32]:

$$G(x,y) = \frac{1}{2\pi\sigma^2}e^{-\frac{x^2+y^2}{2\sigma^2}} \tag{4.4}$$

The expression above corresponds to a 2D Gaussian curve with a mean of (0,0) and a standard deviation (σ) of 1. The 2D Gaussian curve is given in Figure 4.9 [33].

Notice how the Gaussian curve is perfectly symmetrical about the mean at (0,0). The vertical axis, corresponding to $G(x,y)$, can also represent the magnitude or strength of the Gaussian filter. As the filtering strength increases, the resulting image becomes increasingly blurred. This is demonstrated in Figure 4.10 [32].

An extension of Gaussian filtering is the widely used method of bilateral filtering.

4.2.2.8 Bilateral filter

Bilateral filtering is a highly popular denoising method and serves as an extension of Gaussian filtering. Similarly, it is a non-linear denoising technique that focuses on preserving edges while smoothening the image. Instead of directly adjusting a certain pixel value via a 2D Gaussian approximation, however, bilateral filtering takes the weighted average of neighboring pixel intensity values using a Gaussian distribution. The emphasis on radiometric and Euclidean differences gives bilateral filtering edge-preserving properties [34].

As opposed to Gaussian filtering, which applies a uniform gaussian filter on each pixel, the strength of bilateral filtering is adjusted by the similarity between the pixel of interest and neighboring pixels. This is known as weighting the bilateral filter by a similarity measure function. By this measure, if our pixel of interest is similar to neighboring pixels, the Gaussian coefficient is adjusted by a value close to 1 (i.e. it is almost identical to Gaussian filtering). If our pixel of interest is vastly

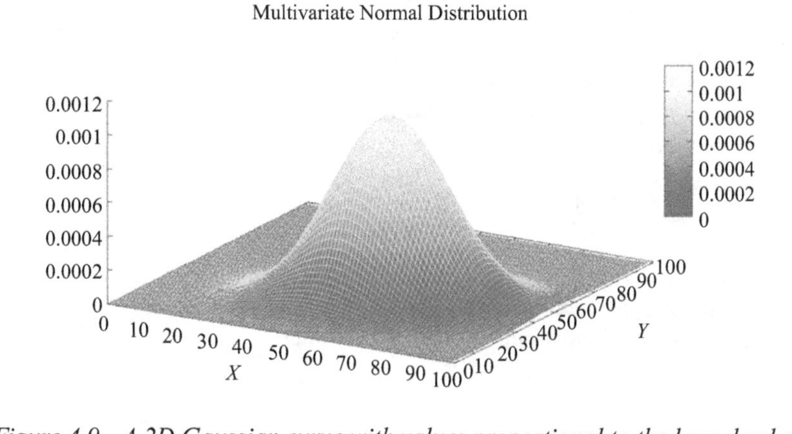

Figure 4.9 A 2D Gaussian curve with values proportional to the kernel values

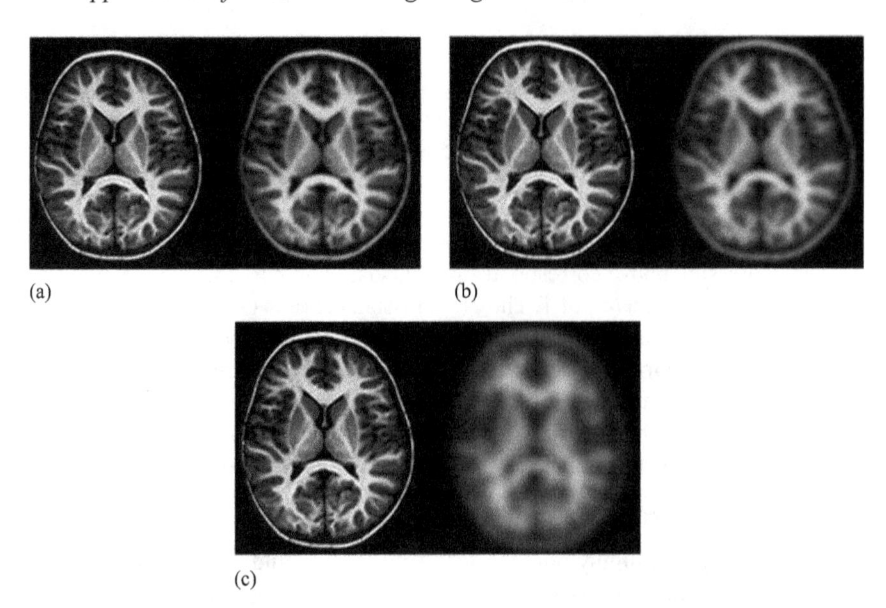

Figure 4.10 Gaussian smoothing applied to the left cranial MRI scan with increasing strengths, (a) $\sigma = 1.25$, (b) $\sigma = 3.0$, and (c) $\sigma = 5.0$

different from neighboring pixels, as in the case of borders or edges, the Gaussian coefficient is adjusted by a value closer to 0 (i.e. filtering is weak or non-existent at that point). This multiplicative property of bilateral filtering enhances overall smoothing while avoiding areas with edges. The expressions below mathematically describe the process of bilateral filtering along with its weighted adjustment based on the similarity function described above:

$$I_f(x) = \frac{1}{W_p} \sum_{x_i \in \Omega} I(x_i) f_r(||I\,(x_i) - I(x)||) g_s(||x_i - x||) \tag{4.5}$$

In the expression above, the input image $I(x_i)$ with a pixel to be filtered x_i is adjusted using the Gaussian function range kernel for smoothing intensity differences between pixels f_r and the spatial kernel for smoothing coordinate differences between pixels g_s, which is weighted by W_p [34]:

$$W_p = \sum_{x_i \in \Omega} f_r(||I(x_i) - I(x)||) g_s(||x_i - x||) \tag{4.6}$$

The expressions above display a clear difference from those of Gaussian filtering. This mathematical difference between Gaussian and bilateral filtering can be observed in Figure 4.11, which displays the application of both filters to a 3D cranial MRI [35].

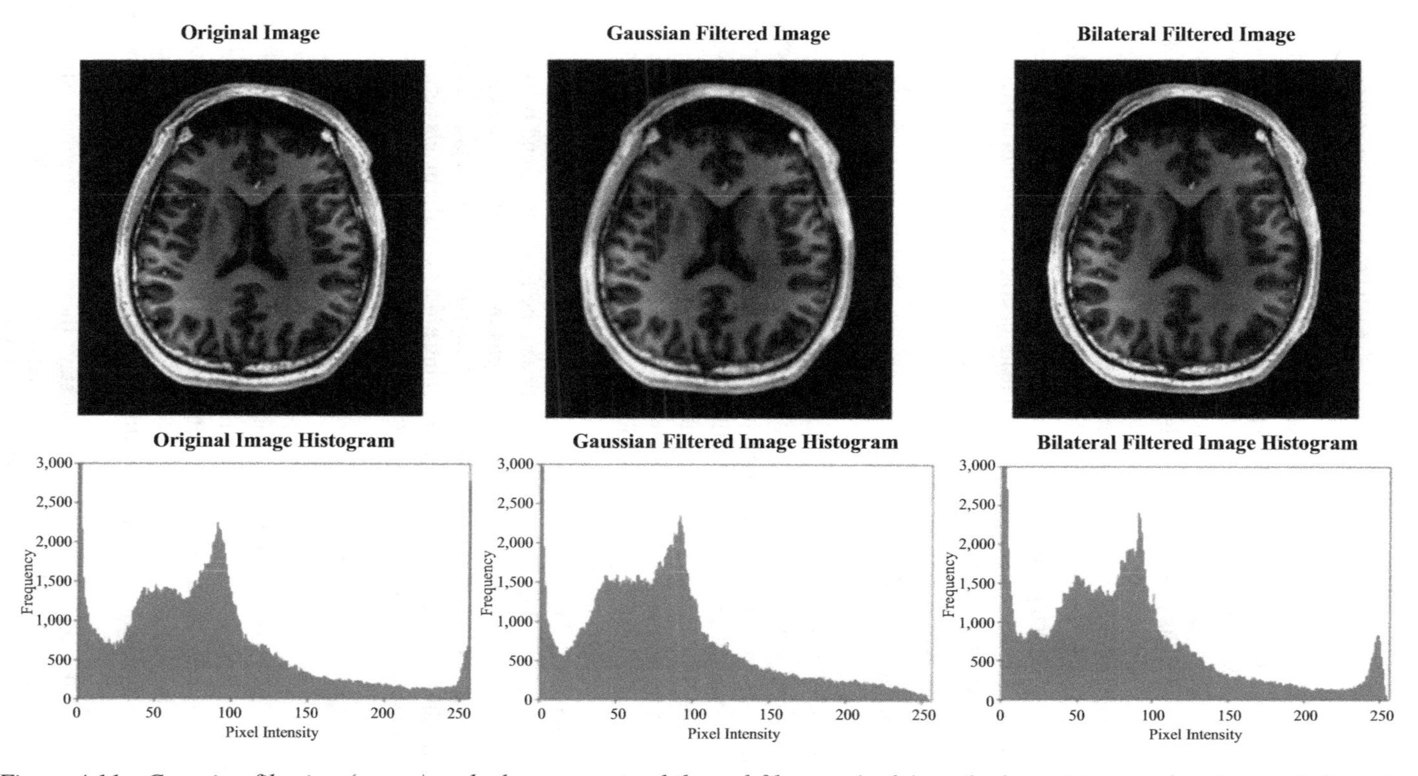

Figure 4.11 Gaussian filtering (center) and edge-preserving bilateral filtering (right) applied to a 3D cranial MRI scan (left) with corresponding intensity histograms (bottom row)

4.2.3 Contrast adjustment

Once global and local noise has been reduced, the second stage of image preprocessing is contrast alteration. Contrast can be defined as the range of intensity values between image pixels. Thus, contrast alteration largely involves strengthening the intensity between the foreground and background. Similar to noise reduction, altering the contrast also consists of both global and local techniques. Global contrast alteration methods adjust image-level characteristics and do not focus on adjusting each pixel relative to its neighboring pixels (i.e. the change in each pixel is uniform and independent of adjacent pixels – the spatial component of each image is ignored). Local contrast alteration methods, on the other hand, consider the spatial component of image pixels (i.e. neighboring pixels influence how the target pixel is altered). Let us begin by first discussing global contrast alteration methods, then shifting our focus to local contrast alteration methods. The two leading contrast alteration methods to adjust global contrast are listed below.

1. Contrast stretching
2. Histogram equalization

4.2.3.1 Contrast stretching

Contrast stretching, or normalization, involves "stretching" the range of pixel intensity values until a desired range is acquired (typically, a desired pixel intensity range is 0–255). As mentioned earlier, contrast defines the dynamic range of intensity values across an image. Contrast can be graphically represented as an intensity histogram, which displays the frequency of each intensity value across the intensity range of an image. For example, take Figure 4.12, which shows a low contrast cranial MRI scan and its corresponding intensity histogram [36].

As we can see the figure, the range of intensities for the given image is quite condensed. Thus, it is difficult to discern between the foreground and background of the scan. Contrast stretching essentially assigns the intensity of the highest and lowest pixels to 255 and 0, respectively, then distributes the remaining pixel intensities between the adjusted range. The contrast stretching process for the pixels between the highest and lowest intensity pixels can be described by the expression below:

$$X_{new} = \frac{X_{input} - X_{\min}}{X_{\max} - X_{\min}} \times 255 \tag{4.7}$$

From the expression above, the input pixel is scaled accordingly and then "stretched" across a desired range. This results in an image where the target features, in this case the chest, becomes accentuated and defined. Figure 4.13 displays contrast stretching and post-processing applied to the previously generated low contrast cranial MRI scan and its corresponding intensity histogram, which now ranges from 0 to 255 [35].

Notice that the shape of the histogram stays the same, it is simply stretched across a new range (i.e. there is a one-to-one relationship between the old and new histogram values). Thus, contrast stretching is analogous to the linear normalization

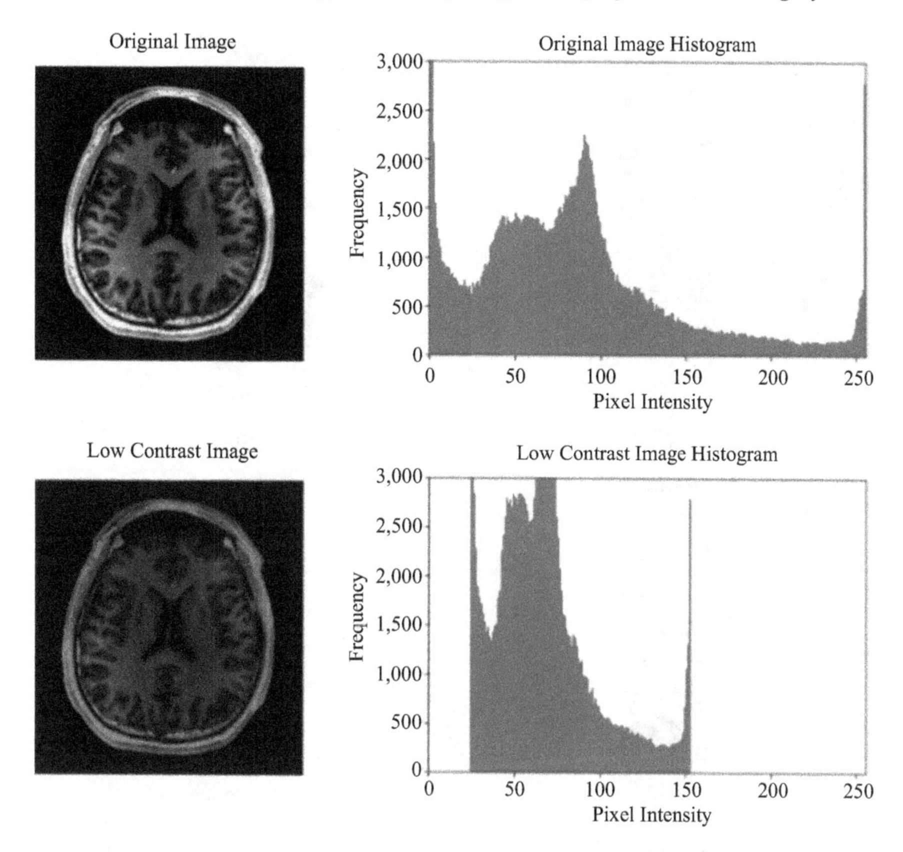

Figure 4.12 A regular cranial MRI scan (top) and low contrast cranial MRI scan (bottom) with corresponding intensity histograms

of pixel intensities. This one-to-one characteristic of contrast stretching also enables us to restore the original histogram from the modified histogram.

Although contrast stretching can enhance the visibility of the foreground and certain target features, the cost is image clarity and granularity. Each pixel has certain characteristics, of which our focus is on intensity. If the range of intensities across all of the pixels is small, then the "stretched" image may appear grainy, as if individual pixels were now grouped into larger voxels (i.e. if the intensity range of image pixels is too condensed, then contrast stretching cannot achieve much better image clarity). In addition to losing image clarity, contrast stretching only focuses on the highest and lowest intensity pixels, which can cause further challenges when dealing with images with an unequal range of pixel intensities in certain regions of the image. For example, if the image contains groups of pixels with intensities of 0 and 255, then contrast stretching may not be efficient since the desired intensity range has already been achieved [37].

In Figure 4.14, notice how the original image is already distributed across the entire intensity range. Thus, upon contrast stretching, the visible difference between

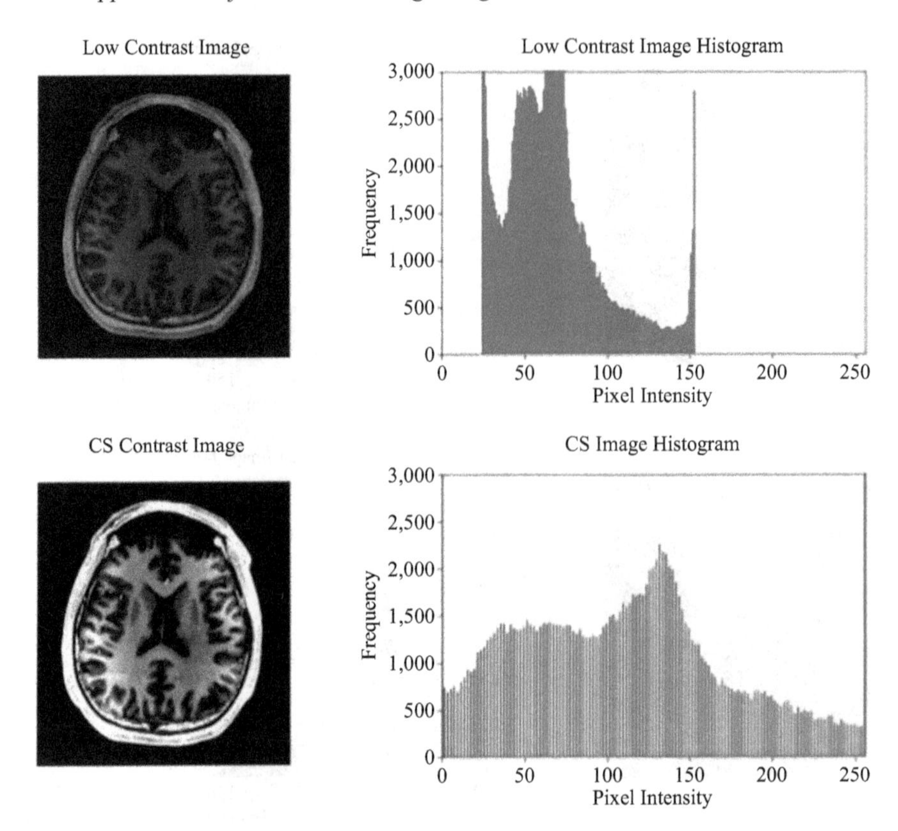

Figure 4.13 A low contrast cranial MRI scan (top) and contrast stretched/processed cranial MRI scan (bottom) with corresponding intensity histograms

the original and modified histogram and image is negligible. It is due to this drawback that the second contrast alteration method, histogram equalization, is more preferred in certain situations.

4.2.3.2 Histogram equalization

Histogram equalization (HE) adjusts the initial intensity histogram by using a target histogram, with the goal of matching the target histogram. Typically, our target histogram, although not limited to, is a flat histogram – the original intensity histogram is essentially "transformed" to fit our flat histogram, which equalizes the intensity range. HE is more complex than simply mapping intensity values over a new range, as done for contrast stretching, and often changes the shape of the original histogram beyond restoration. Thus, histogram equalization is analogous to the non-linear normalization of pixel intensities.

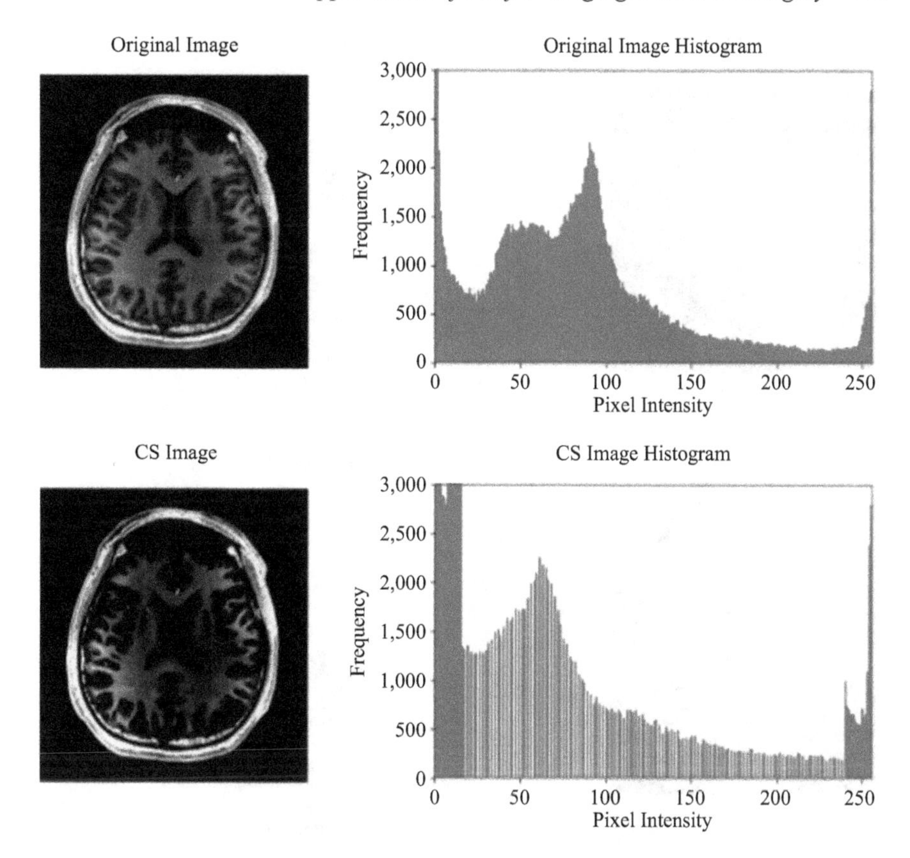

Figure 4.14 Contrast stretching applied to a cranial MRI scan (top), yielding the limited effect contrast stretched image and histogram (bottom)

Mathematically, HE can be thought of as first normalizing the intensity histogram by the total number of pixels then transforming the histogram to create a uniform distribution. In this sense, the histogram can be thought of as a probability density function (PDF). The histogram normalization and equalization process can be described by the following expressions [38]:

$$P_n = \frac{\text{number of pixels with intensity } n}{\text{total number of pixels}}; n = 0, 1, \ldots, L-1 \quad (4.8)$$

$$g_{i,j} = floor\left((L-1)\sum_{n=0}^{f_{i,j}} P_n\right) \quad (4.9)$$

In the expression above, the normalized histogram P_n with a pixel intensity range from 0 to $L-1$ (where $L = 256$) is transformed across the entire image $f_{i,j}$ to yield the equalized image $g_{i,j}$. The expression above is graphically displayed in Figure 4.15. Notice the difference between the following figure and the contrast

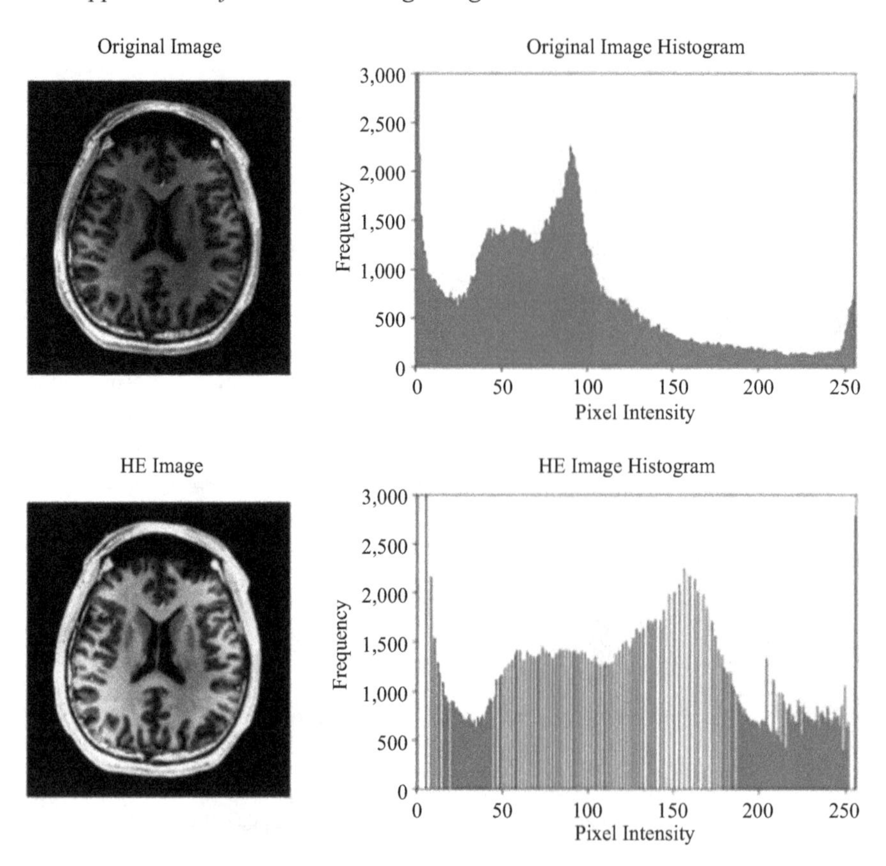

Figure 4.15 Histogram equalization applied to a cranial MRI scan (top), yielding the equalized image and histogram (bottom)

stretched histogram above. The visibility of the original scan has greatly improved upon the application of HE [37].

Although Figure 4.15 demonstrates the usefulness of global HE, there are situations where global contrast alterations may decrease the quality of the image. For example, if a target feature blends into the image as a whole but can be identified when evaluating only a small area of the image, global contrast methods may gloss over this elusive target feature. Specifically, if certain areas of the image have different levels of contrast, it may be more efficient to employ local contrast adjustment methods.

4.2.3.3 Local contrast adjustment

As opposed to global contrast alterations, local contrast alterations do factor in the spatial characteristics of target features – each area is evaluated in comparison to its neighboring areas, and if a local contrast or intensity shift is detected, an area specific contrast alteration may be applied. Similar to above, both contrast stretching and histogram equalization can be applied to adjust local contrast levels, however, slight modifications must be made to account for the spatial component of each pixel (i.e. account for neighboring pixels). Below we briefly discuss two methods for local contrast adjustment. Notice that the first method, adaptive histogram equalization (AHE), is a modified version of global histogram equalization.

1. AHE
2. Unsharp masking (USM)

4.2.3.4 AHE

AHE is mathematically similar to global histogram equalization, however, instead of calculating and equalizing a total image histogram, it creates several histograms corresponding to locally selected regions of the image. This is beneficial when intensity differences are not equally distributed throughout the image (i.e. there are lighter and darker areas). Figure 4.16 graphically displays the AHE process, which uses an adjusted version of the HE equation above on different areas of the image.

As compared to global HE, notice how the AHE conducted above yields a much smoother intensity histogram. This is due to the fact that HE has been conducted to

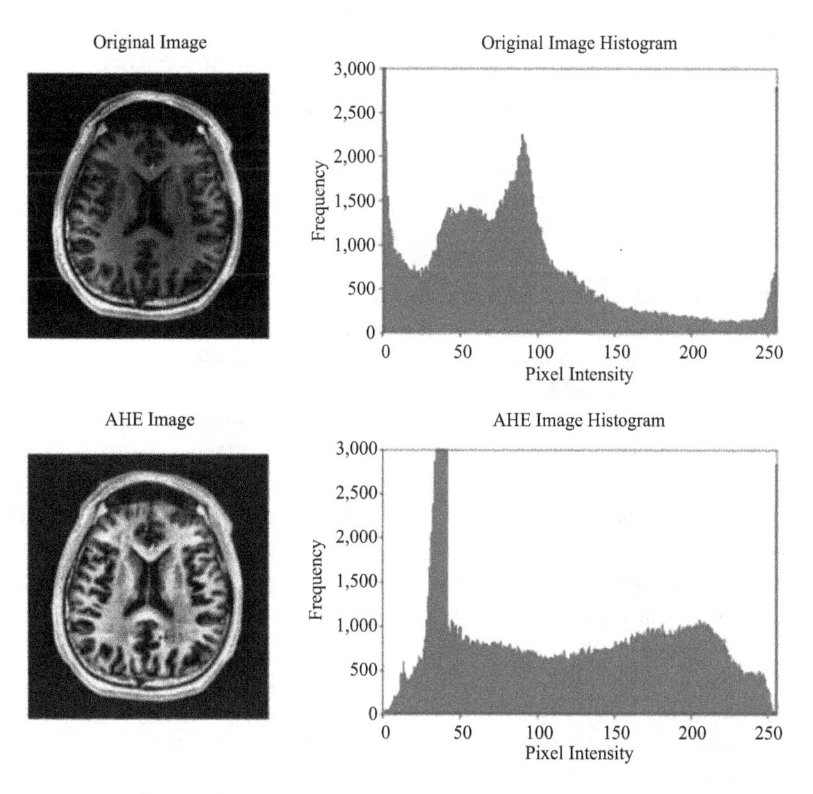

Figure 4.16 Adaptive histogram equalization applied to a cranial MRI scan (top), yielding the equalized image and histogram (bottom)

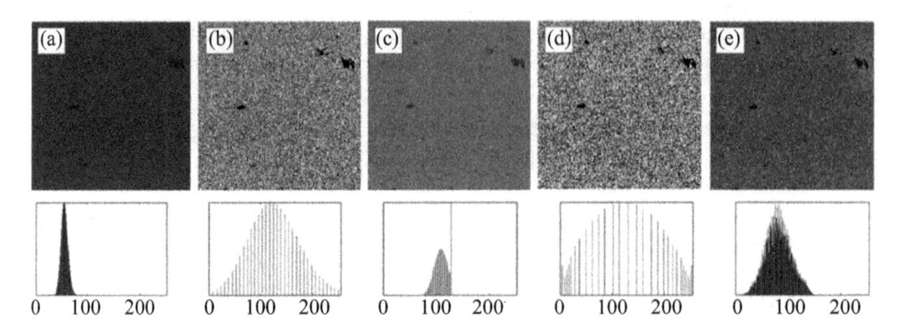

Figure 4.17 Contrast stretching with 1% (b) and 15% (c) saturation limits, histogram equalization (d) and CLAHE (e) applied on the original image (a) with corresponding intensity histograms (bottom row)

different extents at various portions of the image, adjusting lower contrast areas more to create a more uniform histogram. An extension of AHE is CLAHE (contrast-limited adaptive histogram equalization), which is similar to AHE, however, CLAHE limits the extent of contrast adjustment on homogenous regions (i.e. those areas that do not require contrast alterations).

The images in Figure 4.17 display the effects of several global and local contrast alteration techniques on a single reference image [23]. The diagram demonstrates the application of both contrast stretching (with slight modifications) and histogram equalization techniques. Notice, how the application of different local and global contrast alteration methods leads to vastly differing output intensity histograms.

4.2.3.5 USM

USM is a technique to improve the sharpness, or resolution, of an image. The process is analogous to refining an image to enhance its clarity and visibility of finer features. See Figure 4.18, which displays the application of a unique nimble unsharp filter on a 3D mammogram image [39].

In Figure 4.18, notice how after an unsharp mask is applied, the contours, edges, texture, and other subtle features become more apparent. We can say that the image has been "refined" to an extent. Qualitatively, USM subtracts a blurred version of the image from the original image to yield an edge image – an image that quantifies the presence of edges and contours. The contrast of the edge image is increased to "refine" the edges and contours. The edge image is then added back to the original image to yield a sharper image. Note, smoothening an image employs techniques similar to those discussed in the noise reduction section above.

A better intuition of USM can be gained by understanding how certain signals are filtered. Figure 4.19 demonstrates the concept of subtracting a smoothed version of the image (i.e. using a low pass filter to smooth the image) from the original image to yield an edge image (i.e. applying a high pass filter) [40].

In Figure 4.19, notice how the high pass signal, which corresponds to the edges, is isolated. This process can be represented by the following expression, where $g(x, y)$

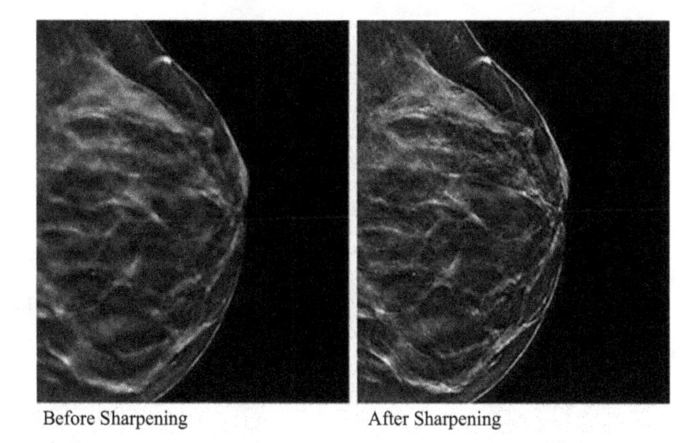

Figure 4.18 Unique nimble unsharp filter applied to a 3D mammogram image (left), resulting in the sharper image (right)

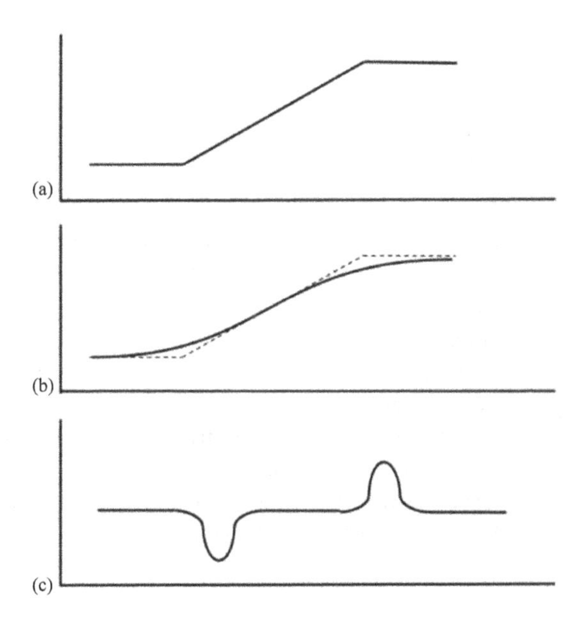

Figure 4.19 The low pass signal is estimated (b) from the original signal (a), yielding only the high pass signal (c)

represents the edge image, $f(x,y)$ represents the original image, and $f_{smooth}(x,y)$ represents the smoothed image [40]:

$$g(x,y) = f(x,y) - f_{smooth}(x,y) \tag{4.10}$$

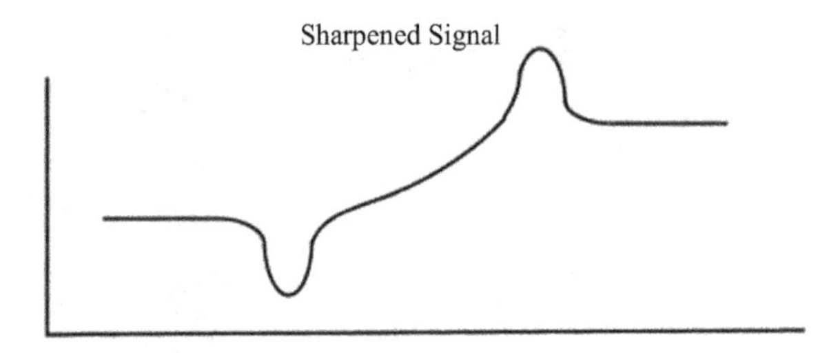

Figure 4.20 The high pass signal from Figure 4.19(c) is weighted and added to the original signal from Figure 4.19(a)

Once the edge image has been processed (via contrast adjustments, filtering, etc.), it can be added back to the original image to yield a sharper image:

$$f_{sharp}(x, y) = f(x, y) + k \times g(x, y) \tag{4.11}$$

In the expression above, k represents a scaling constant, or weight, that adjusts the strength of the applied filter. The final image signal graph is shown in Figure 4.20 [40].

Notice how the edges are now more defined in the signal graph in Figure 4.20, indicating a higher degree of "sharpness."

4.2.4 Preprocessing review

In this section, we reviewed two main image preprocessing steps, noise reduction and contrast alteration, while introducing several examples of both processes. In general, noise reduction and contrast alteration set, the foundation for the remainder of the segmentation process because they facilitates easier target feature detection, which involves identifying the anatomical features of interest and highlighting them using projected overlays or navigational cues. Overall, both the noise reduction and contrast alteration steps of image denoising are rather intuitive – if noise pollutes an image, it is more difficult to discern target features, and if a dark feature is present on a lighter background, it is much easier to identify than a dark feature on a dark background. Let us now review one of the most important components of developing an image-guided system, segmentation.

4.3 Segmentation

Prior to building a visual model, collected images must be partitioned and analyzed to help an observer, or algorithm, distinguish between relevant and irrelevant features. The partitioning of an image to allow the identification of target features is called image segmentation. Segmentation consists of splitting and grouping image pixels based on common features such as intensity, coloring, shapes, and boundaries [41]. There are two overarching types of segmentation that differs in automation: supervised and unsupervised [42]. Both, supervised and unsupervised segmentation, require the use of machine learning, however, the key distinction lies in the presence of ground truth image labels – the true label corresponding to the image. For supervised segmentation, a machine learning model is trained to identify certain image characteristics and features based on a set of labeled images (i.e. the ground truth). Image labeling can be manual, or it can be automated using a feature selection tool. For unsupervised segmentation, the ground truth image labels do not exist, thus, unsupervised machine learning models such as clustering algorithms can be employed to identify similar feature categories and group image pixels of the same class [23]. Although these methods differ in the extent of automation, both can utilize machine learning to aid in the process of identifying and segmenting relevant features.

In this section, we will discuss the segmentation workflow while evaluating key concepts and methodologies that are being harnessed today. The segmentation workflow involves three main steps:

1. Thresholding
2. Post-processing
3. Validation

The table in Figure 4.21 nicely summarizes the three main stages of segmentation. We will also introduce several challenges specific to microsurgery when conducting image segmentation.

4.3.1 Thresholding

Thresholding is similar in principle to the methods discussed in the image preprocessing section – the overarching goal is to strengthen the visibility and identification of target features, but it focuses more on the individual target features rather than simply strengthening the difference between the general foreground and the background. Thresholding can also be categorized as either unsupervised or supervised. Supervised thresholding involves training a model to identify target features, allowing both target detection and labeling. Unsupervised, on the other hand, relies more on separating distinguishable features in the foreground using characteristic-based clustering [43]. Although the differences between these methods are important, our focus will lean towards supervised thresholding with the assumption that anatomical target features are being imaged such that they are manually identifiable. There are two overarching thresholding methods that enhance the selection of target features [44].

1. Region-based thresholding
2. Edge-based thresholding

4.3.2 Region-based thresholding

Region-based thresholding focuses on identifying groups of neighboring pixels with similar intensity values and spatial patterns. Just as was done in the previous contrast alteration stage, region-based thresholding algorithms modify the intensity histogram in various ways – this is why region-based thresholding is also known more broadly

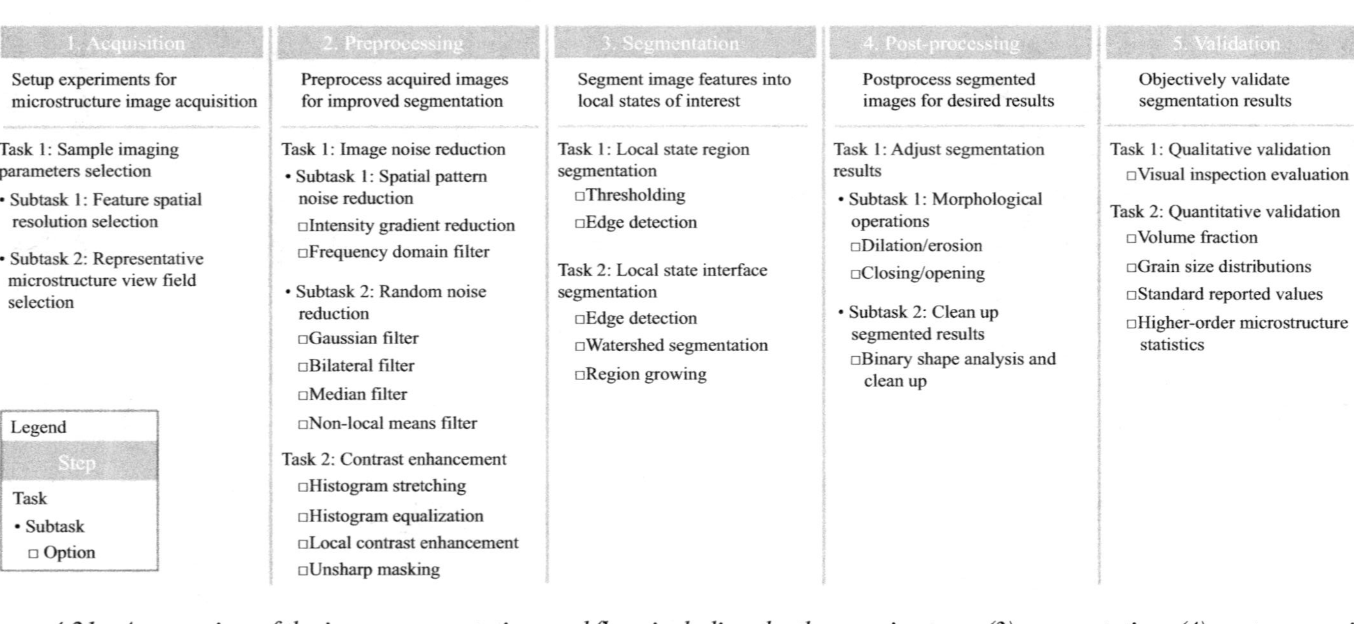

Figure 4.21 An overview of the image segmentation workflow, including the three main steps: (3) segmentation, (4) post-processing, and (5) validation

as histogram-based thresholding. There are a few key supervised and unsupervised region-based thresholding algorithms we will review:

1. Otsu's method
2. Entropy thresholding
3. Gaussian mixture model
4. K-means clustering

In the list above, the former two are supervised methods and the latter two are unsupervised methods. Each one of the aforementioned methods can be adjusted for unique applications, however, we will only cover each method from an overview perspective. In addition, note that as before, thresholding methods can also be global or local. Global image thresholding algorithms apply the same intensity threshold to each pixel, while local image thresholding algorithms apply different intensity thresholds to pixels based on characteristics such as neighboring pixel intensity values and local contrast. The line between global and local thresholding methods is not definite – there are many local thresholding methods can that be extrapolated to the entire image, and vice versa. Thus, we review thresholding from a broader perspective, focusing primarily what each technique aims to accomplish and the methodology behind it. Let us briefly review the region-based thresholding methods listed above, beginning with Otsu's method.

4.3.2.1 Otsu's method

Otsu's method is one of the most popular automatic image thresholding methods. It focuses on the idea of quantifying image features by identifying differences between grouped pixel intensities – the difference between the grouped pixel intensities is established via a selected intensity threshold. Pixels that cross the intensity threshold are now considered the foreground, and those that do not meet the threshold criteria are considered the background. This introduces a subfield of region-based thresholding, intensity-based thresholding, which is reviewed in the box.

Intensity-based thresholding

Intensity-based thresholding algorithms quantify intensity differences between local regions and highlight the regions that cross a certain intensity threshold. The amount a target feature is strengthened is determined by the type of intensity-based thresholding employed. For example, matting is a technique where each threshold is assigned a certain value between 0 and 1. The closer the value is to 1, the more likely a given area or pixel is to be associated with the target foreground. Binary thresholding is similar, but more rigid – it assigns each pixel a 0 or 1 based on an inputted threshold; if the value is 1, it is assigned to the foreground, and if the value is 0, it is assigned to the background [45].

The selection of an intensity threshold can be supervised or unsupervised based on the image intensity distribution. There are a variety of methods for optimizing the intensity threshold when using unsupervised processes. Otsu's method specifically relies on minimizing the intra-class variance (or maximizing inter-class variance) of pixel groups. The minimization of intra-class variance between the foreground and background classes is expressed in the form below [46]:

$$\sigma_w^2 = \omega_0(t)\sigma_0^2(t) + \omega_1(t)\sigma_1^2(t) \tag{4.12}$$

In the expression above, the variances of the two classes, σ_0^2 and σ_1^2, separated by a threshold t are weighted by the class probabilities, ω_0 and ω_1, and summed. The class probabilities are calculated from the intensity histogram with L bins, a process similar to the histogram equalization method in the contrast alteration section [46]:

$$\omega_0(t) = \sum_{i=0}^{t-1} p(i) \tag{4.13}$$

$$\omega_1(t) = \sum_{i=t}^{L-1} p(i) \tag{4.14}$$

Otsu's method conducted on images of cells is shown in Figure 4.22, along with the original image and an image altered via global thresholding. Notice how Otsu's

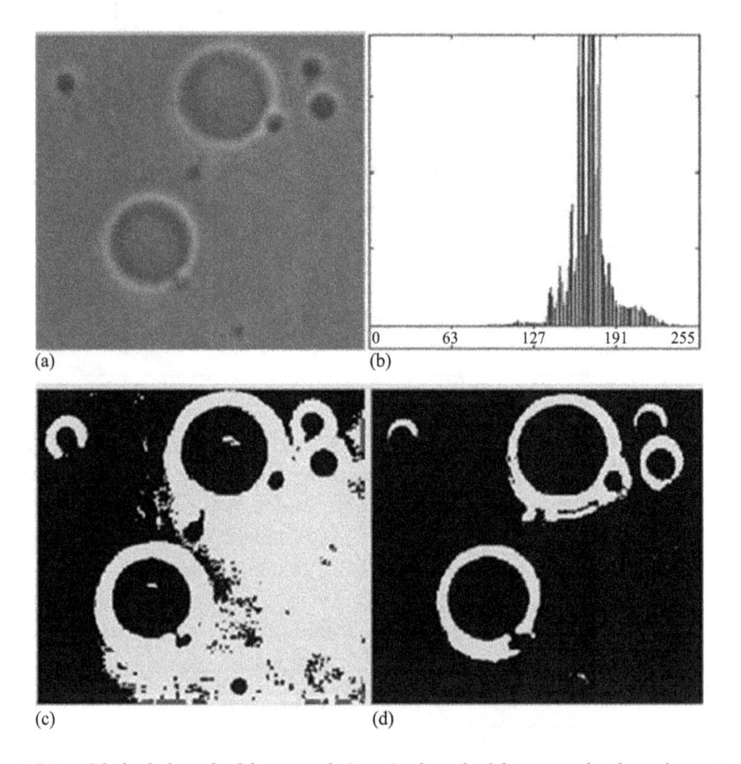

Figure 4.22 Global thresholding and Otsu's thresholding applied to the original image (a) with corresponding histogram (b), yielding images (c) and (d), respectively

method provides a more definite outline of the cells when compared to the image adjusted by global thresholding [47].

As mentioned above for Otsu's method, the selection of an intensity threshold can be manual or automatic, although the former is preferred since automatic thresholding requires specific conditions. In an ideal scenario, the intensity distribution for a given image is bimodal, such that there is a clear distinction between the foreground and background. This enables the algorithm to efficiently choose an optimal intensity threshold. In more advanced cases, however, the appearance of a bimodal intensity histogram is rare, and the threshold selection process must be supervised. The challenge of selecting a threshold in images with a less obvious distinction between the foreground and background is shown in Figure 4.23 [48].

Notice how the coins in Figure 4.23 have distinct boundaries that can be characterized by a bimodal intensity histogram. This is different from the second example showing the segmentation of a laryngeal image, which does not have a perfect bimodal distribution and, therefore, does not result to as pronounced of target features. Notice that the second example is more relevant to image-guided microsurgery – it becomes clear that in surgical settings, ideal scenarios are often not present, and thus, more advanced methods are required to discern the features of an image.

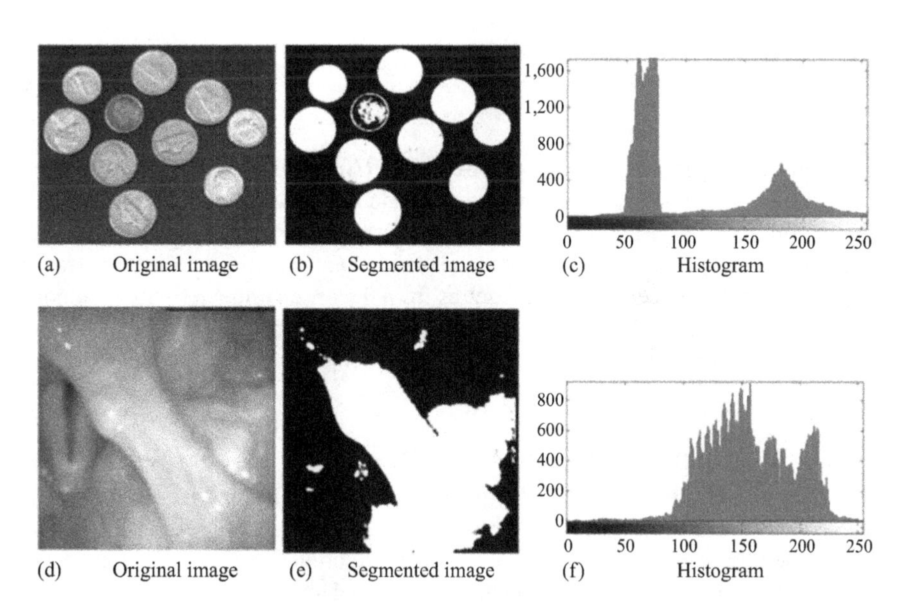

Figure 4.23 Otsu's thresholding applied to an image with a distinct foreground and background (a), yielding a properly segmented image (b) and a corresponding histogram (c). Otsu's thresholding applied to a laryngeal image (d), yielding a poorly segmented image (e) and its corresponding histogram (f).

4.3.2.2 Entropy thresholding

Entropy thresholding is similar to Otsu's method in that it iterates through different thresholds to find an optimal threshold – the difference is in the objective function (the function to be maximized or minimized) used. Many entropy thresholding methods aim to optimize the cross-entropy between two classes separated by a threshold. Entropy is a central concept in information theory and has numerous variations that are beyond the scope of this chapter. From an overview level, entropy can be thought of as the amount of information in an image, which can be represented by its probability distribution. Images with higher entropy values require more information to represent them – these images can be thought of as having a more uniform probability distribution. On the other hand, images with lower entropy require less information to represent them – these images typically have more concentrated, or skewed, probability distributions. Thus, when calculating cross-entropy, it can be thought of as the entropy difference between two histograms with certain, distinct probability distributions [49].

In context to segmenting an image, we can consider entropy thresholding as a method to acquire a target distribution (i.e. a ground truth or labeled parameter) from a given experimental probability distribution. Cross-entropy can also be used between histograms separated by a threshold. The most general model of cross-entropy can be represented by the following expression [49]:

$$H(p,q) = -\sum_{x \in classes} p(x)\log(q(x)) \tag{4.15}$$

In the expression above, the cross-entropy $H(p,q)$ is the entropy difference between our true probability distribution $p(x)$ and our model generated probability distribution $q(x)$. There are numerous variations and sub-methods under this generalized cross-entropy optimization technique, making entropy thresholding a versatile method for extracting certain target features from the original image. This is demonstrated in Figure 4.24, which shows how Shannon entropy can be used to extract the histogram distribution of a nanopore from a microscopy image [50].

4.3.2.3 Gaussian mixture model

The Gaussian mixture model (GMM) thresholding is an unsupervised method that enables us to identify deeper, more obscure, features within an image – this is particularly helpful when no discernable bimodal distribution is present. As seen above, many thresholding algorithms rely on the ability to detect changes in intensity. However, when the intensity histogram does not favorably display a partition, it is difficult to identify such a reasonable value for thresholding. GMM segmentation methods identify separate Gaussian curves within the total intensity histogram, which can then be labeled as either the foreground or background. Figure 4.25 displays how a single-intensity histogram can be decomposed into a set of Gaussian curves [51].

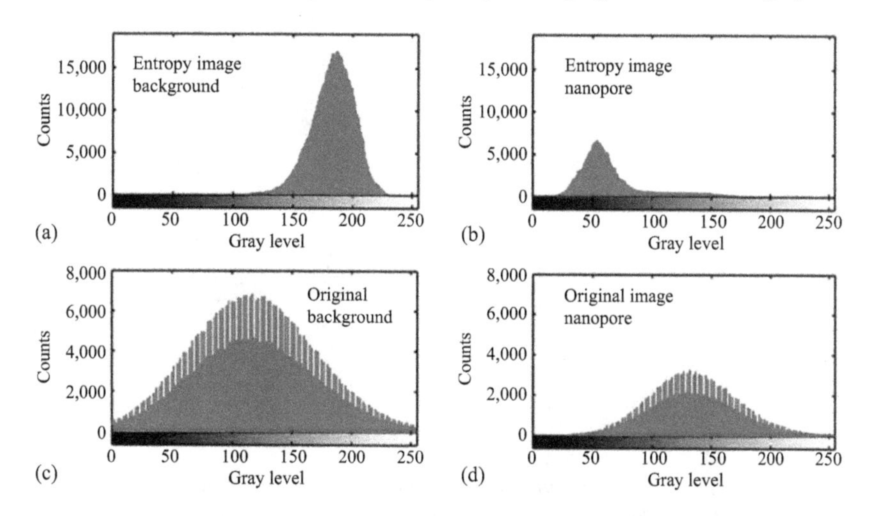

Figure 4.24 Shannon entropy thresholding applied to the background histogram (c) and original nanopore histogram (d), resulting in a filtered background entropy histogram (a) and nanopore entropy histogram (b)

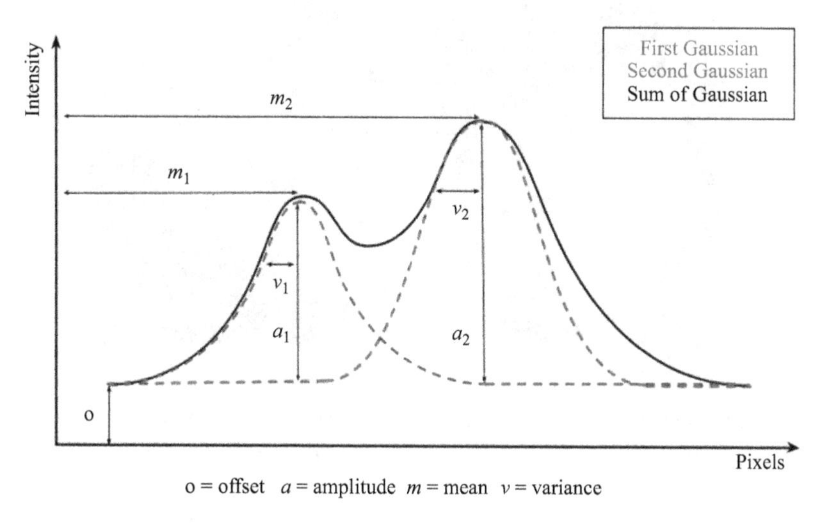

Figure 4.25 GMM used to approximated total curve (solid line) using two individual Gaussian curves (dashed lines)

Similarly, the graphic shown above can be mathematically represented by the summation of component gaussians shown below, with variables corresponding to Figure 4.25 [51]:

$$f(x) = p_1 + \left(p_2 \cdot e^{\frac{(x-p_3)^2}{2 \cdot p_4}} + p_5 \cdot e^{\frac{(x-p_6)^2}{2 \cdot p_7}} \right) \tag{4.16}$$

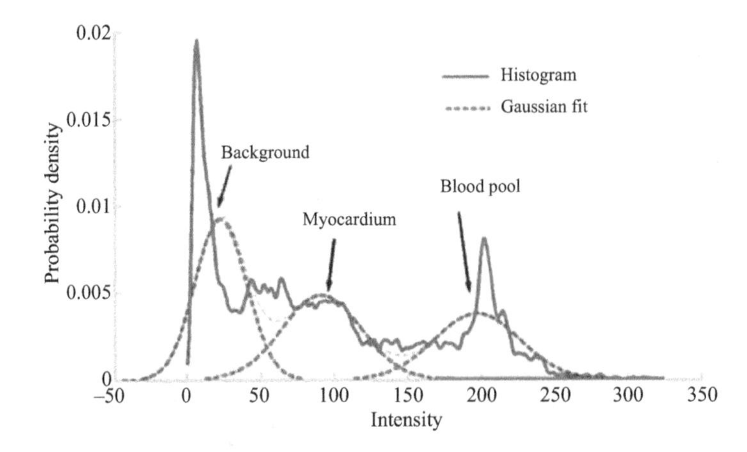

Figure 4.26 GMM approximation on the original histogram (blue line), yielding distinct gaussian curves (red lines) corresponding to the background, myocardium, and blood pool

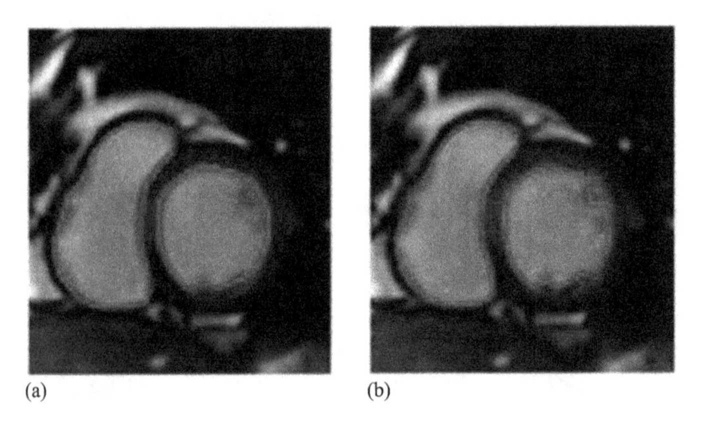

Figure 4.27 Segmentation initialization via GMM clustering on image (a), resulting in a segmented image showing the left and right ventricles (b)

In the context of image-guided systems, GMM segmentation can be applied to discern certain target features from tissue biopsy images. In Figures 4.26 and 4.27, the myocardium can be identified using GMM segmentation on the intensity histogram, which can help outline the left and right ventricles [52].

Notice that although GMM segmentation is discussed as a region-based thresholding method, it is widely applied for edge-based thresholding as well. This demonstrates that even the distinction between region-based and edge-based thresholding methods is not definite. Researchers are constantly modifying existing techniques and extending their applications. Therefore, we advise you not to get bogged down by the category definitions or terminology used for classifying methods (i.e. focus less on the nomenclature and more on the underlying functionality of each technique presented).

4.3.2.4 K-Means clustering

Image segmentation at its core focuses on classifying certain quantified representations of an image (an intensity histogram for example) to separate a desired target feature from the background. This makes clustering-based methods an intuitive choice for segmentation. From a broader machine learning perspective, K-means clustering is an unsupervised machine learning algorithm that aims to group a dataset of n observations into k clusters. Clusters are formed by identifying similarities between quantified features for each observation. In that sense, each observation can be represented as a vector with a certain number of features. Measures such as Euclidean or Manhattan distance can then be used to iteratively compute similarities between observations, thereby, forming a set of clusters [53].

Mathematically, K-means clustering can be broken down into two iterative tasks – generating observation assignments to each cluster and recalculating the centroids (center of the cluster) based on the prior observation assignments. Between these two tasks, observations are adjusted to their respective clusters until a maximum iterations limit is reached or an accuracy threshold is met. This process is also known as an Expectation–Maximization (EM) problem. The expression below shows the general objective function for K-means clustering [54]:

$$J = \sum_{i=1}^{m}\sum_{k=1}^{K} w_{ik}||x^{(i)} - \mu_k||^2 \tag{4.17}$$

In the expression above, we compute a distance measure between all of the points $x^{(i)}$ of cluster k and its centroid μ_k, $||x^{(i)} - \mu_k||^2$. For any points in cluster k, $w_{ik} = 1$, otherwise, $w_{ik} = 0$. This way, the algorithm focuses on one cluster at a time. This iterative process is concisely diagramed in Figure 4.28 [55].

In the context of image segmentation for medical images, K-means clustering can classify an image into distinct clusters based on the intensity values of pixels and the number of clusters assigned, which corresponds to the number of intensity thresholds used. The utility of using K-means clustering for image segmentation is displayed in Figure 4.29, which aims to characterize certain cell types among large cellular populations for enhancing leukemia diagnosis [56].

From Figure 4.29, we can see how K-means clustering can be employed to distinguish target cells from a population of neighboring cells.

4.3.3 Edge-based thresholding

Edge-based thresholding methods focus more on labeling the boundaries between two different pixels. Boundaries, or edges, are identified using a combination of intensity and contrast differences between the pixel of interest and neighboring pixels. When

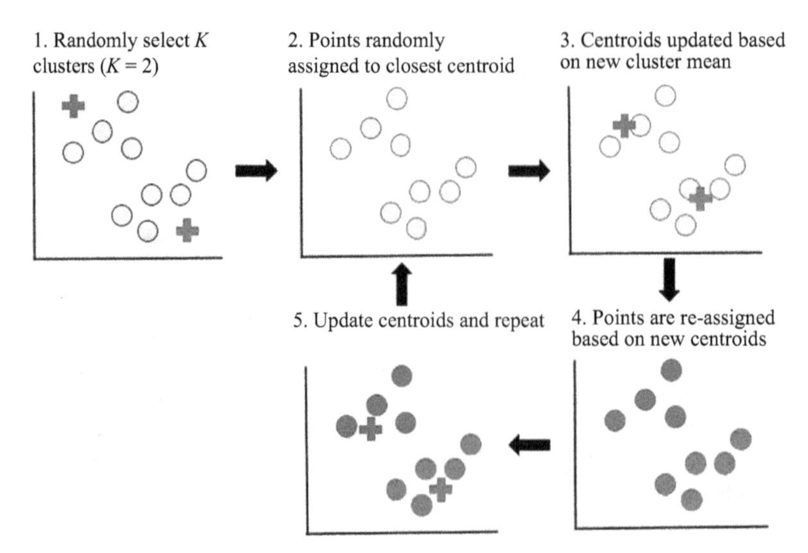

Figure 4.28 K-means clustering workflow

the image is split into pixels, each pixel represents a data point with certain characteristics. These characteristics can be thought of as individual numerical features of that particular data point. Since these characteristics are numerical, statistical methods can be applied to analyze the similarities and difference between neighboring pixels. Most of these statistical methods rely on clustering similar neighboring pixels together – if neighboring pixels are statistically different, however, there is a probability that an edge lies between the border of the two pixels. This approach is similar to many clustering algorithms that group certain data points based on a series of core characteristics (as discussed above). Once edges have been identified, region-based methods can be applied to determine which edge-enclosed area constitutes the foreground or target features [57].

The simplest edge detection model focuses on identifying step edges, which is an area of the image where the intensity rapidly changes when moving from one pixel to an adjacent pixel (i.e. there is a large step, or change, in intensity). Step edge detection models are idealistic, however, they help us gain a baseline understanding of the edge detection process. Mathematically, a simple edge detection model for a 1D image f with an edge at $x = 0$ can be expressed as the following function [58]:

$$f(x) = \frac{I_r - I_l}{2}\left(\text{erf}\left(\frac{x}{\sqrt{2}\sigma}\right) + 1\right) + I_l \tag{4.18}$$

In the expression above, the image f is smoothed via a Gaussian filter, or error function, erf and adjusted using a blur scale σ. The left and right sides of the edges can be represented as the limits $I_l = \lim_{x\to-\infty} f(x)$ and $I_r = \lim_{x\to\infty} f(x)$, respectively. The base model shown above is greatly limited in its application and requires near

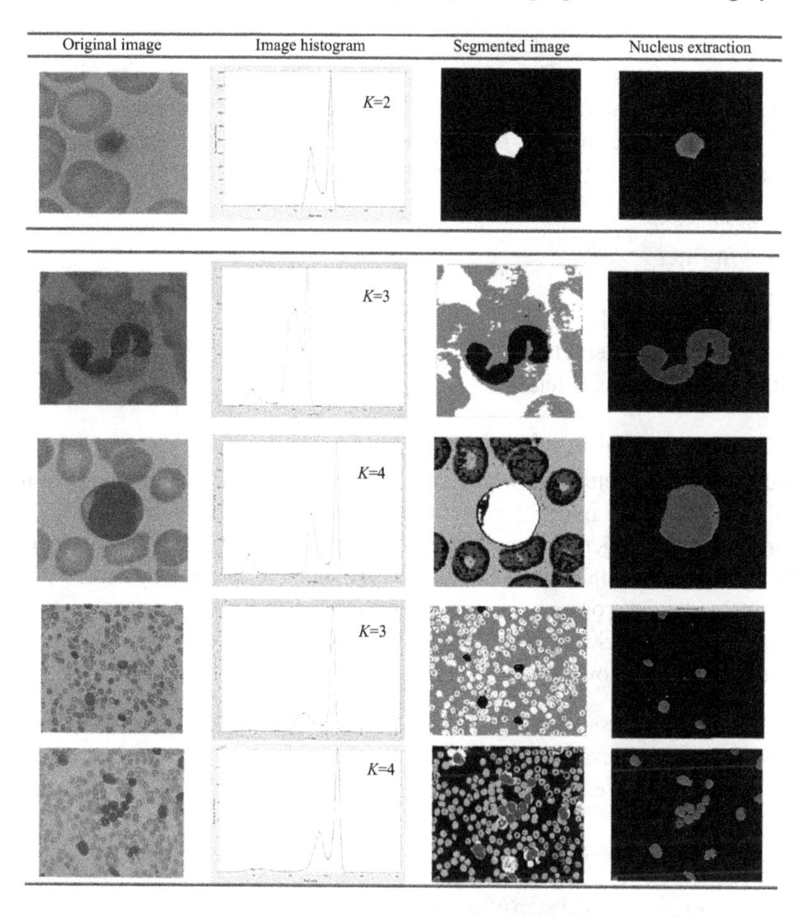

Figure 4.29 K-means clustering with different cluster values (K) applied to leukemia nucleus images, resulting in a segmented and nucleus extracted image

idealistic conditions. Thus, countless variations and extensions have been derived for non-idealistic edge detection scenarios.

Since the realm of edge detection algorithms is immensely vast, we will discuss two broad categories of edge detection methods, while naming a few methods within each category for you to pursue independently.

1. Search-based edge detection
2. Zero-crossing edge detection

4.3.3.1 Search-based edge detection

Search-based edge detection, which is a first-order derivative method, locates an edge via an edge strength measure (gradient or intensity change), then identifies the directional change and orientation of the edge. Therefore, fundamentally, a search-based edge detector is identifying two components of an edge – the intensity gradient magnitude and the direction of the intensity gradient [59]. Typically, search-based edge detectors produce thicker, more pronounced edges. There are numerous search-based edge detection algorithms, the most notable ones are listed below [60]:

1. Roberts filter
2. Prewitt filter
3. Sobel filter
4. Frei-Chen method
5. Kirsch compass kernel
6. Robinson compass filtering

4.3.3.2 Zero-crossing edge detection

Zero-crossing edge detection, which is a second-order derivative method, identifies spatial changes in the image corresponding to an edge. Zero-crossing edge detectors produce finer edges and are typically automated, but are much more complex. Mathematically, the Laplacian is calculated for the image, and an edge corresponds to where the Laplacian crosses zero. Zero-crossing usually occurs in an area with a large and abrupt intensity change [61]. Frequently utilized zero-crossing edge detection methods are listed below [62]:

1. Laplacian of Gaussian (LoG) filter
2. Difference of Gaussians (DoG) filter
3. Determinant of Hessian (DoH) method

4.3.4 Post-processing

Post-processing is the final step of segmentation prior to validating the outputs, which is a verification step focused on checking the quality of the segmented imaging data. The post-processing stage focuses on improving the quality and granularity of the selected features based on the output image's intended usage. For microsurgeries, post-processing is a key component for improving the resolution and strength of microscopic features. Iskakov *et al.* describe post-processing as having two stages: morphological operations and segmentation result cleanup. Morphological operations refer to the process of altering the physical boundaries separating individual target features. The two main morphological operations are dilation and erosion which expand and contract the object boundaries, respectively. Segmentation result cleanup refers to the process of removing extraneous features. The most common method is binary shape analysis, which harnesses geometric features like target feature area and ratio to evaluate whether that object should be retained or excluded. The diagram in Figure 4.30 from the segmentation workflow review by Iskakov *et al.* demonstrates this process. Notice the exclusion of smaller, more abstract highlighted objects to yield a cleaner image [23].

Once the image is properly segmented, it must undergo validation, which ensures the accuracy, quality, and reliability of the resulting output. The decision to pursue

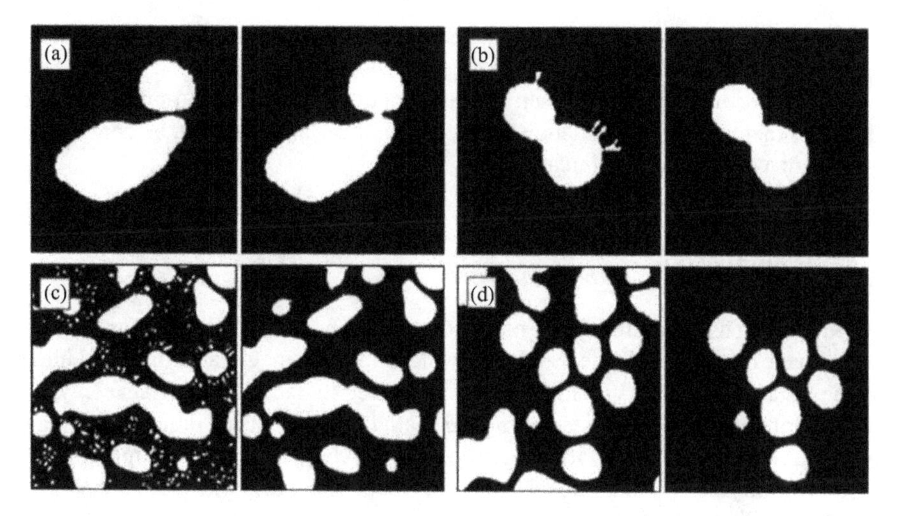

Figure 4.30 *Image (a) increases the border thickness to connect two target features; image (b) disconnects the bridges between the target features and smaller, irrelevant features; image (c) filters out local noise; and image (d) excludes any target features that are only partially in frame*

segmentation validation depends on the quality of the initially captured image, the resolution of the output image, and the target feature of interest.

4.3.5 Validation

Segmentation validation is one of the final post-processing steps and focuses on gauging a segmented image's correspondence to the ground truth. Validation often requires *a priori* knowledge of the microstructures being studied – prior studies provide a ground truth for comparison. However, in most novel studies, ground truth data is unavailable. Thus, the absence of ground truth data and accuracy markers has made validation one of the most difficult stages of image segmentation. Despite this, researchers have developed rigorous methods for validating segmented images, even in the absence of ground truth data. Broadly, these methods can be categorized as either qualitative or quantitative techniques [63].

Qualitative methods involve filtering the imaging data and reviewing the outputs with leading scientists in the field. Although qualitative methods are simpler and less time consuming, for larger datasets, scientists harness quantitative methods to optimize efficiency. Iskakov *et al.* describe quantitative validation methods as those that utilize microstructure statistics, including metrics such as volume fractions, grain size distributions, and spatial indices [23]. A frequently used quantitative method involves having experts review a set of images and develop a rough estimation of the ground truth [64]. Labeled ground truth images can then be quantified (via a matrix or histogram) and compared to the model segmented images via several probabilistic accuracy metrics [65].

Once the imaging data has been validated for quality, the registration process can begin. Without adequate validation, segmented images are unable to be confidently applied in a laboratory or clinical setting.

4.4 Registration

When imaging data is collected, the modalities used, time of collection, and angle often vary. Image registration is the process by which multiple images are combined into a single visualization. The visualization is represented in one coordinate space, which results from integrating both temporal and spatial components of multimodal imaging datasets [66]. The registration workflow ultimately helps bring multiple images, even from varying sensors, into spatial correspondence, which enhances the navigational capabilities of image-guidance systems [67]. This process can be grouped into two broad categories, which are often conducted numerous times to attain the most accurate imaging data [68]:

1. Image labeling, which consists of identifying and marking relevant target features.
2. Transformation, which consists of matching experimental imaging data to reference images.

There are numerous challenges when registering an image, including accuracy, reliability, geometrical correspondence, and image quality. This section will first examine the image registration workflow, then briefly review two most common registration methods that are being applied today. We begin by reviewing the first step of registration, image labeling.

4.4.1 Image labeling

Image labeling data is the first step of image registration. In many cases, this process is automated – the algorithm uses intrinsic information within an image to mark relevant features. If two images differ drastically in content, however, manual labeling may be necessary. Manual image labeling focuses on identifying the extrinsic information present within an image; the user is extrinsically labeling geometrical landmarks to enable image correspondence. The diagram in Figure 4.31 displays how two differing images can be integrated into a single visual representation [69].

Notice how the two images on the left vary in contrast, intensity, and color. Correlating common landmarks in each image enables the second image to be overlayed onto the first, which results in a single visual representation. When the images are similar in content, labeling relies on identifying similar features between multiple images and can be automated. As mentioned above, automated labeling focuses on identifying correspondence between the intrinsic information of multiple images; features that are intrinsic to the image are harnessed to integrate multiple data points. Utilizing

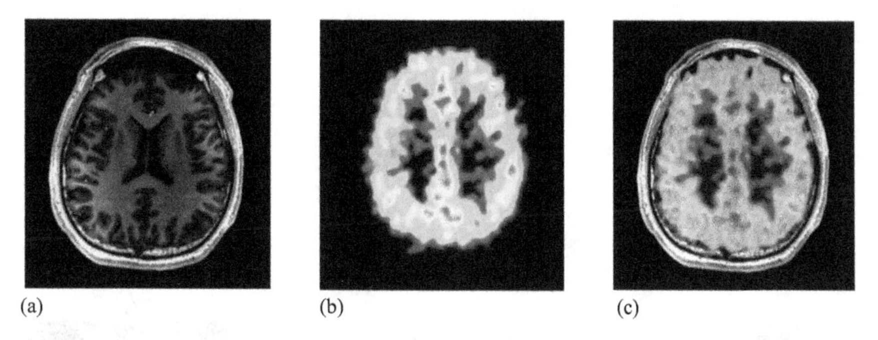

Figure 4.31 Image (a) represents an MRI scan; image (b) represents a SPECT scan; and image (c) displays a combination of images (a) and (b). The red arrows indicate a spatial correlation between images (a) and (b).

intrinsic information requires quantitatively measuring the similarity between multiple images – this is where similarity measures can be applied. Similarity measures are algorithms that map the correlation in characteristics between two images. These algorithms compare intrinsic image features such as intensity, contrast, and spatial characteristics to enable simple visual optimization. For example, the diagram in Figure 4.32 displays two similar images taken from MR and CT scans [70]. The graph on the far right represents a correlation between the intensity of the two images. As the images become more aligned, the correlative relationship becomes more linear, which indicates improve spatial correspondence between the images.

These graphical representations are derived from similarity measures. Although the mathematics behind similarity measures is beyond the scope of this chapter, there are numerous measures ranging in complexity that utilize features such as intensity, pixel size, spatial markers, and relative contrast – you can explore these methods independently using the materials referenced throughout this section. Once the appropriate adjustments have been made, relevant features can be tagged within the resulting image. This methodology can be grouped into two separate techniques as well: feature identification and feature matching. The range of feature identification and matching algorithms is quite expansive – it includes corner detection algorithms, local intensity-comparison functions, and numerous other specialized techniques. Figure 4.33 diagrammatically outlines the workflow and relevant methods for feature identification and matching [71].

Although the workflow above seems daunting, we will aim to simplify the outline by briefly reviewing the two aforementioned image labeling techniques while noting some key methods for achieving identification and matching. Let us begin by discussing feature identification.

4.4.2 Feature identification

Feature detection identifies specific prominent anatomical features within preoperative and intraoperative imaging. Key recognition of these diagnostic features is

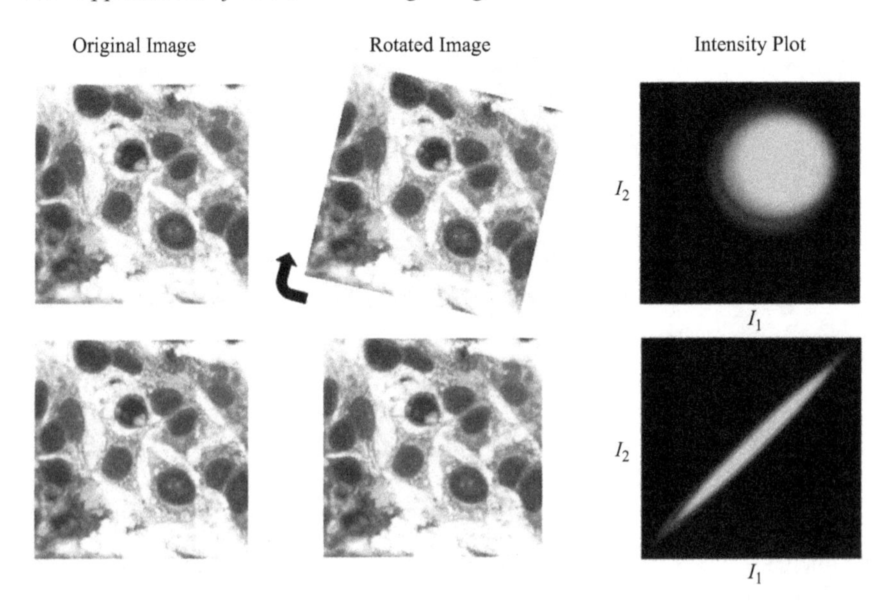

Figure 4.32 Image (a) represents a reference image; image (b) is rotated to fit to reference image; and the two graphs on the far right represent the intensity relationship between images (a) and (b) as image (b) is rotated

essential for accurate surgical procedures. Some examples include an extractable tumor or a round window. The resolution of the imaging modalities directly impacts the validity of feature detection. The main focus of feature identification algorithms is to detect certain characteristics of a target feature (edges, contours, areas, etc.) that can be distinguished from the background. We will discuss two main feature identification techniques:

1. Harris corner detection
2. Scale-invariant feature transform (SIFT)

4.4.2.1 Harris corner detection

Harris corner detection is a feature detection algorithm that aims to identify the corners of our target features. We are familiar with edge-based detection methods, as we reviewed in the segmentation thresholding section – an edge is characterized by a change in intensity or brightness of an image. The intersection or junction between two edges is known as a corner. The difference between an edge and a corner is nicely illustrated in Figure 4.34 [72].

The mathematical process for Harris corner detection relies on identifying intensity changes across groups of pixels. We can begin by taking the sum of squared distances (SSD) between the intensities for a window of pixels $W(x, y)$ and a window

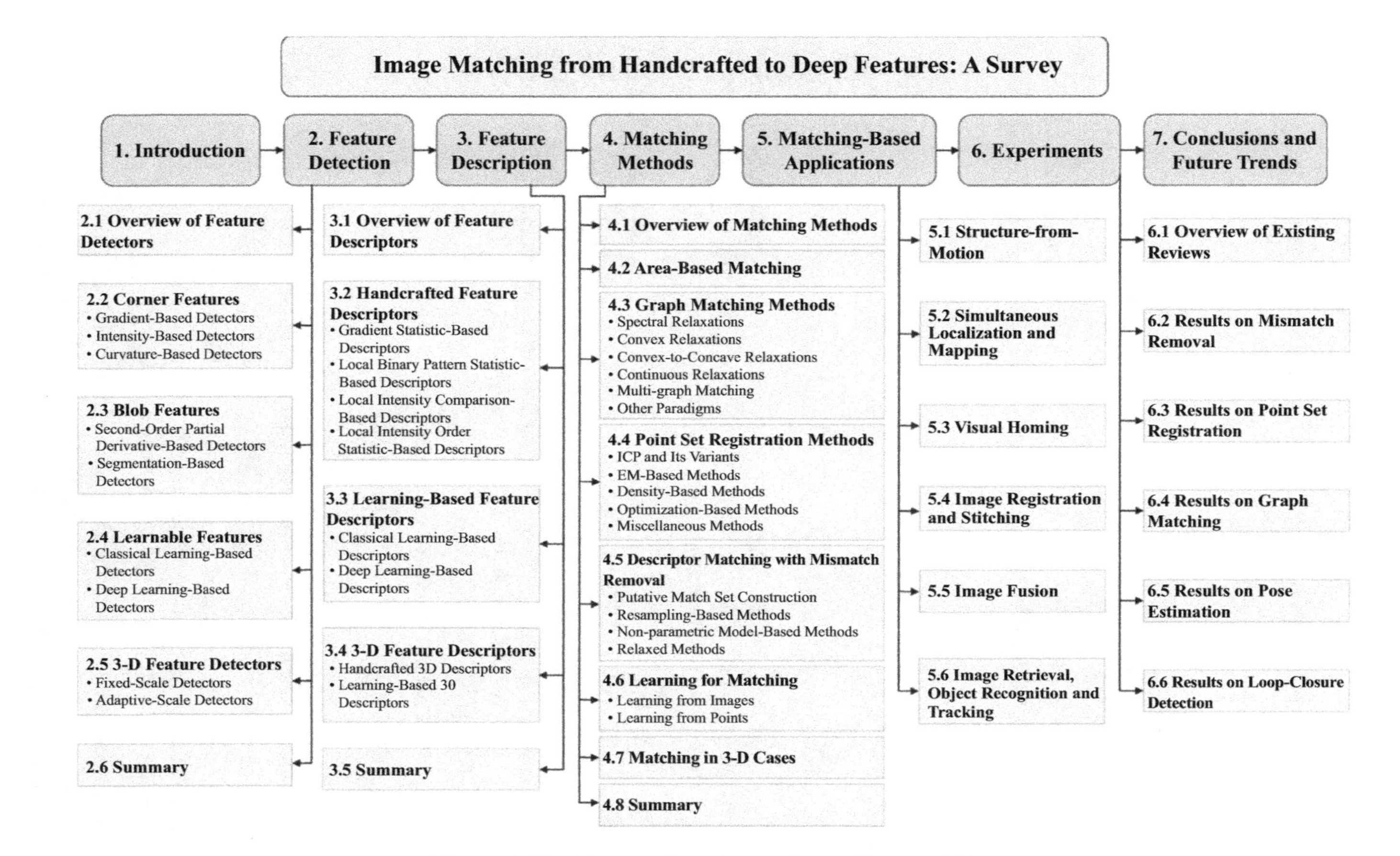

Figure 4.33 Feature detection and matching workflow

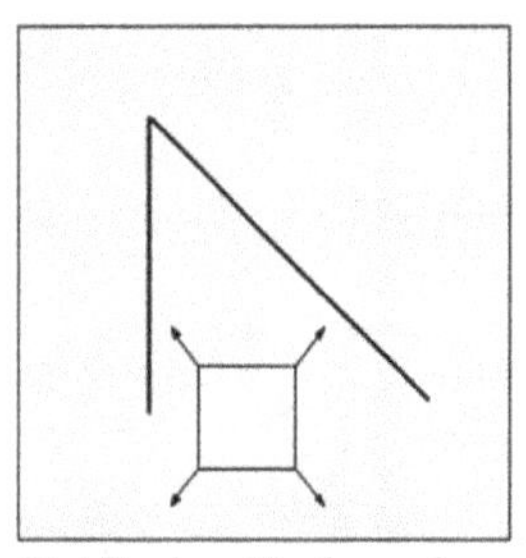

Flat Region: No change in any direction

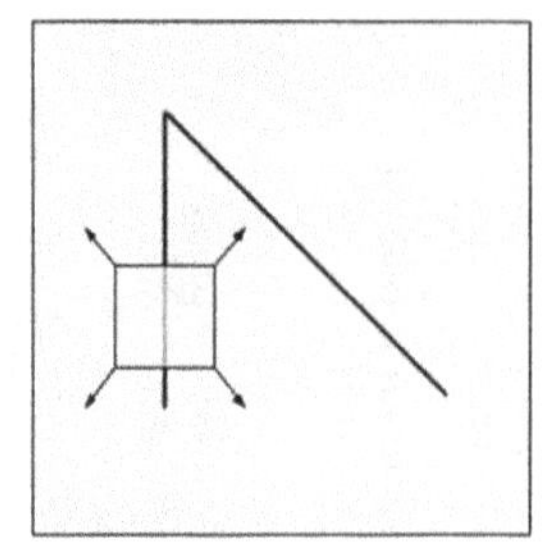

Edge: No change in edge direction

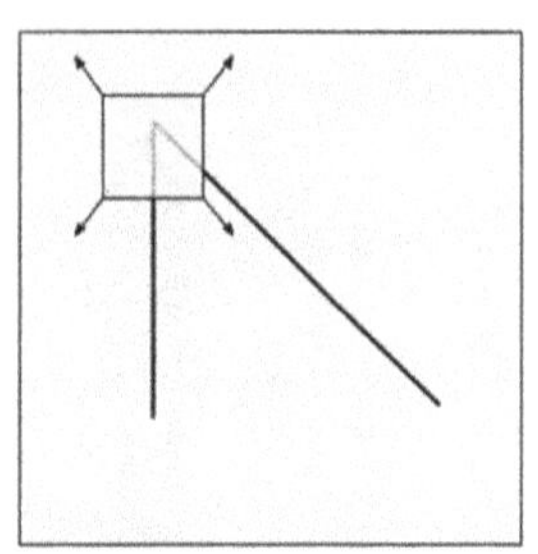

Corner: Significant change in all directions

Figure 4.34 Diagram displaying a flat region (left), edge region (center), and corner region (right)

of pixels shifted by $W(x + \Delta x, y + \Delta y)$. This results in the intensity change expression displayed below:

$$f(\Delta x, \Delta y) = \sum_{(x,y)\in W} [I\,(x + \Delta x, y + \Delta y) - I\,(x, y)\,]^2 \tag{4.19}$$

Upon approximating $I(x + \Delta x, y + \Delta y)$ using a Taylor expansion and converting the expression to a matrix format, we get the following structure tensor M, where I_x and I_y correspond to partial derivatives of our intensity matrix I:

$$M = \sum_{(x,y)\in W} \begin{bmatrix} I_x^2 & I_xI_y \\ I_xI_y & I_y^2 \end{bmatrix} \tag{4.20}$$

The eigenvalues of our structure tensor M enable us to calculate a score R, which indicates how certain shifts in the window correspond to the magnitude of intensity changes. The process of calculating the eigenvalues of M, (λ_1, λ_2), and our score R is shown below [73]:

$$R = detM - k(traceM)^2 \tag{4.21}$$

$$detM = \lambda_1\lambda_2 \tag{4.22}$$

$$traceM = \lambda_1 + \lambda_2 \tag{4.23}$$

The calculated score R can have three ranges of values that correspond to either a flat region, edge, or corner, with respect to the diagram above [74]:

1. If R is small, then that section of the image is a flat region.
2. If $R < 0$, then that section of the image is an edge.
3. If R is large, then that section of the image is a corner.

In general, the Harris corner detection algorithm substantially improved upon one of the earliest corner detection algorithms, the Moravec detector, by removing the invariance to direction and image rotation when identifying a corner. It is still

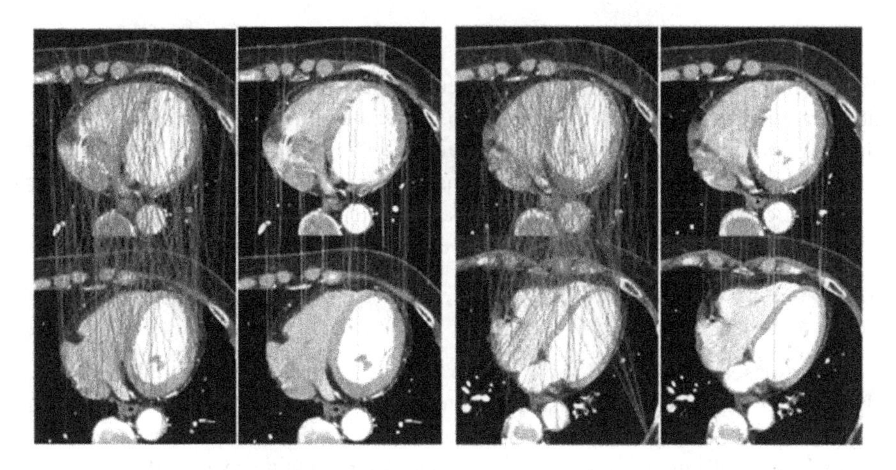

Figure 4.35 Feature identification and matching using SIFT

commonly used in many image processing workflows and has been varied for extended applications in other areas [71].

4.4.2.2 Scale-invariant feature transform

Scale-invariant feature transform (SIFT) is another powerful method for feature detection and can also be applied for feature matching. The SIFT process is quite elaborate and mathematically dense. We will provide a brief qualitative overview of SIFT and its advantages and disadvantages with respect to other detection methods.

SIFT is a feature detection method that identifies points of interest – labeled keypoints. The difference between canonical edge detection methods and SIFT is the size and orientation invariance of the located keypoints produced by SIFT [75]. In addition to identifying more definite match points, SIFT is also highly efficient at feature matching in the presence of noise. Heavy statistical noise, such as gaussian or salt-and-pepper noise, can distort an image, making feature detection and matching difficult [76]. By utilizing a combination of the second-order feature detection, contrast adjustment, and localization methods, SIFT is able to bypass detection and matching interference due to noise. The SIFT detection and matching process is demonstrated in Figure 4.35 [77].

Although SIFT is one of the most accurate methods for feature detection and matching, the complexity of the algorithm decreases its efficiency. Competing methods have been introduced to combat this efficiency barrier, which are listed below [78]:

1. Rotation-invariant feature transform (RIFT)
2. Speeded up robust features (SURF)
3. Gradient location-orientation histogram (GLOH)

4.4.3 *Feature matching*

Feature matching associates the detected features from the preoperative and intraoperative images to one another. The mapping is calculated through every pixel's intensity distribution for both images. In image-guided system applications, the features are matched based on similarity measures among corresponding anatomical and pathological information between preoperative and intraoperative image data sets. Feature matching methods can be grouped into two overarching methodologies [79]:

1. Area-based feature matching
2. Feature-based matching

Area-based feature matching, which is also referred to as correlation-like or dense matching, focuses on identifying similarities between two images based on an assigned pixel window. Rather than evaluating each pixel independently and identifying points of interest, area-based methods compare wide swaths of two images. This enables area-based methods to compare images rapidly, however, in many cases, matching precision is diminished [80]. Feature-based matching employs numerous previously discussed techniques, including SIFT, Harris corner detection, and multiple edge detection algorithms. Since the main task is to identify points or edges with correspondence between a model and experimental image, feature matching methods go hand-in-hand with techniques discussed in the thresholding and feature detection sections [81,82].

Ultimately, the goal of feature matching is to optimize similarity measures to achieve temporal and spatial correspondence. In these applications, precise transfer of corresponding information between preoperative image and intraoperative patient image is a crucial but challenging task. Poor matching can result in alignment errors, leading corresponding information to be transferred to the wrong position. This can potentially have serious clinical consequences for the patient [83,84]. Therefore, proper image labeling is a necessity. This propels us into the second stage of image registration: transformation, which allows overlay of the matched features.

4.4.4 *Transformation*

Transformation is the physical process of geometrically merging two images to generate a single visual representation mapped onto one coordinate plane. For microsurgeries in particular, this process relies heavily on the quality and accuracy of the similarity measures discussed above. Transformation is another process that requires optimization, with the goal of aligning two images with the least number of physical conversions [85]. The similarity measures between two images can similarly be mapped to derive a topological representation. This is shown in Figure 4.36 – the topological plane is an arbitrary representation demonstrating the relationship between a coordinate plane and a similarity measure. Based on whether the similarity measure needs to be minimized or maximized and the complexity of the transformation, the shortest path to match an image to a reference can be mapped. In the case below, the similarity measure needs to be minimized, thus the red arrows map the shortest path of transformation.

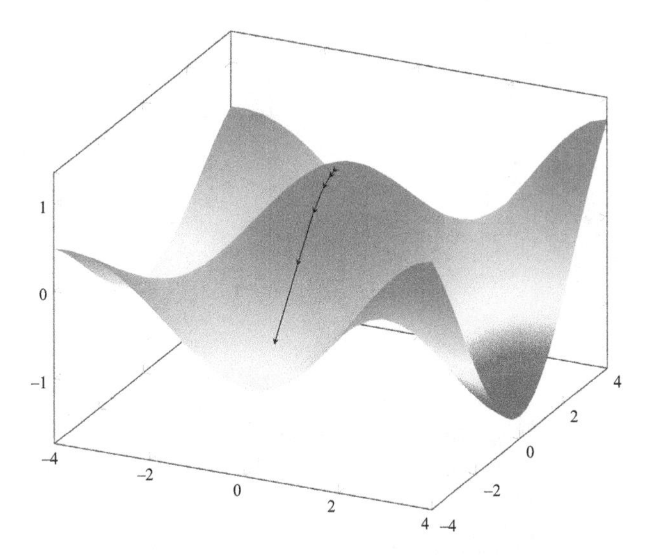

Figure 4.36 Topological representation displaying the relationship between two arbitrary similarity measures. The red arrows are mathematically derived visuals to identify the shortest path between a reference and experimental image.

Notice that the representation above has multiple local minimums, which detracts emphasis from the global minimum – this presents a large barrier to optimization. In addition, as more degrees of freedom are factored in, the model increases in dimensionality, introducing further difficulty when attempting to identify a global minimum. The optimization process is beyond the scope of this chapter; however, an entire branch of mathematics and computation is dedicated towards creating algorithms to determine the shortest path of transformation [86].

There are two broad types of transformations: rigid and nonrigid. Rigid transformations are linear and preserve the shape and size of the original image. This includes rotation, translations, and reflections, which preserve the Euclidean distance between pixels. Rigid transformations also include shearing and scaling, which preserve parallelism. In general, rigid transformations are global – the quantitative change applied to each pixel is uniform. Optimization of rigid transformations is a relatively intuitive process that can be conceptualized without mathematical expressions.

Nonrigid transformations, in contrast, are nonlinear transformations that alter the shape and/or size of the image – the Euclidean distance between pixels is altered. Some nonrigid transformations include stretching, shrinking, and curving. Nonrigid transformations are more local – point-specific deformations are identified and shifted accordingly to match the reference image. The process of identifying deformations is known as interpolation – this involves shift estimations between two points rather than beyond two points, which is known as extrapolation. An example of local nonrigid transformation is demonstrated in Figure 4.37 [87]. Notice how the nonrigid deformations alter how the red outline is mapped onto the blue outline.

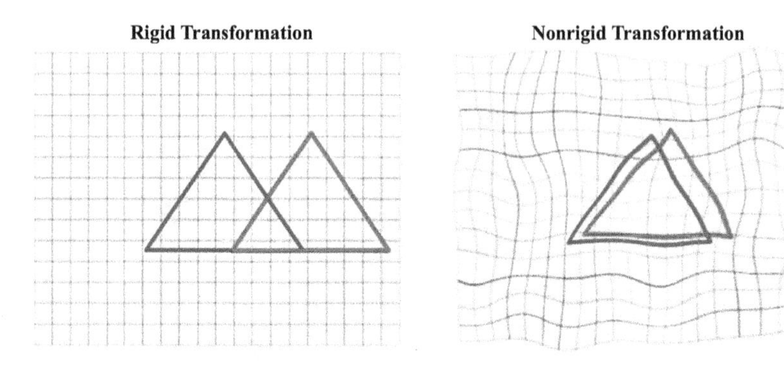

Figure 4.37 The top row displays rigid transformation while the bottom row displays nonrigid transformations. Nonrigid transformations alter the space of the image, distorting the blue and red lines to reach alignment.

Although more difficult, since the parameters of nonrigid transformation functions are very flexible, transformational versatility is greatly enhanced. This added versatility, however, comes at the price of accuracy and resolution. This can be analogized in the form of shrinking and stretching images in any text processor – as you stretch and deform the image, it loses resolution, ultimately becoming blurrier. For microsurgeries in particular, the loss of resolution and accuracy can actively hinder the reliability of an image-guided system. Mapping microsurgical images and features like nerve endings and arteries requires millimeter accuracy, thus interpolation must be meticulously conducted in order to retain the most information within an image. In order to achieve this, researchers use a technique called transformation resampling.

The process of image resampling geometrically transforms the coordinates of the source image into the target image using a mapping function. For image-guided systems, several types of images from a patient are obtained (i.e. both preoperative and intraoperative images with varying resolutions and coordinate systems are all collected). The process of resampling brings all of them into a common coordinate frame. In registration, resampling uses transformation function, thus, information in both images is properly mapped and a more informative registered image is generated.

Broadly, transformation merges multiple images into one spatial representation. The process is highly detailed and requires careful attention to detail and resolution. Note that the transformation and image labeling steps are interchangeable based on the quality and resolution of the collected imaging data. For example, if a collected image closely matches the reference, then image labeling can be conducted without any complex transformation steps. Thus, the interdependent nature of image-labeling and transformation often creates a complicated registration workflow that differs between cases. Despite this complexity, registration can still be broadly grouped into two categories:

1. Feature-based registration
2. Intensity-based registration

4.4.4.1 Feature-based registration

Feature-based registration recognizes and marks complex parts of an image such as curves, edges, points, and lines. These are identified by comparing the pixel values coordinates between images and checking for certain cues such as shapes, textures, and colors. This registration process depends on mapping the features (such as a blood vessel outline) from the preoperative image to the intraoperative images. Neurosurgeons often apply this registration technique for intricate procedures. The transformation between images is efficiently processed because it depends on geometric values. Feature-based registration can be divided into three categories: surface-based registration, curved-based registration, and point-based registration. Surface-based registration would not be suitable for structures with large deformation between imaging. Curve-based registration anticipates these deformations and elastically contours the image to fit the target image. Point-based registration utilizes exterior or interior anatomical landmarks for coordination. Point-based registration is the most widely adapted registration in commercial navigation systems because of its tested error minimized properties. Challenges still arise while using an instrument navigated by multiple landmarks. Thus, modifications to the aforementioned methods are continuously being made to improve navigational accuracy.

4.4.4.2 Intensity-based registration

Intensity-based registration examines scalar values of pixels without regard to feature-based landmarks. This is the most popular registration process and aims to maximize similarities between the preoperative and intraoperative images. The matching parameters are essential for proper synchronization. These parameters include cross correlation, correlation coefficient, normalized mutual information, normalized correlation, mean squared difference and the sum of absolute differences. The transformation for intensity-based registration is conducted iteratively. During each iterative cycle, the image pixel value is compared by position. The intensity-based registration shows more accurate results because more image data is processed. Further, intensity-based registration is best when comparing different imaging modalities and dimensions. Also, it does not require much pre-operative user image manipulation.

Segmentation-based registration maps correlating features from source and target segmented images. The registration can be applied for rigid or deformed structures. Segmentation methods may also be employed for calculating tissue capacity, localizing pathology, or guiding interventions. When the images are missing anatomical parts, segmentation-based registration is better fit than intensity-based registration or feature-based registration. Additionally, this registration supports multi-modalities registration and depends on the accuracy of the segmentation process.

Now that we have reviewed the individual steps and broader methods to accomplish registration, we will review the final step in developing image-guided systems for microsurgical procedures, visualization.

4.5 Visualization

Visualization focuses on using the processed imaging data to support microsurgical procedures. The visual model that is rendered is the component of image-guided systems that interfaces with the clinical environment the most. Visualization strategies range from augmented-reality (AR) overlay models to simple navigational cues being projected onto the lens of a microscope. Unlike traditional surgical procedures, microsurgical procedures require visualization techniques that can be adapted to the microscope scale of the operation. This introduces many challenges, including accuracy, field of view, and resolution. In this section, we will discuss the range of visualization techniques in the context of microsurgery while simultaneously addressing relevant challenges. Prior to reviewing specific visualization modalities, however, we must review a few general visualization concepts and strategies.

4.5.1 Real-time motion tracking

Real-time motion tracking is the method by which incoming visual feed is processed to provide navigational cues during a procedure. This requires preoperative calibration between the position of the patient and the position of the microscope. Once the microscope's position is registered in reference to its placement in the body, its movements can be tracked relative to landmark anatomical features. Often, tracking sensors are placed in the operating room to map the position of the microscope relative to the patient. Microscopes will contain LED markers that enable the tracking sensors to identify it. Emerging methods, however, are enabling preoperative registration to spatially align the position of the microscope and relevant patient anatomy. Using motion trackers attached to the microscope itself, each movement is measured relative to a particular target feature [88]. Although this method does simplify the tracking process, gradual increase in noise may distort the estimation of relative position. At the microscale, this can lead to large diversions from the area of focus, thus, much effort is being put towards improving the accuracy and precision of motion tracking systems.

4.5.2 Overlaying

Overlaying is another method employed during visualization that uses holographic projections to highlight target anatomical features. Previously, we learned about the various stages of developing an image-guidance system. The resulting output of the image processing workflow is a computational visualization of the target anatomical feature within the human body. By feeding a set of images through machine learning classifiers, we can enable the computer to capture real-time imaging feed, process it, and identify the position of the target feature. Then, visual cues or lighter holographic projections can be displayed on the microscope lens to guide the surgeon. Figure 4.38 displays what a surgeon may see with holographic projections.

Overlaying is particularly helpful when target anatomical features tend to blend in with the surrounding environment. Using computational power to identify subtle nuances in intensity, shape, and contrast, the blended features can be outlined, greatly enhancing the surgeon's navigational capabilities.

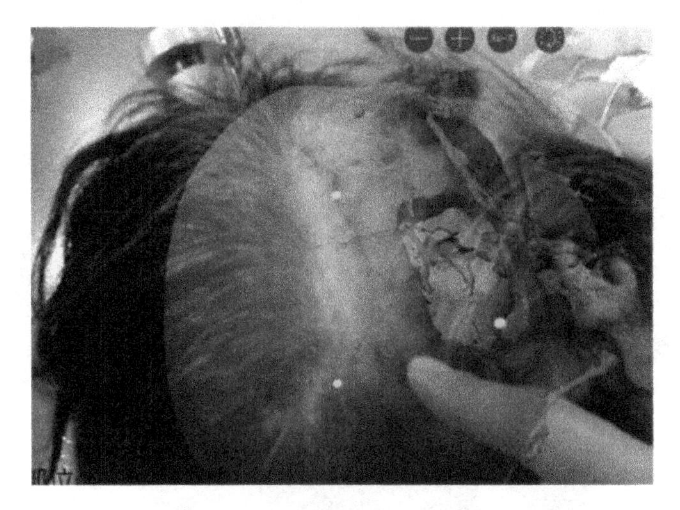

Figure 4.38 A virtual craniotomy developed via the Amira imaging program projects a tumor overlay onto the patient's head

4.5.3 Image-guided microscopic surgery system

With a brief review of both real-time motion tracking and overlay strategies, we can delve into an emerging microscopic image-guidance system. One method that we will review focuses on projecting navigational cues onto the lens of a microscope and overlaying, or outlining, target anatomical features to direct a surgeon during a procedure. The development of this system is aptly described in a study conducted by Nobuhiko Hata. Hata *et al.* use histogram-based alignment models to overlay 3D renderings with visual feeds from microscopes [89]. By receiving video feed from microscopes during the surgery, the tracking algorithm localizes certain target features and presents navigational cues directly onto the microscope lens. This not only simplifies the surgical procedure but also minimizes its invasiveness nature. The principle of overlaying navigational cues in real-time has led to the development of numerous other microsurgical systems, including those based on augmented-reality technologies.

4.5.4 Augmented-reality-based microsurgical systems

Rather than using 2D visual cues projected onto a microscope lens, augmented-reality-based microsurgical systems add another spatial component to provide surgeons with a 3D profile of the patient's anatomical environment. Using holographic headsets, such as the Microsoft Hololens, surgeons can easily identify target features using holographic projections that highlight areas of interest. The process of using augmented-reality-based headsets is described further in reviews by Andrews *et al.* and Southworth *et al.* [90,91]. Figure 4.39 displays how augmented-reality technologies provide a more resolved topological understanding of the patient's anatomy [92].

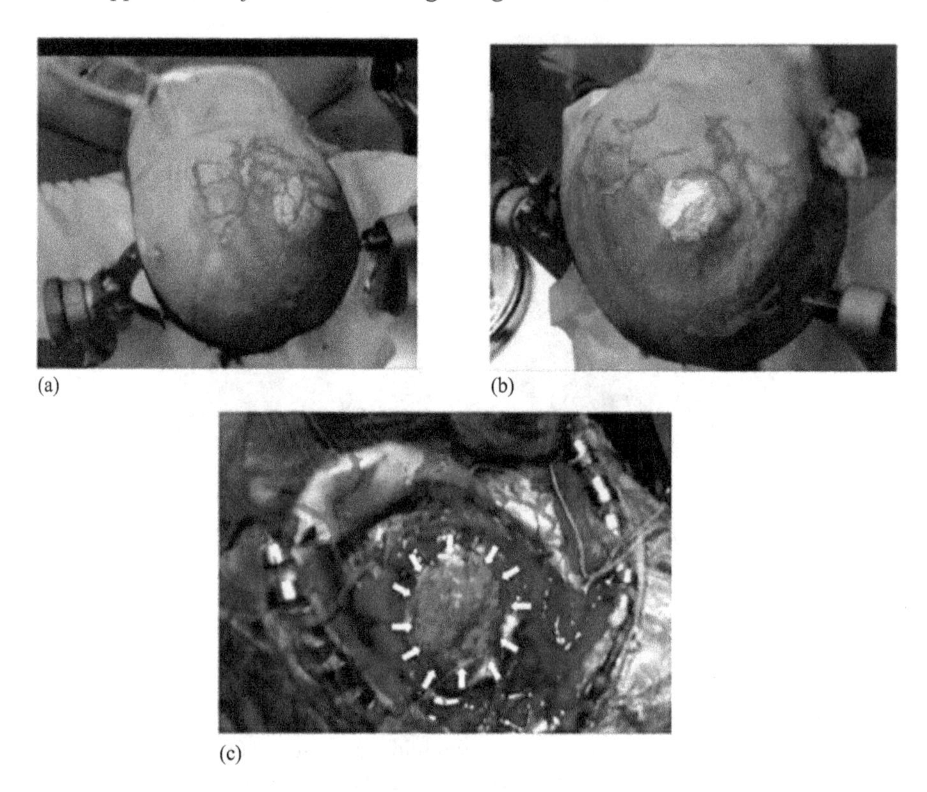

Figure 4.39 Virtual scans of the tumor and bridging veins are displayed in images (a) and (b). The image is projected onto a video feed of the surgery (c).

Augmented-reality-based microsurgery is a rapidly growing field with numerous clinical benefits, however, the presence of technical challenges such as cost and infrastructure shortages is still limiting the technology's applicability. Companies and startups around the world are still working to alleviate technical and practical barriers for augmented-reality-based microsurgical systems – a few of which are mentioned in the following section.

4.6 Challenges

Throughout this chapter, we have largely discussed the vast number of technical methods used for developing an image-guided system for microsurgical procedures. The breadth of these methods may give the false impression that image-guided systems are rapidly making their way towards the clinical environment. Although image-guided microsurgical systems provide a promising path towards improving procedural safety and outcomes, there are several key practical barriers that remain to be addressed. These practical challenges primarily revolve around infrastructure, safety, and cost.

4.6.1 Infrastructure challenges

Integrating augmented-reality-based surgical procedures with the current clinical workflow is logistically very challenging. The practical challenges can be subdivided into software and hardware challenges. Since most AR platforms operate through either Bluetooth or Wi-Fi, lack of connectivity could seriously hinder a procedure. Furthermore, the software created for processing AR data needs to mesh well with all of the software currently being utilized in a particular hospital. Therefore, a single AR device must be compatible with EMR databases, image collection tools, and other healthcare software. Considering that each hospital utilizes different software, the challenge of complete software adaptability, or interpolation, still persists.

The physical tools required for performing image-guided microsurgery also hold their own challenges. First, the use of bulky goggles or headsets during a surgical procedure can perturb a surgeon's ability to operate. An alternative to bulky headsets involves strategically placing cameras around the operating room to track a surgeon's movements. However, this method also becomes challenging, because the visualization of the procedure in the absence of a headset requires large monitors and other scanning devices that tend to crowd the operating room. Second, since there is often a large disconnect between engineers and medical practitioners, image-guided systems can be difficult to adopt in the clinic due to their inherent complexity of use. Therefore, the cost and resources for training the surgeons on how to use the devices is also a barrier for image-guided systems.

4.6.2 Safety challenges

Another broad area limiting the penetration of image-guidance systems into the clinical environment is patient safety – for both the patient physically and of their data. Out of the near 3,000+ studies conducted on virtual reality systems, only roughly 600 examine the applicability of these systems within the clinic. Furthermore, even fewer studies have been conducted that directly analyze the efficacy and usability of image-guided systems by surgeons. Therefore, the confidence in the underlying technology and validity image-guided systems still remains to be completely affirmed. This lack of studies supporting the use of image-guided systems for microsurgery directly fuels regulatory concerns and privacy issues as well. Image-guided systems offer a novel way to visualize and analyze patient data. The availability of new healthcare data opens up conversations regarding accessibility to this new data, specifically, whether healthcare insurance companies should be able to use this data to determine whether a procedure will be insured.

4.6.3 Cost challenges

The final practical limiting area is cost. Most of the equipment and technology that constitute AR or image-guided surgery is expensive, particularly when being implemented within the surgical setting. Hospitals must weigh the option of saving money over a long-term period due to reduced costs of surgery via AR and image-guided models versus the implementation and usage costs of these systems. Furthermore, developing AR platforms requires an immense amount of funding. For just one system software, the development expenditures can range from 300K to upwards of a million dollars. Therefore, the cost of producing and implementing AR and image-guided surgical tools is a significant barrier to entering the market.

4.7 Chapter review

Image-guided microsurgery is a potential solution for numerous clinical challenges involving difficult procedures. Despite the barriers that must be overcome to achieve complete clinical adoptability, the benefits of image-guided microsurgery systems outweigh the aforementioned hurdles. In this chapter, we have become familiar with the workflow and methodology to create algorithms and technologies for image-guided microsurgery systems from an overview perspective. We covered, first, data collection and preoperative imaging; second, preprocessing the raw images to yield higher quality images; third, segmenting imaging data in to processable components; fourth, registering and adjusting images for learning algorithms; and fifth, displaying and visualizing the images for clinical applications.

Although the five fundamental steps covered in this chapter lay the foundation for novel methodologies, it is only the surface of what has been achieved. Beyond these four steps lies a complex network of strategies, algorithms, and technologies that are advancing the capabilities of image-guided microsurgery systems. The engine of this complex network is the interdependence between medicine and computation – a partnership that is elaborated on throughout this chapter and book.

References

[1] "Microsurgery." American Society of Plastic Surgeons. Accessed January 16, 2023. https://www.plasticsurgery.org/reconstructive-procedures/microsurgery.

[2] Mavrogenis, A.F., K. Markatos, T. Saranteas, *et al.* "The history of microsurgery." *European Journal of Orthopaedic Surgery & Traumatology* 29, no. 2 (2019): 247–54. https://doi.org/10.1007/s00590-019-02378-7.

[3] "Computed Tomography (CT)." National Institute of Biomedical Imaging and Bioengineering. U.S. Department of Health and Human Services, June 2022. https://www.nibib.nih.gov/science-education/science-topics/computed-tomography-ct.

[4] Berger, A. "How does it work?: magnetic resonance imaging." *BMJ* 324, no. 7328 (2002): 35–35. https://doi.org/10.1136/bmj.324.7328.35.

[5] Alder, M.E., S. Thomas Deahl, and S.R. Matteson. "Clinical usefulness of two-dimensional reformatted and three-dimensionally rendered computerized tomographic images." *Journal of Oral and Maxillofacial Surgery* 53, no. 4 (1995): 375–86. https://doi.org/10.1016/0278-2391(95)90707-6.

[6] Hildebolt, C.F., M.W. Vannier, and R.H. Knapp. “Validation study of skull three-dimensional computerized tomography measurements.” *American Journal of Physical Anthropology* 82, no. 3 (1990): 283–94. https://doi.org/10.1002/ajpa.1330820307.

[7] Bayer, S., A. Maier, M. Ostermeier, and R. Fahrig. “Intraoperative imaging modalities and compensation for brain shift in tumor resection surgery.” *International Journal of Biomedical Imaging* 2017 (2017): 1–18. https://doi.org/10.1155/2017/6028645.

[8] Golby, A.J. *Image-Guided Neurosurgery*. New York, NY: Elsevier Science Publish, 2015.

[9] Ng, A. and J. Swanevelder. “Resolution in ultrasound imaging.” *Continuing Education in Anaesthesia Critical Care & Pain* 11, no. 5 (2011): 186–92. https://doi.org/10.1093/bjaceaccp/mkr030.

[10] Laughlin, S. and W. Montanera. “Central nervous system imaging.” *Postgraduate Medicine* 104, no. 5 (1998): 73–88. https://doi.org/10.3810/pgm.1998.11.402.

[11] Mahesh, M. *MDCT Physics: The Basics – Technology, Image Quality and Radiation Dose*. LWW, 2012.

[12] Van Reeth, E., I.W. Tham, C.H. Tan, and C.L. Poh. “Super-resolution in magnetic resonance imaging: a review.” *Concepts in Magnetic Resonance Part A* 40A, no. 6 (2012): 306–25. https://doi.org/10.1002/cmr.a.21249.

[13] Pandit, P., S.M. Johnston, Y. Qi, J. Story, R. Nelson, and G. Allan Johnson. “The utility of micro-CT and MRI in the assessment of longitudinal growth of liver metastases in a preclinical model of colon carcinoma.” *Academic Radiology* 20, no. 4 (2013): 430–39. https://doi.org/10.1016/j.acra.2012.09.030.

[14] Fabiszewska, E., K. Pasicz, I. Grabska, W. Skrzyński, W. Ślusarczyk-Kacprzyk, and W. Bulski. “Evaluation of imaging parameters of ultrasound scanners: baseline for future testing.” *Polish Journal of Radiology* 82 (2017): 773–82. https://doi.org/10.12659/pjr.904135.

[15] Lankenau, E.M., M. Krug, S. Oelckers, N. Schrage, T. Just, and G. Hüttmann. “IOCT with surgical microscopes: a new imaging during microsurgery.” *Advanced Optical Technologies* 2, no. 3 (2013): 233–39. https://doi.org/10.1515/aot-2013-0011.

[16] Hüttmann, G., J. Probst, T. Just, *et al.* “Real-time volumetric optical coherence tomography OCT imaging with a surgical microscope.” *Head & Neck Oncology* 2, no. S1 (2010): 1. https://doi.org/10.1186/1758-3284-2-s1-o8.

[17] Wieser, W., W. Draxinger, T. Klein, S. Karpf, T. Pfeiffer, and R. Huber. “High definition live 3D-Oct in vivo: design and evaluation of a 4D OCT engine with 1 Gvoxel/s.” *Biomedical Optics Express* 5, no. 9 (2014): 2963. https://doi.org/10.1364/boe.5.002963.

[18] Storey, D. “3D View of My Skull.” Flickr. Yahoo!, December 20, 2006. https://www.flickr.com/photos/duanestorey/328383705.

[19] “File: PTPR Brain.jpg – Wikimedia Commons.” Wikimedia, November 30, 2006. https://commons.wikimedia.org/wiki/File:PTPR_MRI.jpg.

[20] "File: Radiografía Pulmones Francisca Lorca.cropped.jpg." Wikimedia, May 11, 2010. https://commons.wikimedia.org/wiki/File:Radiograf%C3%ADa_pulmones_Francisca_Lorca.cropped.jpg.
[21] "File: CRL Crown Rump Length 12 Weeks Ecografia Dr. Wolfgang Moroder.jpg ..." Wikimedia, February 2, 2012. https://commons.wikimedia.org/wiki/File:CRL_Crown_rump_length_12_weeks_ecografia_Dr._Wolfgang_Moroder.jpg.
[22] Coste, A. "Project 1 : histograms." Image Processing, September 5, 2012. http://www.sci.utah.edu/~acoste/uou/Image/project1/Arthur_COSTE_Project_1_report.html.
[23] Iskakov, A. and S.R. Kalidindi. "A framework for the systematic design of segmentation workflows." *Integrating Materials and Manufacturing Innovation* 9, no. 1 (2020): 70–88. https://doi.org/10.1007/s40192-019-00166-z.
[24] Swain, A. "Noise in digital image processing." Medium. Image Vision, August 9, 2020. https://medium.com/image-vision/noise-in-digital-image-processing-55357c9fab71.
[25] Damerjian, V., O. Tankyevych, N. Souag, and E. Petit. "Speckle characterization methods in ultrasound images – a review." *IRBM* 35, no. 4 (2014): 202–13. https://doi.org/10.1016/j.irbm.2014.05.003.
[26] Basavarajaiah, M. "6 Basic things to know about convolution." Medium. March 29, 2022. https://medium.com/@bdhuma/6-basic-things-to-know-about-convolution-daef5e1bc411.
[27] Manansala, J. "Image processing with python: image effects using convolutional filters and kernels." Medium. The Startup, February 14, 2021. https://medium.com/swlh/image-processing-with-python-convolutional-filters-and-kernels-b9884d91a8fd.
[28] "Kernel (Image Processing)." Wikipedia. Wikimedia Foundation, January 13, 2023. https://en.wikipedia.org/wiki/Kernel_(image_processing).
[29] "Matlab Image and Video Processing." Matlab Tutorial: Digital Image Processing 6 – Smoothing: Low Pass Filter – 2020. Accessed January 17, 2023. https://www.bogotobogo.com/Matlab/Matlab_Tutorial_Digital_Image_Processing_6_Filter_Smoothing_Low_Pass_fspecial_filter2.php#:~:text=Low%20pass%20filtering%20(aka%20smoothing,noise%20from%20a%20digital%20image.&text=The%20operator%20moves%20over%20the,make%20an%20image%20appear%20sharper.
[30] Swain, A. "Noise filtering in digital image processing." Medium. Image Vision, July 16, 2020. https://medium.com/image-vision/noise-filtering-in-digital-image-processing-d12b5266847c.
[31] Ozcan, A., A. Bilenca, A. E. Desjardins, B.E. Bouma, and G.J. Tearney. "Speckle reduction in optical coherence tomography images using digital filtering." *Journal of the Optical Society of America A* 24, no. 7 (2007): 1901. https://doi.org/10.1364/josaa.24.001901.
[32] "File: Multivariate Gaussian Inequality Demonstration.svg." Wikimedia Commons, October 9, 2006. https://commons.wikimedia.org/wiki/File:Multivariate_Gaussian.png.

[33] Usman, O.L., R.C. Muniyandi, K. Omar, and M. Mohamad. "Gaussian smoothing and modified histogram normalization methods to improve neural-biomarker interpretations for dyslexia classification mechanism." *PLoS One* 16, no. 2 (2021): e0245579. https://doi.org/10.1371/journal.pone.0245579.

[34] "Bilateral Filter." Wikipedia. Wikimedia Foundation, March 18, 2022. https://en.wikipedia.org/wiki/Bilateral_filter.

[35] "GPU Implementations of 3D Image Denoising Filters." LBNL Visualization Group, April 2010. https://dav.lbl.gov/archive/Vignettes/ImageDenoising/.

[36] "File: MRI Brain.jpg – Wikimedia Commons." Wikimedia, June 15, 2022. https://commons.wikimedia.org/wiki/File:MRI_brain.jpg.

[37] "Contrast Enhancement and Smoothing of CT Images for Diagnosis." ResearchGate, March 2015. https://www.researchgate.net/publication/280003458_Contrast_enhancement_ and_smoothing_of_CT_images_for_diagnosis.

[38] "Histogram Equalization." UCI iCAMP, March 10, 2011. https://www.math.uci.edu/icamp/courses/math77c/lab_11w/.

[39] Al-Ameen, Z. "Sharpness improvement for medical images using a new nimble filter." *3D Research* 9, no. 2 (2018): Article no. 12. https://doi.org/10.1007/s13319-018-0164-0.

[40] "Unsharp Filter." Spatial Filters – Unsharp Filter. Accessed January 17, 2023. https://homepages.inf.ed.ac.uk/rbf/HIPR2/unsharp.htm.

[41] Yuheng, S. and Y. Hao. "Image segmentation algorithms overview." arXiv.org, July 7, 2017. https://arxiv.org/abs/1707.02051.

[42] Baby, D., S.J. Devaraj, S. Mathew, M. M. Anishin Raj, and B. Karthikeyan. "A performance comparison of supervised and unsupervised image segmentation methods." *SN Computer Science* 1, no. 3 (2020): 122. https://doi.org/10.1007/s42979-020-00136-9.

[43] Poole, A.D. "Real-time image segmentation for augmented reality by combining multi-channel thresholds." *Computer Science*, vol. 1: 5–7, 2017.

[44] Nirpjeet, E. "A review on various methods of image thresholding." *Computer Science*, vol. 3. no. 10: 3441–3442, 2011.

[45] "Thresholding." Point Operations – Thresholding. Accessed January 17, 2023. https://homepages.inf.ed.ac.uk/rbf/HIPR2/threshld.htm.

[46] "Otsu's Method." Wikipedia. Wikimedia Foundation, November 6, 2022. https://en.wikipedia.org/wiki/Otsu%27s_method.

[47] Norouzi, A., M.S. Rahim, A. Altameem, *et al*. "Medical image segmentation methods, algorithms, and applications." *IETE Technical Review* 31, no. 3 (2014): 199–213. https://doi.org/10.1080/ 02564602.2014.906861.

[48] Andrade Miranda and Gustavo Xavier. "Analyzing of the vocal fold dynamics using laryngeal videos." *ResearchGate*, June 2017. https://doi.org/10.20868/upm.thesis.47122.

[49] Brownlee, J. "A gentle introduction to cross-entropy for machine learning." MachineLearningMastery.com, December 22, 2020. https://machinelearningmastery.com/cross-entropy-for-machine-learning/.

[50] Wojcik, T.R. and D. Krapf. "Solid-state nanopore recognition and measurement using Shannon entropy." *IEEE Photonics Journal* 3, no. 3 (2011): 337–43. https://doi.org/10.1109/jphot.2011.2129503.

[51] Gustavsson, T., R. Abu-Gharbieh, G. Hamarneh, and Q. Liang. "Implementation and comparison of four different boundary detection algorithms for quantitative ultrasonic measurements of the human carotid artery." *Computers in Cardiology*, 24 (1997): 69–72. https://doi.org/10.1109/cic.1997.647832.

[52] Dharanibai, G. and J. Raina. "Gaussian mixture model based level set technique for automated segmentation of cardiac mr images." *International Journal of Engineering Science and Technology* 3 (2011): 170–80.

[53] Dabbura, I. "K-means clustering: algorithm, applications, evaluation methods, and drawbacks." Medium. Towards Data Science, September 27, 2022. https://towardsdatascience.com/k-means-clustering-algorithm-applications-evaluation-methods-and-drawbacks-aa03e644b48a.

[54] Dhanachandra, N., K. Manglem, and Y.J. Chanu. "Image segmentation using K-means clustering algorithm and subtractive clustering algorithm." *Procedia Computer Science* 54 (2015): 764–71. https://doi.org/10.1016/j.procs.2015.06.090.

[55] Gupta, A. "X-means algorithm-a complement to the K-means algorithm." Medium. Geek Culture, June 3, 2021. https://medium.com/geekculture/x-means-algorithm-a-complement-to-the-k-means-algorithm-b087ae88cf88.

[56] Inbarani H.H., A.T. Azar, and G. Jothi. "Leukemia image segmentation using a hybrid histogram-based soft covering rough k-means clustering algorithm." *Electronics* 9, no. 1 (2020): 188. https://doi.org/10.3390/electronics9010188.

[57] Padmapriya, B., T. Kesavamurthi, and H. Wassim Ferose. "Edge based image segmentation technique for detection and estimation of the bladder wall thickness." *Procedia Engineering* 30 (2012): 828–35. https://doi.org/10.1016/j.proeng.2012.01.934.

[58] "Edge Detection." Wikipedia. Wikimedia Foundation, November 6, 2022. https://en.wikipedia.org/wiki/Edge_detection.

[59] Atul, K. and K. Atul. "First-Order Derivative Kernels for Edge Detection." TheAILearner, September 24, 2019. https://theailearner.com/2019/05/24/first-order-derivative-kernels-for-edge-detection/.

[60] "Image Edge Detection Operators in Digital Image Processing." GeeksforGeeks, May 17, 2020. https://www.geeksforgeeks.org/image-edge-detection-operators-in-digital-image-processing/.

[61] "Zero Crossing Detector." Feature Detectors – Zero Crossing Detector. Accessed January 17, 2023. https://homepages.inf.ed.ac.uk/rbf/HIPR2/zeros.htm.

[62] "Edge Detection and Zero Crossing." Accessed January 17, 2023. http://portal.survey.ntua.gr/main/labs/rsens/DeCETI/IRIT/GIS-IMPROVING/node18.html.

[63] Frounchi, K., L.C. Briand, L. Grady, Y. Labiche, and R. Subramanyan. "Automating image segmentation verification and validation by learning test

oracles." *Information and Software Technology* 53, no. 12 (2011): 1337–48. https://doi.org/10.1016/j.infsof.2011.06.009.

[64] Warfield, S.K., K.H. Zou, and W.M. Wells. "Validation of image segmentation by estimating rater bias and variance." *Philosophical Transactions of the Royal Society A: Mathematical, Physical and Engineering Sciences* 366, no. 1874 (2008): 2361–75. https://doi.org/10.1098/rsta.2008.0040.

[65] Zou, K.H., W.M. Wells, R. Kikinis, and S.K. Warfield. "Three validation metrics for automated probabilistic image segmentation of brain tumours." *Statistics in Medicine* 23, no. 8 (2004): 1259–82. https://doi.org/10.1002/sim.1723.

[66] "Image Registration." Wikipedia. Wikimedia Foundation, September 21, 2022. https://en.wikipedia.org/wiki/Image_registration.

[67] "Correspondence Problem." Wikipedia. Wikimedia Foundation, December 10, 2022. https://en.wikipedia.org/wiki/Correspondence_problem.

[68] Narayan, R., C. Hellmich, D. Mantovani, A. Wong, W.Z. Rymer, and L. Hargrove. *Encyclopedia of Biomedical Engineering*. New York, NY: Elsevier, 2019.

[69] Studholme, C. "Image registration in medical imaging." Lecture, January 17, 2023.

[70] Meijering, E. "World Molecular Imaging Congress," n.d.

[71] Ma, J., X. Jiang, A. Fan, J. Jiang, and J. Yan. "Image matching from handcrafted to deep features: a survey." *International Journal of Computer Vision* 129, no. 1 (2020): 23–79. https://doi.org/10.1007/s11263-020-01359-2.

[72] Tyagi, D. "Introduction to Harris Corner Detector." Medium. Data Breach, April 7, 2020. https://medium.com/data-breach/introduction-to-harris-corner-detector-32a88850b3f6.

[73] Vendra, S. "Addressing Corner Detection Issues for Machine Vision Based UAV Aerial Refueling," 2006.

[74] "Harris Corner Detector." Wikipedia. Wikimedia Foundation, January 11, 2023. https://en.wikipedia.org/wiki/Harris_corner_detector.

[75] Singh, A. "SIFT: how to use SIFT for image matching in python." Analytics Vidhya, December 23, 2020. https://www.analyticsvidhya.com/blog/2019/10/detailed-guide-powerful-sift-technique-image-matching-python/.

[76] Karami, E., Siva Prasad, and M. Shehata. "Image matching using SIFT, surf, brief and Orb: performance comparison for distorted images." arXiv.org, October 7, 2017. https://arxiv.org/abs/1710.02726.

[77] Zhang, J., L. Chen, X. Wang, *et al.* "Compounding local invariant features and global deformable geometry for medical image registration." *PLoS One* 9, no. 8 (2014): Article no. 76. https://doi.org/10.1371/journal.pone.0105815.

[78] "Scale-Invariant Feature Transform." Wikipedia. Wikimedia Foundation, January 15, 2023. https://en.wikipedia.org/wiki/Scale-invariant_feature_transform.

[79] Tyagi, D. "Introduction to feature detection and matching." Medium. Data Breach, April 7, 2020. https://medium.com/data-breach/introduction-to-feature-detection-and-matching-65e27179885d.

[80] Yixin, C. and J.Z. Wang. "A region-based fuzzy feature matching approach to content-based image retrieval." *IEEE Transactions on Pattern Analysis and Machine Intelligence* 24, no. 9 (2002): 1252–67. https://doi.org/10.1109/tpami.2002.1033216.

[81] Gandhi, V.H., and S.R. Panchal. "Feature based image registration techniques: an introductory survey: semantic scholar." *International Journal of Engineering Development and Research*, 2(1): 368–372, January 1, 1970. https://www.semanticscholar.org/paper/Feature-Based-Image-Registration-Techniques%3A-An-Gandhi-Panchal/c51e04a6b32048bc7303277c29a8f0b5dccfa68e.

[82] Somaraju B. "Feature-based image registration," Master's thesis, National Institute of Technology, Rourkela, 2009.

[83] Zitová, B. and J. Flusser. "Image registration methods: a survey." *Image and Vision Computing* 21, no. 11 (2003): 977–1000. https://doi.org/10.1016/s0262-8856(03)00137-9.

[84] Guan, S.-Y., T.-M. Wang, C. Meng, and J.-C. Wang. "A review of point feature based medical image registration." *Chinese Journal of Mechanical Engineering* 31, no. 1 (2018): 76. https://doi.org/10.1186/s10033-018-0275-9.

[85] Harmouche, R., F. Cheriet, H. Labelle, and J. Dansereau. "Multimodal image registration of the scoliotic torso for surgical planning." *BMC Medical Imaging* 13, no. 1 (2013): 723–37. https://doi.org/10.1186/1471-2342-13-1.

[86] Ikonen, L., P. Toivanen, and J. Tuominen. "Shortest route on gray-level map using distance transform on curved space." *Image Analysis*, (2003): 305–10. https://doi.org/10.1007/3-540-45103-x_42.

[87] Zhang, X., H. Dong, D. Gao, and X. Zhao. "A comparative study for non-rigid image registration and rigid image registration." arXiv.org, January 12, 2020. https://doi.org/10.48550/arXiv.2001.03831.

[88] Chen, L., S. Bai, G. Li, *et al.* "Accuracy of real-time respiratory motion tracking and time delay of gating radiotherapy based on optical surface imaging technique." *Radiation Oncology* 15, no. 1 (2020): Article no. 170. https://doi.org/10.1186/s13014-020-01611-6.

[89] Hata, N., W.M. Wells, M. Halle, *et al.* "Image guided microscopic surgery system using mutual-information based registration." *Lecture Notes in Computer Science*, (1996): 317–26. https://doi.org/10.1007/bfb0046969.

[90] Andrews, C., M.K. Southworth, J.N. Silva, and J.R. Silva. "Extended reality in medical practice." *Current Treatment Options in Cardiovascular Medicine* 21, no. 4 (2019): 1–12. https://doi.org/10.1007/s11936-019-0722-7.

[91] Southworth, M.K., J.R. Silva, and J.N. Silva. "Use of extended realities in cardiology." *Trends in Cardiovascular Medicine* 30, no. 3 (2020): 143–48. https://doi.org/10.1016/j.tcm.2019.04.005.

[92] Watanabe, E., M. Satoh, T. Konno, M. Hirai, and T. Yamaguchi. "The trans-visible navigator: a see-through neuronavigation system using augmented reality." *World Neurosurgery* 87 (2016): 399–405. https://doi.org/10.1016/j.wneu.2015.11.084.

Chapter 5

Electrophysiology and consciousness: a review

Ioanna Stavraki,[1] Nicoletta Nicolaou[2] and Marcela Paola Vizcaychipi[3]

5.1 Introduction

Objective clinical analysis of brain activity and more complex brain functions, such as consciousness/arousal, is a challenging task. There is no single definition of consciousness, but one definition could be as the act of remembering; therefore, one could assume that the opposite could be called unconsciousness. Unconsciousness could describe a physiological state, such as sleep, or an acquired state following a disease or pharmacological intervention, such as stroke-induced coma or administration of an anesthetic agent respectively. The brain physiology is based on biological circuitry signs within and between the brain structures. The brain produces dynamic patterns of electric potentials on the scalp called the electroencephalogram (EEG). These complex brain functions of consciousness/unconsciousness are not comprised solely of mental components, but there are also associated physical components to consider. The mental components relate to emotions that involve cognition and awareness. These cause so-called feelings, which in turn lead to actions and associated physical changes, such as tachycardia, hypertension and sweating.

One of the anatomical structures involved in these processes is the limbic system, formerly known as the rhinencephalum. The limbic system has multiple layers and connections to the hippocampus, mamillary bodies which are connected to the anterior nuclei of the thalamus. The anterior nuclei of the thalamus project to the cingulate cortex and from there back to the hypothalamus, completing a complex closed circle knows as the Papez circuit. The connections of the limbic system to the neocortex seem to modify emotional behavior, and vice versa. However, emotions cannot be simply turned on/off, meaning that there is also influence by more autonomous structures, such as the autonomic nervous system (ANS).

The ANS is responsible for the delivery of the motor impulses from the central nervous system (CNS) to the visceral portions of the body and transmitted differently

[1] Independent Researcher, UK
[2] Department of Basic & Clinical Sciences, University of Nicosia Medical School, Cyprus
[3] Department of Surgery & Cancer, Imperial College London, UK

from those to the skeletal muscles. It is organized on the basis of the reflex arc and has two different structures that slow down or accelerate more primitive physiological responses, such us feeding, sexual, emotions, and sleep.

Sleep and the anesthetic-induced state of unconsciousness can be monitored through surrogate makers or somatic and autonomic responses, or subjective measures prompting the presence of recall events. Objective assessment of (un)consciousness is yet to be understood, despite great advances in the field of neuroscience and better understanding of physiological response to disease and medical interventions, such as anesthesia.

In this chapter, we outline the physiology of the central and ANSs, as well as their interaction, and how the electrical activity that is generated by these systems can be captured and studied. The understanding of the electrical activity of the brain is recognized as the window of the mind and gives cognitive scientists, engineers and healthcare professionals an integrated and 3-D visualization of these complex functions.

5.2 Nervous system signals

5.2.1 Central nervous system

The central nervous system (CNS), comprising of the brain and spinal cord, is critical for our daily functioning and physiological coordination. It is responsible for various key roles including our thoughts and cognition, movement, body temperature, breathing/heart regulations, hormone secretion, and emotion regulation. To protect the CNS due to its crucial role, both the brain and the spine are wrapped in three layers of meninges membranes: the dura mater, arachnoid mater and the most inner layer called the pia mater [1]. Additionally, as a further protective measure, hard layers of bone encase the skull to protect the brain while the spinal cord innervates with vertebrae. As many bodily operations are dependent on the proper functionality of the CNS, any alterations/malfunctions such as neurodegeneration and age-related declines have been documented to lead to dysregulations in behavior, neuroendocrine metabolic rhythms, cognition, and our overall physical functioning [2]. This truly highlights the widespread reach and key role of the CNS as the body's processing center that enables us to perform our day-to-day activities.

Through technological advancements, we are now able to capture physiological signals via an array of mechanisms that further our neurophysiological understanding of our bodily functions and provide neurorehabilitation guidance. Such areas include motor disorders, aging, sleep, cardiovascular, and mental health. When investigating motor disorders, there are several tools used with the most prominent ones being EEG, transcranial magnetic stimulation (TMS), transcranial direct current stimulation (tDCS), transcranial alternating current stimulation (tACS), and reflexology studies. Researchers can investigate the excitability of spinal reflexes using indexes such as F waves and H reflexes, look at the startle, jaw jerk and blink reflexes, or combine EEG and TMS for a more multimodal approach [3]. Regarding aging, MRI, diffusor tensor imaging (DTI), TMS and EMG are frequently utilized for investigating conditions such as Alzheimer's (AD) and mild cognitive impairment (MCI) as well as normal aging CNS alterations. Various findings have been uncovered such as an increased motor cortex excitability in AD/MCI patients [4], and changes in mean diffusivity of frontal white matter when using DTI was observed with aging [5]. Hence, such neuroimaging and neurophysiological tools can help us better understand the physiological CNS mechanisms and assist in delineating normal vs disorder aging patterns for more accurate diagnoses. While the ability to utilize such techniques can be still limited due to the expensive nature and invasiveness of some machinery, further discoveries about our physiology and new tools to uncover them are always on the horizon.

Sleep and the role of the CNS has also been investigated in regards to various sleep disorders such as apnea, narcolepsy, insomnia, the role of neuroendocrine secretions in sleep regulation as well as general investigations into the evolutionary mechanisms of sleep in humans and other animals. As previously discussed, the CNS is responsible for body temperature modulation and cardiovascular regulation. Any dysregulations on either mechanism can, thus, lead to cardio-physiological implications, as well as obstructions in the sleep/wake cycle. Polysomnographic recordings have uncovered such sleep state dependent bidirectional relationship between the CNS and cardiorespiratory networks in the forms of heart and respiration rates [6]. This in conjunction with evidence of sleep being highly important for not only typical memory consolidation but also immune system and T Cell immunological memory formation [7], further highlights the key endocrinological and cardiorespiratory role that the CNS has in our sleep hygiene and functioning. Lastly, to further elaborate on the brain–heart physiological mechanisms, it is paramount to consider the homeostatic role of the CNS. It regulates cardiorespiratory systems through two main types of regulation, feed-forward (anticipatory system) and feedback (reactionary system). Examples of when such systems are used are e.g. when we need to respond to a perceived threat. Signals reach the amygdala either via the hippocampus or the thalamus, and cardiorespiratory and cardiovascular responses are then elicited through amygdala pathways resulting in e.g. elevated heart rate, and increased respiration [8]. Dysregulation of such systems can thus cause severe physiological disruptions. Some conditions which can lead to this are mental health disorders such as anxiety/depression, pharmacological reactions and cardiovascular disorders [9].

5.2.2 ANS

The ANS is responsible for homeostatic changes and comprises two main branches, the sympathetic (SNS) and parasympathetic (PSNS) branch. SNS is responsible for the "fight or flight" or the stress response, while the PSNS is responsible for the "rest and digest" response, i.e. allows the body to recover after the "fight." The importance of the ANS system for cardiovascular health has instigated great interest in investigating the use of non-invasive measurements that represent both the SNS branch and the vagus nerve activity as a proxy for studying physiological, as well as pathophysiological, mechanisms of cardiac function.

The ANS innervates with various organs including tissues and glands making it highly influential in homeostatic functions such as thermoregulation, blood regulation, and bladder contractions [10]. Similarly, the sympathetic nerve chain clusters running down from the head all the way towards the spine's tailbone have ample cardiovascular physiological responses from increasing heart rate, airflow and skeletomuscular blood supplies [11]. The adrenal medulla is another gland that has a crucial role, this time in the fight-or-flight physiological response. This is due to its responsibility for neurotransmitter release of adrenaline and norepinephrine [12]. In healthy subjects, after the release of such hormones, a resting state of the body followed while hormonal regulations ensued. However, frequent activations of this system, often caused by various pathologies of the ANS and the adrenal medulla, can result in ramifications and a dysregulated stress response. Some of the most common illnesses can arise in both intrinsic and extrinsic manifestations such as diabetes, adverse drug effects, inflammation, cardiac disease, and neurological conditions [13]. Some of the most prevalent ones include those of neoplasms (neuroblastomas and ganglioneuromas) and tumors (such as a phaeochromocytoma). The body can thus end up with dysregulated immune responses, blood vessel contractions, and fluid homeostasis [14].

The long-term impacts of a dysregulated ANS have been widely documented across various population groups. Chronic stress conditions can have detrimental cardiovascular effects with vagal nerve withdrawals, a blunted baroreflex sensitivity and overall higher arterial pressure in healthy subjects when compared to controls [15]. In population groups who suffer from chronic stress, a direct impact was not only observed on physiological responses but on a cognitive level too, with hindered performances on various mental tasks and increased reported cortisol levels [16]. This has direct real-life implications which can further drive individuals into a worsening psychological state, in turn further impacting their physiological capabilities creating thus a vicious cycle. In patients suffering with diagnosed mental illnesses such as depression and anxiety, which often correlate with poor cardiovascular health, such effects are even more prevalent. Specifically, they were found to be at an increased risk of cardiac morbidity and can exhibit several abnormal physiological signals such as a lower heart rate variability (HRV), dysregulated responses to stressors and high ventricular repolarization variability [17]. Sleep disorders can also arise leading to a decrease in sleep quality and an overall decline in health-related outcomes. Specifically, SNS neurotransmitter and cortisol dysregulations and/or PSNS cardiovascular variability are some of the main ways ANS directly impacts sleep hygiene [18].

When discussing the implications of ANS and physiology, it is key to consider the clinical implications of such interactions, how modern medicine is addressing them and what gaps persist in the field of bioengineering tools. Adverse manifestations of ANS dysregulation may present themselves in either a chronic or acute or nature. One of the biggest challenges is discerning the symptomatology as symptoms of ANS dysfunction e.g. gastrointestinal difficulties or cardiovascular hyperactivity can often be attributed to various conditions as well. Comorbidity with other dysfunctions e.g. sleep dysregulations may also arise especially in patients with long-term ANS conditions. This becomes of extreme importance when considering the fact that heart disease is a leading mortality cause [19] and ANS physiological disruptions can lead to severe cardiological episodes such as a cardiac arrest. With the advancement of technology, we are now able to capture multi-modal recordings, to aid our understanding of a patient's physiological underpinnings. The most frequently utilized tools include capturing blood pressure, body temperature, EEG/ECG/EMG recordings, MRI, and microneurography (MSNA) [20]. Through such approaches, we aim to bridge the gap in our understanding between ANS caused physiological dysregulations, clinical applications and therapeutic outcomes.

5.2.3 CNS–ANS connection in physiological mechanisms

The ANS as part of the PNS is responsible for promoting plasticity through neuronal regeneration. In contrast, the CNS due to its maintenance responsibility of the neural network stability is unable to perform such regeneration, which can greatly and poorly impact the prognosis of CNS injuries as seen in this article by [21]. Figure 5.1 provides a schematic of the two nervous systems and their function.

A key interrelationship between these two systems lies in the bidirectional gut-brain axis connection. Research indicated the presence of a relationship between stressors and conditions such as IBS or a temporary alteration in bowel movements (such

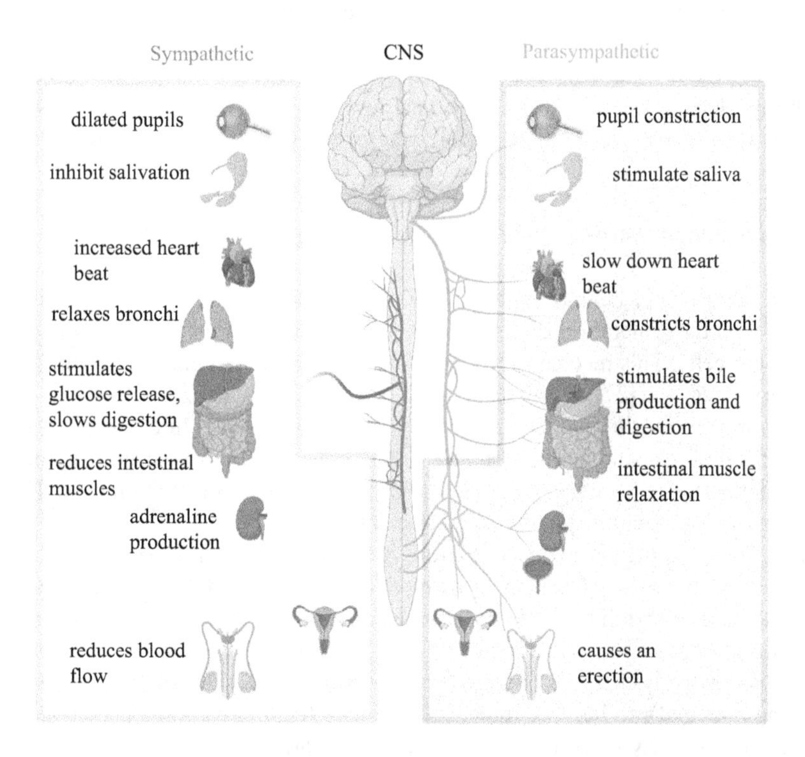

Figure 5.1 Illustration of CNS (brain and spinal cord) and ANS function (sympathetic and parasympathetic branches). Created with BioRender.com.

as diarrhea or constipation) as long as the stressors are present [22]. Hypothalamic–pituitary–adrenal or (HPA axis) alterations) have also been documented in association with elevated cortisol levels in IBS patients and cellular alterations of the immune response [23]. The HPA axis is one of the neuroendocrine pathways regulated by the CNS further documenting the inter physiological relationship between the gut-brain axis and the CNS–ANS relationship. Dysregulation in either nervous system can lead to documented gastrointestinal disorders and conditions of the pancreas. Specifically, hepato-biliary (CNS controlled) and gastrointestinal hormone (ANS controlled) secretions are necessary for the normal daily function of the enteric system. Disruptions in the form of e.g. parasympathetic rebounds can lead to gastritis, reflux esophagitis, pancreatitis, and biliary dyskinesia [24].

Sleep has a crucial role in regulating our conscious–unconscious bodily state, having a direct impact on our daily function. The interconnected interaction between cortical and cardiac activation is key in the implementation and sustainability of sleep/wake cycles; highlighting thus the strong relationships between the CNS and ANS in sleep functionality. Hence, irregularities in one or both of these systems can directly impact sleep homeostasis and our circadian rhythm leading to various abnormalities such as insomnia, sleep apnea, and irregular sleep/wake patterns [25]. Research indicates that a strong coupling is present during sleep between these two systems with the CNS regulating REM cycle timings and the ANS regulating breathing and cardiovascular responses. This relationship generates various cortico-cardiac pathways that are key in maintaining a healthy sleep hygiene and can be impacted by factors such as age, sleep habits and sleep disorders [26].

5.3 Neurophysiological signal recording

The basic principle behind the ability to capture the nervous system activity from the body is the generation of electrical pulses due to the movement of ions in and out of our cells. This mechanism is described in detail in a number of Neuroscience textbooks (e.g. see [27]), but we provide a brief summary here.

As ions move through the intracellular and extracellular fluid, electrical pulses are generated, causing action potentials that propagate from neuron to neuron, or through muscle cells (e.g. cardiac muscle). The primary ions that are involved in this mechanism are sodium, potassium, and calcium (Na^+, K^+, Ca^{2+}). These ions are usually found at difference concentrations in the intracellular and extracellular space (Na^+ and K^+ are mostly found in the extracellular and intracellular space respectively) and maintain cell equilibrium in the absence of stimuli. When a stimulus of adequate intensity arrives, it initiates a series of events that cause the internal cell potential, which is negatively charged compared to the outside of the cell in absence of a stimulus (resting potential), to become transiently positive (depolarization). This is achieved by the opening of voltage-gated Na^+ channels, allowing an influx of Na^+ ions into the intracellular space. The depolarization causes the K^+ voltage-gated channels to open, leading to K^+ efflux into the extracellular space. This causes call repolarization, which is the decrease of the intracellular potential aiming at restoring the membrane resting potential. The membrane potential becomes more negative than the resting potential, i.e. there is an overshoot of the resting potential (hyperpolarization), but this is quickly followed by restoration to the cell's membrane resting potential. This completes one cycle of the action potential generation. These waves of depolarization–repolarization and action potential propagation cause an electric current, which can be captured by electrodes placed on the surface of the body. The electrodes capture the voltage difference caused by the propagation of this electrical current along the body.

5.3.1 Recording the electroencephalogram (EEG)

The first time that any electrical activity was recorded from humans dates back to 1924 by the German psychiatrist Hans Berger, who in 1929 recorded for the first time also the electrical activity from the scalp surface. He named these low-frequency oscillations *alpha* waves. The EEG is a record of such oscillations that are observed simultaneously and at different frequency ranges. The oscillations that are captured by the EEG reflect the summation of synchronous excitatory and inhibitory post-synaptic potentials, that is the potentials originating from ion movement when the action potential is propagated from one neuron (pre-synaptic) to the next (postsynaptic). The EEG is obtained from electrodes placed on the surface of the head at pre-determined positions, as recommended by the American EEG Society [28]. This electrode configuration is called the International 10–20 system, as the distances between the electrodes are either at 10% or 20% of the distance from nasion to inion or from left to right side of the head (Figure 5.2). The standard configuration allows for 21 electrodes but this has been extended to accommodate an even larger number of electrodes, e.g. 10–10 or 10–5 systems (distances between the electrodes

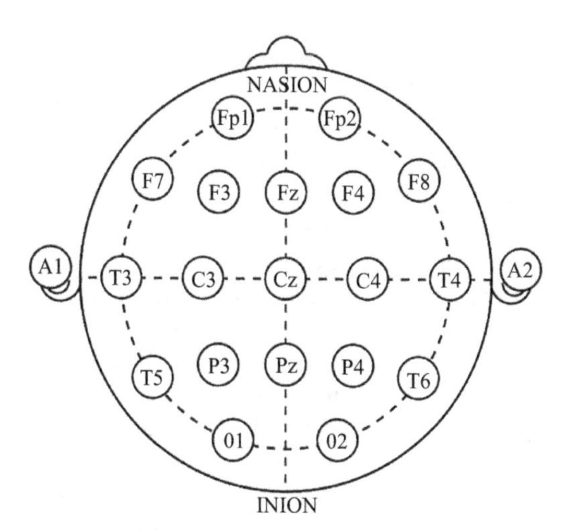

Figure 5.2 International 10–20 EEG electrode placement system (public domain image)

are smaller). The electrodes are conventionally named after the cortical area located directly beneath them, e.g. "F" for "Frontal," "O" for "Occipital," and using numbers to indicate left-right hemisphere (odd for left hemisphere, even for right hemisphere) and increasing from midline towards the side of the head (e.g. electrodes "C3" and "C5" are both placed over the cortex on the left hemisphere but "C3" is nearer the midline compared to electrode "C5"). Equally important are the "ground" and "reference" electrodes. The former is an electrode that serves as the reference for the recording equipment (placing the subject and equipment at the same reference electrical potential). The latter should be placed at "infinity" from the source generators of interest, which practically means any location that does not record activity from those sources of interest. In EEG, this is difficult to achieve as we do not know where the specific brain sources are located, therefore there is no such thing as an "ideal reference." Common reference strategies include a single scalp reference electrode (usually at location FCz), the linked ears or linked mastoids, the average reference. For the interested reader, the reference problem is discussed in detail in Nunez and Srinivasan [29]. There are also two main types of recording montages: the monopolar or unipolar montage, which is recording the voltage difference between the reference electrode and an EEG electrode, and the bipolar montage, which is the voltage difference between two EEG electrodes that are spatially close. The bipolar montage is, in essence, the voltage difference between two unipolarly recorded EEG electrodes. The conversion from unipolar to bipolar montage can easily be done post-recording by simple subtraction, but if the montage is set to bipolar from the start, then it is not possible to convert to unipolar montage post-recording.

To improve the quality of the recorded signal and the signal-to-noise ratio, the impedance of the electrodes is lowered by preparing and cleaning the scalp via an abrasive gel and placing the electrodes, which are commonly made from gold or silver, on the scalp using conductive gel. The EEG signal originates from different brain sources and is attenuated through brain tissue, cerebrospinal fluid and the skull. It is difficult to record activity from deep brain sources, e.g. thalamus and brainstem, from surface EEG, but it is not impossible provided these sources produce powerful electric fields that are consistent across a very large number of trials that, when averaged, will provide a sufficient signal-to-noise ratio. The EEG is estimated to be dominated by neuronal activity mostly originating from approximately 10,000–50,000 neurons at superficial cortical layers [30]. Due to this volume conduction and attenuation, the surface EEG signal is very small (μV) compared to activity from other physiological sources, hence to record it requires a special configuration of amplifiers to reduce the effect of environmental noise. Instead of amplifying the signal at each electrode individually, the potential difference is amplified instead, as this first cancels out the common environmental noise and then amplifies the remaining signal that corresponds largely to the weak brain biopotential [31].

Regarding frequencies of interest, the EEG oscillatory brain rhythms can be grouped into δ (1–4 Hz), θ (3.5–7.5 Hz), α (7.5–13 Hz), β (14–30 Hz), and γ (30–100 Hz) range rhythms, with the lower and upper limits to these ranges varying slightly between different references. Oscillations that are lower than 0.1 Hz are difficult to measure with EEG due to limitations of EEG recording equipment. Similarly, activity

> 100 Hz is of low power and may be difficult to distinguish from background noise. With these in mind, as well as the Nyquist theorem, the optimum sampling rate for EEG recording should be at least 256 Hz. In practice, sampling rates from 500 Hz to 2 kHz are commonly used, with appropriate bandpass filtering to reduce aliasing phenomena. The higher sampling frequencies provide a better signal-to-noise ratio, with EEG data usually down-sampled prior to analysis.

5.3.2 Recording the ECG

The paramount cardiovascular event of interest that is required for maintaining a constant flow of blood throughout the body is the cardiac output (output of the heart per minute). This is achieved through a regular cycle of contraction–relaxation of the heart atria and ventricles, regulated by a series of complex electrophysiological events orchestrated by the cardiac cells. The heart has its own "pacemaker," i.e. the ability to initiate electrical potentials spontaneously by the sinoatrial node (SA) in the right atrium. The spontaneous impulse (depolarization) generated by the SA initiates a cardiac cycle, which propagates along from the atria to the ventricles and causes the contraction (and subsequent relaxation through repolarization) of the cardiac muscle (myocardium).

This cyclic activity of contraction–relaxation is captured by the ECG via surface electrodes placed on pre-determined locations on the chest area. Each cardiac cycle causes a deflection of the ECG isoelectric line that has a specific shape and distinct features (Figure 5.3). The P wave, which is the first (positive) deflection, represents atrial muscle cell depolarization. This is followed by the Q wave (negative deflection), reflecting septal depolarization (however, the Q wave is small and not always detectable in the ECG). The main wave, the QRS complex, combines the Q, R, and S waves into what represents ventricular muscle cells depolarization. Finally, the T wave represents the repolarization of the ventricular muscle.

The first recording of the heart activity was obtained in 1901 by William Einthoven, a Dutch physiologist, using a galvanometer. He discovered that the shape and amplitude of the tracings varied depending on where he placed the positive and negative electrodes, and proceeded to describe three lead locations (including the reference electrode) that formed a triangle around the heart. Today this is known as Einthoven's triangle (Figure 5.4), and these lead arrangements form the primary ECG leads (I, II, and III). Lead I records electrical current from right to left arm and, thus, reflects activity of the left ventricle lateral wall. Leads II and III reflect activity from the bottom part of the heart, with Lead I recording electrical current from right arm to left leg and Lead III from left arm to left leg. The primary leads have been expanded with three augmented leads (augmented voltage right arm, left arm, and feet–AVR, AVL, and AVF) and six chest leads (V1, V2, V3, V4, V5, V6) to the 12-lead ECG that is commonly used for clinical purposes today. Augmented leads AVR, AVL, and AVF reflect cardiac activity from the right surface of the heart, the left lateral part of the heart, and the inferior surface of the heart respectively.

A simple, inexpensive, and non-invasive alternative method for monitoring heart rate (HR) from the skin surface is photoplythesmography (PPG). PPG uses optics to

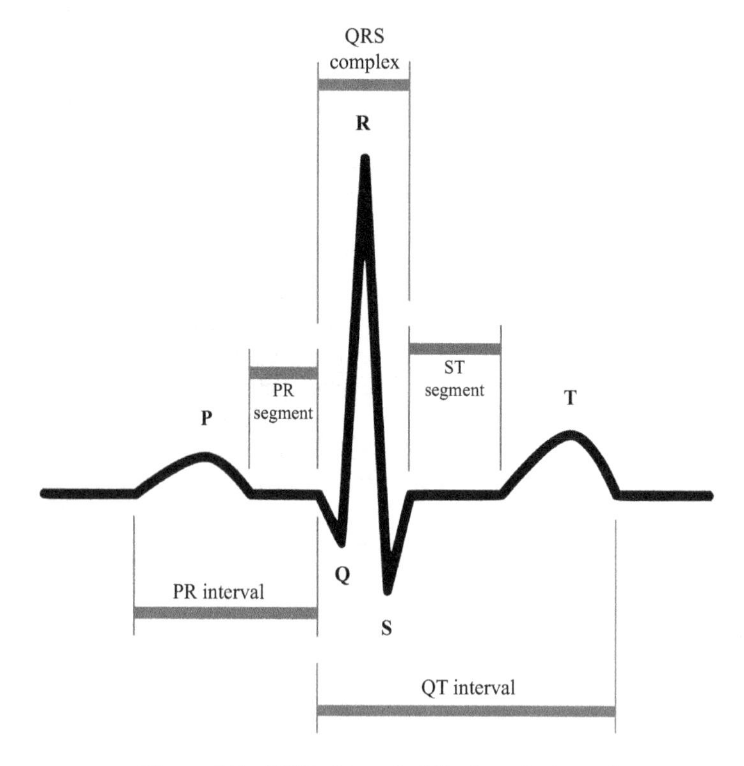

Figure 5.3 ECG shape (public domain image)

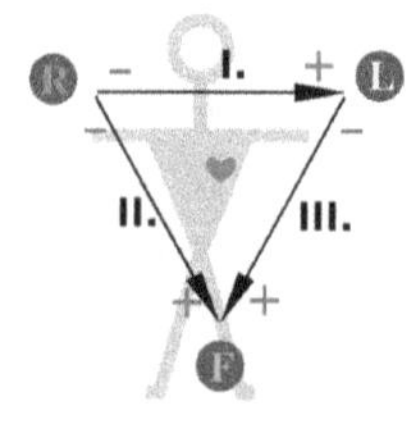

Figure 5.4 Eindhoven's triangle (this file is licensed under the Creative Commons Attribution-Share Alike 3.0 Unported license)

measure the rate of blood flow or the variations of blood volume in the body that occur during heart activity. While ECG uses electrical signals that are produced directly by the heart activity, PPG uses electrical signals that are derived from the reflected light due to the changes in blood volume: the changes in blood volume are correlated with the intensity of the reflected light on the skin. The PPG signal is a composite waveform comprising an AC component (synchronous cardiac changes in blood volume) superimposed on a DC component (baseline, low frequency components). The waveforms represent interactions between the cardiovascular, respiratory and autonomic systems.

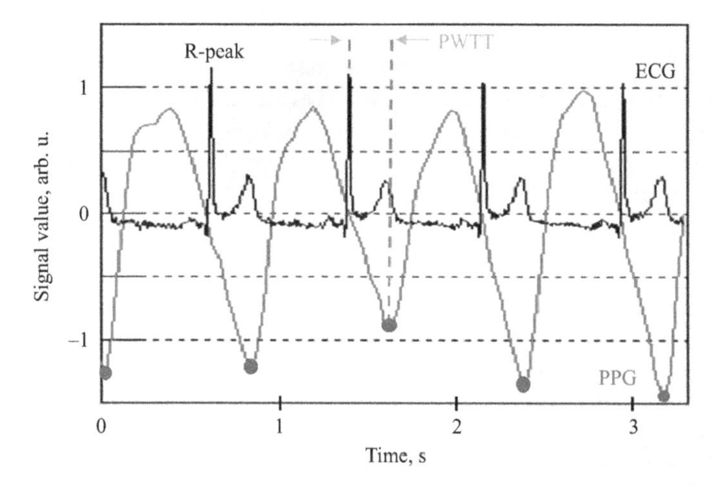

Figure 5.5 Simultaneously recorded ECG and PPG. PWTT stands for pulse wave transit time, which is the time taken for the arterial pulse pressure wave to travel from the aortic valve to the peripheral site of the body. (Figure from [33]; content from this work may be used under the terms of the Creative Commons Attribution 3.0 license.)

The most common PPG sensors are pulse oximeters, which are placed on the finger and use similar technology to measure both our HR, as well as oxygen saturation (the latter measurement is based on differences in the light absorption properties of oxygenated versus non-oxygenated hemoglobin). Pulse oximeters utilize a light emitting diode (LED) to shine light on the skin of the finger and measure the intensity of the reflected light, which correlates with changes in the volume of blood in the arteries right underneath the skin. As the volume of blood changes with the cardiac cycle, so does the light intensity, hence, this mechanism can be employed to measure the HR. While the ECG-derived HR is estimated from the ECG wave R–R intervals, the PPG-derived HR is estimated from the pulse period (P–P intervals) (Figure 5.5). It has been shown that, despite the difference in underlying mechanisms, the two HR derivations are highly correlated [32]. Thus, even though ECG is the "gold standard" for accurate beat-to-beat HR monitoring, PPG is suitable for average HR measurements and has a number of advantages over ECG, such as it can be embedded in a wearable device (e.g. pulse oximeter, wearable watch) and it does not require conductive gel or special electrodes. However, it is also susceptible to factors such as ambient light and motion artifacts.

5.4 Applications of biopotentials in health and disease

For many years, biopotentials have been used as a means of characterization of different states, including health and cognition, as well as indicators of diseases/disorders.

This is more straight-forward for ECG activity, and less so for EEG. Even though the causal involvement of EEG in cognitive processes is non-trivial to prove, the use of transcranial magnetic stimulation has provided evidence for the role of specific oscillations in cognition [34]. Despite the difficulty of proving causality, EEG in particular has provided robust characterizations of various physiological states, some examples of which we refer to below.

5.4.1 Neurodegeneration

Neurodegeneration is associated with specific changes in both the central and the ANSs, each reflected by EEG and ECG activity respectively.

Changes in CNS activity are evidenced from EEG activity commonly recorded under rest conditions (resting state) [35]. Neurodegenerative-related EEG changes in patients with mild cognitive impairment (MCI) and AD compared with age-matched healthy controls include [36]: (a) frequency-specific changes in power content (increased power in lower frequency ranges 0.5–8 Hz and decreased power in higher frequencies 8–30 Hz); (b) decrease in complexity and increase in EEG regularity; (c) perturbations in resting-stage EEG synchrony; and (d) decrease in small world properties of synchrony-derived brain networks. The observed EEG changes are not associated solely with loss of neurons, but also with changes in functional or anatomical coupling, which may result in disruption of temporal information coordination in the brain and, ultimately, loss of normal cognitive processing.

With regards to the ANS, neurodegeneration likely causes associated ANS dysfunction. Despite this, the particular direction of research is not commonly pursued and a limited amount of studies exist, with contrasting findings. Some initial studies conducted a number of years ago report changes in sympathetic response to stress in AD patients [37,38]. More recently, increase in sympathetic dominance and suppression in parasympathetic activities, which are the two branches of the ANS, have been reported in AD patients (the latter is likely associated with cholinergic deficiency) [39]. In contrast, Allan *et al.* report normal autonomic function in AD patients [40]. Considering some evidence that points to a cholinergic deficit in cognitively impaired patients related to EEG slowing [41], we expect that neurodegeneration would be characterized by an overall reduced PSNS function. A novel way to investigate this is via the ANS and specifically the parasympathetic (PSNS) branch, which is entirely cholinergic, as opposed to the sympathetic (SNS) branch. We have recently shown that the activity of PSNS and SNS can be estimated from complexity analysis (Poincaré analysis) of ECG-derived HRV [42].

5.4.2 Anesthesia

Anesthesia is a reversible state of drug-induced unconsciousness in which the patient neither perceives nor recalls noxious stimulation. The administration of the anesthetic drugs is achieved either intravenously or via inhalation of gaseous anesthetics, together with administration of a neuromuscular blocking agent to prevent patient movements during surgery. Anesthesia is considered more of an art than science as there are no exact "recipes" for dosage of the various anesthetic agents; the anesthetist relies more on past experience and patient characteristics to administer the correct amount of anesthetic. Hence, it has been known for patients to experience awareness and recall due to anesthetic underdose or, in the most severe of cases, even death due to anesthetic overdose. The percentage of people who become aware during anesthesia depends on the type of anesthetic agent used and generally ranges from 0.2% to 2%, while it also depends on the type of surgery, e.g. 0.4–1.1% for cardiac surgery and up to 20% for trauma surgery [43]. As a result, it has long been identified that anesthesia is a highly risky procedure, which could benefit from some form of a monitoring system that would alert anesthetists in cases of emergency. Such a device would make it possible to optimize the delivery of anesthetic agents to each individual patient, and to guarantee an adequate depth of anesthesia (DOA), and loss of awareness and recall [44]. A few commercial systems have already been developed for this purpose, however, these systems suffer from a number of unresolved issues. Their operation concentrates on measures of DOA obtained from the electrical brain activity (EEG). The nature of the changes observed in the EEG during anesthesia suggests that an alternative method to monitor DOA would be to utilize a measure representative of the deeper interactions within the brain itself, which in turn give rise to the observed EEG. A plausible candidate describing such underlying mechanisms is the emerging patterns of phase-synchronization (PS).

There are a number of models that describe the effects of anesthetics on neural connectivity and model the direct anesthetic effects on EEG: (1) the neural-mass model [45]; (2) the oscillator-based model [46], and (3) the thalamocortical model [47]. These models use different mechanisms for describing neural behaviors under different conditions and simulating the effects of anesthesia in such models can be achieved through modifying specific model parameters, such as $GABA_A$ conductance. The individual spikes are propagated through the network via inhibitory and excitatory connections, the summations of which result in the simulated EEG.

Current clinical practice during surgery allows the routine monitoring of critical autonomic factors, such as heart rate, to assess whether the patient is regaining consciousness during surgery. However, these autonomic signs are not accurate and, thus, measures extracted from the EEG of individuals under general anesthesia have recently been utilized for monitoring DOA. The use of the EEG is motivated by the observation that the administration of anesthetic agents causes observable changes in the EEG that are dependent on the type of anesthetic used [48]. For example, intravenous agents such as Propofol cause a slowing and non-reactive EEG, something particularly observed over anterior regions [34].

A number of different measures extracted from the EEG have been utilized for monitoring DOA so far, based on a number of approaches including network science, information transfer and integration, nonlinear dynamics, and artificial intelligence [49]. In addition, measures extracted from (mainly auditory) evoked responses have also been used; e.g. it has been observed that loss of awareness is associated with increase in the amplitude of particular peaks of the auditory evoked potential (N_a and P_a) that is correlated with the latency decrease of N_b under Sevoflurane and Propofol anesthesia [50]. Currently there are very few commercially available systems for monitoring DOA based on measures extracted from the EEG: BIS (Aspect Medical

Systems, USA; most widely used commercial system), Narcotrend™, and Datex-Ohmeda Entropy Sensor (GE, Germany) [51]. Despite the existence of such systems, they have still not become part of routine anesthetic practice as they suffer from robustness-related issues. The BIS has received criticism of not being responsive to some anesthetic agents, such as Nitrous Oxide (N_2O) which does not alter BIS index values when used either as a sole agent or as an adjunct to anesthesia [52], and not being robust across patients or time. The BIS index is estimated as the weighted sum of three features derived from spectral, bispectral and time-domain features of the EEG and, thus, its composite nature makes it difficult to pinpoint the exact cause of the criticisms it has received. Entropy values are also not altered by N_2O [53]. It has also been found that myorelaxant administration influences some measures of DOA in light anesthesia, such as the BIS index and response entropy [54]. In addition, different measures can predict different features, e.g. the BIS index may provide a better indication of recovery of consciousness at the end of anesthesia, whereas measures based on evoked potentials may better detect the transition from consciousness to unconsciousness [55]. The majority of studies performed so far mainly investigate the performance of DOA measures for the use of single anesthetic agents acting separately; however, their performance when a combination of agents is used should also be investigated, as anesthesia induction using a combination of agents is the general practice. A review of measures used for monitoring of DOA conducted in 1996 concludes that none seems to have the sensitivity and specificity to allow the clinician to draw conclusions about DOA in the individual patients for whom they treat [56].

Despite that not much has changed in terms of more sensitive and more specific measures for routine DOA monitoring in the operating theatre, one of the main reasons why the EEG has the potential to provide an extremely useful tool for monitoring of DOA is due to the characteristic changes observed in the EEG during anesthesia. A zero-line is observed in the EEG, prior to which a burst-suppression phase may appear, as during anesthesia the neo-cortical neurons fall silent. This is due to the fact that the activity of the neo-cortical neurons is increasingly synchronized by the oscillating activity of neurons from the thalamus, while at the same time, the interactions between the neo-cortical neurons decrease as a result of their decreased excitability [57]. This may be summarized by the hypothesis that as the depth of anesthesia increases, the highly complex neuronal interactions during the awake state are being replaced by simpler oscillations. This also implies that the degrees of freedom of human brain activity is lowered, nonetheless one should bear in mind that this may still be large and not accessible from an EEG time series. The nature of the changes observed in the EEG during anesthesia, namely the changes in the observed synchronization, support the hypothesis that in order to best utilize EEG as a monitoring tool for DOA a measure representative of the deeper interactions within the brain itself, which in turn give rise to the observed EEG, should be utilized. In this way the anesthetic-induced EEG changes will be captured resulting in a reliable monitor for anesthetic practice. A plausible candidate describing the mechanisms involved in the integration of incoming information and endogenous brain activity is via the emergence of transient dynamic connections between distributed local networks of phase-synchronized neurons (here, phase-synchronization is defined as a constant difference between the instantaneous phases of two signals for a few milliseconds; the instantaneous phase of a signal can be obtained via various methods, e.g. Hilbert transform, wavelets). The nominal work by Varela and colleagues that presents evidence obtained from the EEG of subjects during a number of cognitive tasks suggests that the emergence of cognitive acts is mediated by large-scale phase synchronization (PS) of different brain structures [58]. The utilization of such measures that encompass the deeper operation of the brain and potentially lie behind the generation of the recorded EEG waves could produce a measure that could be utilized in a robust manner for continuous and real-time monitoring of DOA.

5.4.3 Peri-operative stress

Physiological changes to emotional stimuli are strongly related to changes in (para)sympathetic system (de)activation, a key component of which is the vagal tone that regulates the body's internal organ systems, such as the heart and lungs. Other physiological signals of interest include EEG, heart rate, blood pressure, etc. These signals are often treated as independent entities and independent changes in response to different emotional stimuli, e.g. signal amplitude, are reported. The perioperative setting represents a host of conditions, including increased stress, post-operative trauma and other affective and cognitive disorders (e.g. delirium), in which the brain dynamics are significantly affected.

Non-pharmacologic interventions have been increasingly proposed to have beneficial effects in cognitive settings while at the same time minimizing the dose of medications needed and potentially reducing associated undesired effects in a perioperative setting [59]. One such non-pharmacological intervention is the non-invasive delivery of music stimuli, as music entrains body rhythms to decrease psychological stress and affects physiological activity (heart rate, blood pressure, respiratory rate, oxygen consumption, epinephrine levels, sweat gland activity and brain activity) [60–62]. For example, acoustic noise played during sleep leaves significant memory traces [63]; this is encouraging for intraoperative delivery of music due to the similarities in neural pathways between sleep and anesthesia. The anxiolytic effects of music may be due to multiple modes of music interaction – affecting cognitive/affective processing via distraction from stressful stimuli; entrainment of complex brain dynamics via modulation of mutual synchronization of oscillatory brain rhythms; and physiological effects via release of endorphins (audioanalgesia) [64]. The music parameters commonly associated with changes in emotional state are the tempo, volume and tone. The tempo in particular is suggested to play a key role in actively modulating listener emotions, including anxiety. Fast (slow) tempo is associated with increased (decreased) arousal, with measurable physiological effects (e.g. changes in heart rate and entrainment of brain activity oscillations at tempo-related frequencies). However, how to manipulate the music to achieve the desired effects, while at the same time respecting its "musicality," is an open question. Moreover, the research on anxiolytic effects of passive music listening in a perioperative setting is inconclusive, as some studies report benefits (reduction in anesthetic doses, higher levels of patient satisfaction, reduced pain and changes in physiological markers associated with sympathetic nervous system indicators of wellbeing), while others find no such effects [65]. This variability might be in part due to lack of adjustment of musical properties to ongoing changes in the patient state.

5.5 Analysis tools

5.5.1 ECG analysis

5.5.1.1 HRV

The physiological variation of the time interval between consecutive heart beats is called the HRV. It is usually measured as the time duration, in milliseconds, of the R–R interval, i.e. the interval between two consecutive R waves in the ECG (Figure 5.6). The reason why HRV is important is that it is regulated by the ANS, representing both the sympathetic and parasympathetic branches, and, hence, can be used as a non-invasive marker of the non-linear ANS dynamics. The sympathetic branch is the "fight or flight" or the stress response and, as such, activates stress hormones, which in turn increase the heart cardiac output (heart contraction rate). This translates to a reduction in HRV (as there is less time between consecutive heart beats). On the other hand, the parasympathetic branch is the "rest and digest" response, i.e. allows the body to recover after the "fight". Its effect is opposite to the sympathetic branch, so following a stressful event, the parasympathetic branch restores homeostasis by slowing down the heart rate, which in turn increases the HRV. Thus, in healthy situations the HRV should increase when relaxing activities take place, and decrease during stressful activities. This balance between the sympathetic and the parasympathetic function allows the heart to be ready for quick responses to different circumstances.

By monitoring the HRV, we can get a picture of unhealthy situations, such as chronic stress during which the natural interplay between the two ANS branches is disrupted and the sympathetic "fight or flight" state dominates (high stress hormones, low HRV) despite the person being in a rest state. We can also get a picture of healthy situations, which are characterized by a generally higher HRV (high HRV level is associated with reduced morbidity, healthy heart and improved psychological well-being). One can also improve their HRV level through exercise and increased fitness.

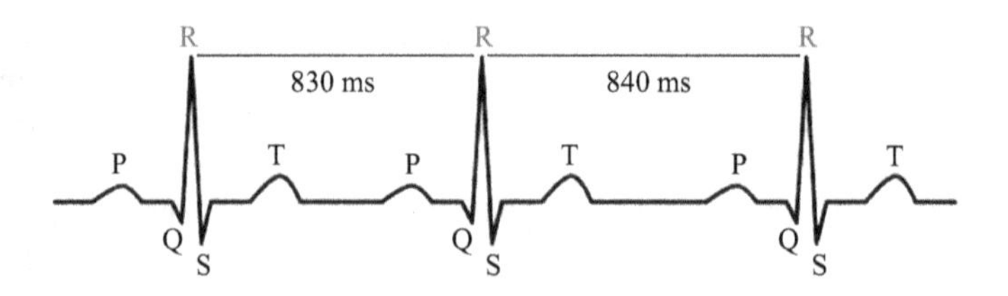

Figure 5.6 HRV measured as the duration between consecutive R–R intervals. In this specific example, HRV is (840–830) ms = 10 ms.

Prior to any ECG analysis, it is usually necessary to apply a high-pass filter (0.5–0.6 Hz) to remove low-frequency baseline drift (wander) in the ECG recording. Commonly, HRV analysis can be conducted in both the time-domain and/or the frequency-domain. The former methods quantify the amount of variability that exists in measurements of the R–R interval (also known as inter-beat interval), while the latter provides information on the distribution of power in predefined frequency bands. Both provide important information on the ANS function and overall cardiac health. An overview of HRV analysis methods and metrics can be found in [66]. We provide here a summary of some commonly used methods.

The most common time-domain analysis method is the Root Mean Square of Successive Differences (RMSSD) between each successive heartbeat. That is, it is estimated by calculating the differences in successive heart beats (R–R intervals), squaring this difference, and obtaining the square root of the average squared differences (5.1). It reflects the short-term beat-to-beat variance in HR and represents changes mediated by the vagus nerve, as reflected in the HRV:

$$RMSSD = \sqrt{\frac{1}{N-1}\sum_{i=1}^{N-1}(RR_{i+1} - RR_i)^2} \tag{5.1}$$

Other popular time-domain measures are estimated from the R–R interval of normal beats, known as the N–N interval. This is the same as the R–R interval but with abnormal beats removed, e.g. ectopic beats or abnormal beats due to artifacts. Example measures include the standard deviation of the N–N intervals (SDNN), the standard deviation of the averages of N–N intervals in a 24-h ECG recording estimated over 5-min segments across the entire recording (SDANN), and the percentage of differences between N–N intervals that differ by more than 50 ms from the previous N–N interval (pNN50).

The most popular frequency-domain analysis method involves power spectral density analysis of the ECG and extraction of power in two frequency ranges: (a) the high frequency (HF) range, 0.15–0.4 Hz, includes oscillations associated with baroreceptor reflexes and represents the parasympathetic activity, (b) the low frequency (LF) range, 0.04–0.14 Hz, includes respiratory sinus arrhythmia (fluctuations in heartbeat intervals linked with respiration) and represents the activity of both the sympathetic and parasympathetic branches, as well as (c) the ratio of LF/HF power, which represents the balance between the sympathetic and parasympathetic branches of the ANS (sympathovagal balance).

Some of these measures are correlated. For example, RMSSD is correlated with HF band power, while SDNN is correlated with LF band power. The 24-h SDNN in particular is considered the "gold standard" in cardiac risk categorization: patients with SDNN values above 50 ms are categorized as "unhealthy", 50–100 ms as "compromised health" and above 100 ms as "healthy" [67].

5.5.1.2 HRV complexity analysis

Despite the popularity of time–frequency HRV analysis, it has been shown that this is susceptible to high levels of respiratory noise [68]. On the contrary, Poincaré plots provide a non-linear, geometrical representation of the HRV dynamics over a period of time, and are not as susceptible to respiratory noise compared to other methods of HRV analysis, while correlating directly with spectral data [69]. Poincaré plots are, essentially, a scatterplot of consecutive R–R intervals and represent the relationship between two consecutive heartbeats, thus providing a visual representation of beat-to-beat variability over time (Figure 5.7).

The shape of the Poincaré plots can provide information on the non-linear dynamics of the ANS function. The plot is a cluster of points along the line of identity ($x = y$) and deviations from this line indicate changes in the heart rate: deviations above the line indicate acceleration in heart rate form one beat to the next, while deviations below the line indicate a deceleration. The movement of points along the line represents long-term changes in heart rate. A wide and long plot indicates high overall variability, which is indicative of a high level of autonomic tone, whereas narrow, bullet-shaped plots indicate low HRV and are typical of patients with a low level of autonomic function [42]. The plots can be analyzed both visually and geometrically. Visually, narrow and long plot indicate a prevalence of sympathetic activity, while large fan-shaped plots indicate a prevalence of parasympathetic activity. Geometrically, they can be quantified by fitting an ellipse to the plots and measuring the width and length of the points distribution along the line of identity (Figure 5.7). Changes in the plot shape are captured by these geometrical measures and reflect change in the ANS function. Specifically, the width (SD2) of the Poincaré plot reflects parasympathetic activation, while its length (SD1) reflects sympathetic antagonism to vagal tone. Furthermore, the SD1/SD2 ratio is analogous to the spectral measure of LF/HF ratio, indicating sympathovagal balance.

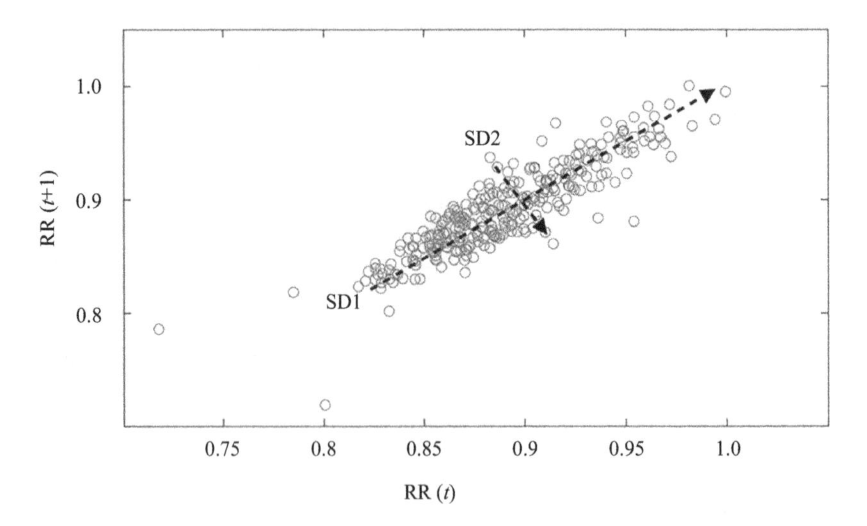

Figure 5.7 An example of a Poincaré plot of consecutive R–R intervals, also showing the geometrical measures of SD1 and SD2

Piskorski and Guzik provide a more detailed description of Poincaré plots and the mathematical estimates of the geometrical SD1 and SD2 measures, as well as their MATLAB® implementations [70].

5.5.2 EEG analysis methods

The brain is considered a chaotic dynamical system and, as such, the EEG is a signal that exhibits significant non-stationary, dynamic and non-linear complex behavior, with its amplitude changing randomly with respect to time. Analysis methods are generally divided into two main categories: linear and nonlinear. Linear methods include time-domain and frequency-domain analysis, such as Fourier and wavelet analysis, and autoregressive models. Even though linear measures have been used widely in EEG analysis, the amount of information they can provide is limited as they do not capture the underlying non-linear dynamics of the EEG activity. Hence, nonlinear measures, such as measures of complexity, are becoming increasingly popular, as they reveal additional dimensions that cannot be captured by linear measures alone. In the next sections, we concentrate on some commonly-used linear and non-linear analysis methods.

5.5.2.1 Artifact removal

One of the main issues in EEG analysis is the contamination of the signals with artefacts (other biological signals of no direct interest and signals of non-biological origin). It is imperative that prior to processing any artifact contamination is removed. Line noise can be easily removed through either a notch filter (centered either at 50 Hz or 60 Hz, depending on the region where the EEG was recorded) or a low-pass filter (commonly set at 40 Hz). The low-pass filter is usually preferred; however, it does not allow for analysis of frequencies higher than 40 Hz, which represent the EEG gamma range and which is identified as important for higher cognitive processes. Common methods of artifact removal utilize independent component analysis (ICA) to separate the recorded EEG in independent components. The ICs can then be inspected (either visually or using automatic methods) and those identified as of artifactual origin can be removed. An example of an automatic artifact removal based on ICA is clustering of the ICs based on the non-linear temporal dynamics; the clean EEG signals are then reconstructed by excluding the components that fall in clusters identified as artefacts [71].

5.5.2.2 Synchronization and causality

In the last few decades, there has been growing evidence concerning *large-scale functional integration* of information in the brain, which amounts to transient integration of numerous neuronal ensembles that are widely distributed through the brain [72]. These functional couplings are: (1) transient – their duration is of the order of a few milliseconds; (2) dynamic – coupling strength between two cortical regions varies in time; and (3) frequency-specific – the oscillation of the neuronal groups occurs in specific bands according to their precise phase relationship [46]. Synchronization can be generalized, a state in which functional dependence between the systems exists, or phase-synchronized, a state in which the phases of the systems are correlated whereas their amplitudes may not be.

In practice, there are three fundamental steps for the estimation of the phase synchrony [73]:

(1) Filtering: since brain signals typically contain multiple oscillatory components, a bandpass filter is applied centered around the frequency of interest.
(2) Estimation of the phase: instantaneous phase is estimated either via the Hilbert Transform or complex wavelets. Both methods can separate phases and amplitudes with a comparable time-frequency resolution.
(3) Quantification of the degree of phase-locking: several measures have been proposed. In the presence of noise, a statistical approach is required for identification of the phases and phase synchrony can be understood as the appearance of a peak in the probability distribution of the phase differences within a small time window.

One popular way of estimating synchronization is via Phase Synchrony (PS), defined as:

$$\Delta\phi = \left|n\phi_i - m\phi_j\right| < \varepsilon \tag{5.2}$$

where ϕ_i (ϕ_j) is the instantaneous phase of signal i (j), n and m are positive integers, ε is a small number, and $\Delta\phi$ is the difference between the instantaneous phases of signals i and j. When this relationship holds, at least for a short period of time (few milliseconds), then the two signals, i and j, are said to be phase synchronized. Due to the existence of various methods of estimating the instantaneous phase (e.g. Hilbert transform, analytic signal) and the PS values (e.g. Phase-Locking value (PLV)), different ways of estimating the instantaneous phase and the phase-locking between two signals have been proposed in the literature. The main methods of estimating instantaneous phase and PS are via the analytic signal and Hilbert transform, the wavelet transform and the two-dimensional projection of the phase-space. There are also different measures that have been developed to describe the PS. The main ones are the PLV, mean phase coherence and nearest neighbor phase synchronization. Published comparisons of the different methods conclude that the method of choice is very much dependent on the specific application.

Obtaining a global picture of synchronization in multichannel data is non-trivial. A promising approach involves detecting common instantaneous frequencies (estimated from the derivative of the instantaneous phases) using wavelets among groups of recording sites. This reveals transient periods of ensemble synchronization in the time–frequency domain and provides detailed information on the dynamic evolution of multiple synchronized clusters that are present in multichannel data [72]. However, analysis of synchronization alone does not address the question of causality, i.e. “who drives whom” – the influence one neuronal system exerts on another.

While phase synchrony measures capture synchronization between pairs of electrodes, the measure of global field synchrony (GFS) is a method that captures global functional connectivity at a specific frequency, by quantifying synchronization between multichannel data as a function of phase [74,75]. GFS estimation is based on the intuitive idea that, if there is significant functional connectivity over a specific frequency range, the individual phases of the observed multichannel data will be dominated by a single-phase angle at that frequency range. Transforming each time series into the frequency domain and creating a scatterplot of the individual channel phases could provide information about this single-phase angle: more elongated clouds indicate a greater likelihood of the scatterplot being dominated by a single-phase angle. As the cloud spreads, the likelihood of more EEG-generating processes being active grows, implying a lower degree of global phase angle synchrony.

Estimating GFS involves transformation of the data the frequency domain via a fast Fourier transform (FFT). The FFT of a multichannel EEG time series results in sine and cosine coefficients for each electrode and frequency point. Following the idea introduced by Lehman and Michel, these coefficients can be displayed as a sine–cosine diagram (scatterplot), whereby each electrode is represented as a point in a two-dimensional imaginary (sine) vs real (cosine) diagram at the frequency of interest [76]. The length of a vector drawn from the origin to a point in the figure represents the contribution of the frequency of interest to the signal at the given electrode, while the angle formed by this vector and the x-axis represents the equivalent phase angle. The dots in real EEG data sine-cosine diagrams tend to form an extended ellipsoid constellation; this ellipsoid shape indicates the fact that the observed EEG is created by a number of sinusoidal processes with varying amplitudes and phases, rather than a single underlying process. If we characterize multichannel data synchrony in terms of individual phase differences between all pairs of electrodes, we would get an exponential amount of estimated phase differences, that is, $N \times (N-1) \times (N-2)/2$ phase angle differences for N electrodes. Principal component analysis (PCA), which estimates and projects data onto the axes of maximal variance, yields one such best-fit straight-line approximation. As a result, the GFS can be estimated using the resulting PCA projection:

$$GFS(f) = \frac{|\lambda_1 - \lambda_2|}{\lambda_1 + \lambda_2} \tag{5.3}$$

where $\lambda_{1,2}$ are the PCA-estimated eigenvalues, representing the contribution of each variance direction to the single multichannel phase. GFS takes values between [0,1]. If the observed time series' phases are strongly dependent, the Fourier transforms will form a straight line in the complex plane. As a result, the estimated normalized eigenvalues will be close to 1 and 0; thus, $\lambda_1 \approx \lambda_2 \approx 0.5$, and GFS will tend to 1. If the time series have unrelated phases, GFS will tend to zero. In summary, higher GFS values indicate higher global functional connectivity, whereas the absence of a dominant phase results in lower GFS values and, consequently, lower global functional connectivity. The main benefits of GFS are that it is a reference-independent method and that no implicit/explicit model of brain sources is required for interpretation of the results [74].

Measures such as partial coherence and Granger causality provide such information. Granger causality (GC) is a bidirectional measure of causal connectivity. It was initially introduced by Wiener [77], and later formalized by Granger [78]. It is used to infer the directionality and dynamics of influence between two (or more) sources.

In its simplest definition, a process, X, is said to Granger-cause another process, Y, if the past of X assists in the prediction of the present of Y beyond the degree by which the present of Y is predicted by the past of Y alone. The traditional formulation relies on the use of autoregressive (AR) models of order p for the linear prediction of each variable, $X = [x(1), x(2), \ldots, x(T)]$, and $Y = [y(1), y(2), \ldots, y(T)]$ (where T: number of time-series samples) using information from its own past only (univariate AR – (5.4)) or using information from the past of both variables (bivariate AR – (5.5)). The GC is then estimated by comparing the variance of the residuals from the two models (5.6):

$$x(t) = \sum_{i=1}^{p} a_i x(t-i) + \varepsilon_x(t) \tag{5.4}$$

$$x(t) = \sum_{i=1}^{p} a_i x(t-i) + \sum_{i=1}^{p} b_i y(t-i) + \varepsilon_{xy}(t) \tag{5.5}$$

$$GC(y \to x) = ln \frac{var(\varepsilon_x)}{var(\varepsilon_{xy})} \tag{5.6}$$

There are some limitations/considerations relating to GC. First, the traditional GC definition is linear and pairwise. When performing pairwise GC analysis, it is impossible to distinguish between direct and indirect causal relationships. This is related to the issue of spurious causality, which can appear between two processes when both are influenced by unaccounted-for external sources [79]. When the interdependence between two time series cannot be fully explained by their interactions, the covariance of the noise terms in the estimated AR models can be examined to capture the remaining interdependence [80]. Conditional GC, which conditions the estimated GC onto external sources, is another solution [81]. To infer a more precise structural causality, all sources of influence must be considered. However, in practice, measuring all sources of influence relevant to a given problem is impossible, so conditional methods will always be provisional [82]. Second, the selection of an appropriate order for the AR model used is not easy: if the order is too low, the signal's properties are not captured; however, if the order is too high, any measurement noise or inaccuracies are also represented as a result of overfitting, and the resulting model is not a reliable representation of the signal. Popular methods of estimating the appropriate model order are the Akaike information criterion (AIC) and the Bayesian information criterion (BIC). The issue of stationarity is another factor to consider. Because AR models assume a stationary process, nonstationary EEG signals must be analyzed in short-term windows. For segments less than 20 s, it is widely accepted that EEG has stationary properties [83]. As a result, EEG analysis in short segments is common practice.

From the GC definition, it can be seen that GC is always positive and represents the amount by which the past of one variable improves the prediction of another variable. If there is significant GC, then it is said that "Y is causal to X," and vice versa if in the above equations the roles of X and Y are switched. GC becomes zero if there is no improvement in prediction. In this traditional definition of GC, it can be seen that causal effects resulting either from direct or indirect relationships with other processes are not considered. Thus, the traditional definition of pairwise GC has been extended to include multivariate AR models that include observations from additional variables, as well as non-parametric models that capture non-linear relationships [84]. GC is a popular directional connectivity measure in neuroscience applications [80].

5.5.2.3 Entropy-based complexity measures

Permutation entropy (PE) measures the complexity of the data by searching for patterns contained within the data itself, based on comparison between neighboring values [85]. Thus, it is a way of quantifying the relative occurrence of the different motifs. A symbolic mapping of the continuous time series captures the relationship between present values and a fixed number of equidistant values at a given past time. This mapping is accomplished by dividing the time series into segments containing m samples (where m is known as the embedding dimension) separated by τ samples and overlapping by $(m-1)$ samples. There will be $m!$ possible permutations for a given embedding dimension (motifs). If each permutation is thought of as a symbol, the embedded time vectors can be represented by a symbol sequence, j, each with its own probability distribution, p_j. Thus, using the Shannon entropy definition, the normalized PE, H_p, of a given time series is defined as:

$$H_p(m) = -\frac{1}{\ln(m!)}\sum_{j=1}^{J} p_j lnp_j \tag{5.7}$$

where J is the distinct number of symbols for a given embedding dimension ($J \leq m!$). PE measures a time series' deviation from complete randomness: the lower the value of PE, the more regular the time series. According to Olofsen *et al.*, "*When the EEG signal is dominated by high frequencies, there will be almost equal numbers of each motif in each EEG segment analyzed. For EEG, when the signal consists of slow delta waves, there will be relatively more of the 'slope' motifs (210 and 012) and fewer of the 'peak' (120, 021) and 'trough' (201, 102) motifs and the entropy decreases. The PE of a signal consisting of a single motif (such as one very long 'up-slope') is zero. [...] The PE tends to decrease as the frequency decreases, whereas the value of the spectral entropy is completely independent of frequency per se, but only measures the sharpness of the frequency peak.*" [48].

A similar complexity measure is that of approximate entropy (ApEn), a nonlinear dynamics measure that quantifies the unpredictability or randomness of a signal [87]. It is also estimated by first dividing the N-dimensional time series into n-dimensional segments, x_i, so that the predictability of current samples can be estimated based on knowledge of the previous n samples. The number of close pairs of n-dimensional segments, *count*, is estimated using a distance function such that *count* = number of $\boldsymbol{x}_j$ for which $d|x_j, x_i| \leq r$. d[...] is the maximum absolute difference between the corresponding scalar components of each segment, while r is a variable that specifies the tolerance in the 'closeness' of the vectors (a good choice is usually $r = k\sigma$, where

$k \in [0.1, 0.25]$ and σ is the signal's standard deviation). $C_i^n(r) = count/(N - m + 1)$ is then defined using the vector proximity, and ApEn is then estimated as:

$$ApEn = \Phi^n(r) - \Phi^{n+1}(r) \tag{5.8}$$

where $\Phi^n(r) = \frac{1}{N-n+1} \sum_{i=1}^{N-n+1} lnC_i^n(r)$.

ApEn represents the difference between the occurrence of patterns of length n and more complex patterns of length $n+1$. In other words, the greater the similarity between $n-$ and $(n+1)$-dimensional segments, the more regular the time series. Thus, the ApEn value of a perfectly regular time series is 0. The ApEn increases as the irregularity of the time series increases.

5.5.2.4 Surrogate data

When dealing with data where the "ground truth" is not known a priori, statistical testing for significance of extracted patterns identified during different states is achieved via the use of surrogate data created after random shuffling of the samples of the original signals. In general, surrogate data testing attempts to find the least interesting explanation that cannot be ruled out based on the data. The measure of interest, for example, PLV, is estimated between pairs of surrogate data – this is repeated a number of times depending on the desired statistical significance level and a different surrogate pair is generated each time. The measure of interest between the original pair of EEG signals is considered statistically significant at a given significance level when its estimated value exceeds the maximum value obtained from the surrogate data. Using PLV as an example, the PLV between pairs of surrogate data is estimated – this is repeated a number of times depending on the desired statistical significance level and a different surrogate pair is generated each time. The PLV between the original pair of EEG signals is considered statistically significant at a given significance level if its PLV value exceeds the maximum PLV value obtained from the surrogate data. Once the significance of individual values is assessed, the statistical significance of the observed patterns between different states and participant groups can be asserted via standard statistical significance methods (e.g. t-test, ANOVA, Wilcoxon ranksum).

5.5.3 Machine learning methods

Technological advancements and increased computer processing power in the last few decades, as well as the increased availability of patient health records in digital form, have facilitated the development of machine learning (ML) methods and their application in bioengineering areas, including the analysis of complex CNS-ANS activity. ML is a branch of Artificial Intelligence that focuses on the use of data and algorithms to learn and uncover relationships and patterns in the data, with the aim of improving the accuracy of recognition between different tasks/states/conditions/groups [87]. There are three main categories of ML: supervised, unsupervised, and reinforcement learning. They are distinguished by whether algorithm training and learning is achieved through the use of labeled datasets, unlabeled datasets with subsequent discovery of latent patterns in the data, or through trial and error, respectively. Commonly used ML methods include neural networks (including deep learning), regression (linear, logistic, non-linear), linear discriminant analysis, clustering, decision trees, random forests, support vector machines [87].

The application of ML to EEG analysis has enabled automated analysis and recognition in a variety of clinical areas [88], including epileptic seizure prediction and detection [89], automated sleep staging [90], emotion recognition [91], and brain-computer interfaces [92]. Furthermore, it allowed for neurological diagnostics of disorders such as neurodegeneration [93], depression [94,95], autism spectrum disorders [96], strokes [97], and disorders of consciousness [98], and of EEG pathology in general [99]. A review of ML methods applied to EEG signal processing is provided by Hosseini and colleagues [100].

ECG analysis has also benefited from advancements in ML methods for patient screening and risk stratification, as well as diagnosis and prediction of cardiovascular diseases, in combination with computer modelling methods that can aid in the interpretation of key physiologically-relevant ML-derived ECG biomarkers. ML has been utilized for classification of normal and abnormal heartbeats, diagnosis of cardiovascular diseases, sleep disorders, and real-time detection of ECG abnormalities [101–106].

The application of ML to EEG and ECG analysis has been a welcome development, as ML methods can make sense of these complex signals and detect underlying differences that the human eye may find difficult to detect. Particularly for EEGs, which are complex multivariate signals whose visual analyses for clinical diagnoses is a non-trivial and time-consuming task requiring intensive training, while also including some degree of subjectivity from the assessor (moderate inter-rater agreement). The development of algorithms for automated EEG diagnostics in particular could aid clinicians in screening EEGs by reducing clinician workload, as well as allow for earlier disease detection and treatment, potentially improving patient care. Moreover, they could provide high-quality EEG interpretation and classification to patients who are unable to travel to specialized centers. However, bringing novel techniques to the clinic necessitates both technological advancements and practical solutions to issues, such as regulatory approval and inclusion in clinical protocols, both of which necessitate collaborations between different sectors.

5.6 Conclusion

Computational measurements and interpretation of consciousness and unconsciousness states have evolved over the years. The computational representational understanding of mind hypothesis has yet to be tested as there are still multiple surrogate variables used for understanding how the CNS structures process data to generate outcomes. ML helps to measure and test physiological signals. However, the outcome is subject to interpretation in context. Nevertheless, discrimination between physiological states, whether these are healthy states, neurological disorders or pharmacologically induced states, seems to be better understood with the use of ML algorithms, but a multimodal approach to acquire neurological outcome is required for more objective measurement of neurophysiological responses. The field of ML of neurocognitive science is transformative and has great potential, but is still in its embryonic phase.

References

[1] Brodal, P. (2010). *The Central Nervous System: Structure and Function* (4th ed.). Oxford: Oxford University Press.

[2] Smith, R., Betancourt, L., & Sun, Y. (2005). Molecular endocrinology and physiology of the aging central nervous sys. *Endocrine Review*, *26*(2), 203–250.

[3] Rothwell, J., Antal, A., Burke, D., *et al.* (2021). Central nervous system physiology. *Clinical Neurophysiology*, *132*(12), 3043–3083.

[4] Mimura, Y., Nishida, H., Nakajima, S., *et al.* (2021). Neurophysiological biomarkers using transcranial magnetic stimulation in Alzheimer's disease and mild cognitive impairment: A systematic review and meta-analysis. *Neuroscience & Biobehavioral Reviews*, *121*, 47–59.

[5] Abe, O., Aoki, S., Hayashi, N., *et al.* (2002). Normal aging in the central nervous system: Quantitative MR diffusion-tensor analysis. *Neurobiology of Aging*, *23*(3), 433–441.

[6] Yang, J., Pan, Y., Wang, T., Zhang, X., Wen, J., & Luo, Y. (2020). Sleep-dependent directional interactions of the central nervous system-cardiorespiratory network. *IEEE Transactions on Biomedical Engineering*, *68*(2), 639–649.

[7] Lange, T., Born, J., & Westermann, J. (2019). Sleep matters: CD4+ T cell memory formation and the central nervous system. *Trends in Immunology*, *40*(8), 674–686.

[8] Dampney, R. (2016). Central neural control of the cardiovascular system: Current perspectives. *Advances in Physiology Education*, *40*(3), 283–296.

[9] Grippo, A., & Johnson, A. (2009). Stress, depression and cardiovascular dysregulation: A review of neurobiological mechanisms and the integration of research from preclinical disease models. *Stress*, *12*(1), 1–21.

[10] McCorry, L. (2007). Physiology of the autonomic nervous system. *American Journal of Pharmaceutical Education*, *71*(4), 78.

[11] Craven, J. (2011). The autonomic nervous system, sympathetic chain and stellate ganglion. *Anesthesia & Intensive Care Medicine*, *12*(2), 55–57.

[12] Ungar, A. & Phillips, J. (1983). Regulation of the adrenal medulla. *Physiological Reviews*, *63*(3), 787–843.

[13] Goldberger, J., Arora, R., Buckley, U., & Shivkumar, K. (2019). Autonomic nervous system dysfunction: JACC focus seminar. *Journal of the American College of Cardiology*, *73*(10), 1189–1206.

[14] Fung, M., Viveros, O., & O'Connor, D. (2008). Diseases of the adrenal medulla. *Acta Physiologica*, *192*(2), 325–335.

[15] Lucini, D., Di Fede, G., Parati, G., & Pagani, M. (2005). Impact of chronic psychosocial stress on autonomic cardiovascular regulation in otherwise healthy subjects. *Hypertension*, *46*(5), 1201–1206.
[16] Teixeira, R., Díaz, M., Santos, T., *et al.* (2015). Chronic stress induces a hyporeactivity of the autonomic nervous system in response to acute mental stressor and impairs cognitive performance in business executives. *PLoS ONE*, *10*(3), e0119025.
[17] Carney, R., Freedland, K., & Veith, R. (2005). Depression, the autonomic nervous system, and coronary heart disease. *Psychosomatic Medicine*, *67*, S29–S33.
[18] Grimaldi, D., Reid, K., Papalambros, N., *et al.* (2021). Autonomic dysregulation and sleep homeostasis in insomnia. *Sleep*, *44*(6), zsaa274.
[19] Benjamin, E., Muntner, P., Alonso, A., *et al.* (2019). Heart disease and stroke statistics-2019 update: A report from the American Heart Association. *Circulation*, *139*, e56–e528.
[20] Silvani, A., Calandra-Buonaura, G., Dampney, R., & Cortelli, P. (2016). Brain–heart interactions: Physiology and clinical implications. *Philosophical Transactions of the Royal Society A: Mathematical, Physical and Engineering Sciences*, *374*(2067), 20150181.
[21] Nagappan, P., Chen, H., & Wang, D. (2020). Neuroregeneration and plasticity: A review of the physiological mechanisms for achieving functional recovery postinjury. *Military Medical Research*, *7*(1), 1–16.
[22] Jones, M., Dilley, J., Drossman, D., & Crowell, M. (2006). Brain–gut connections in functional GI disorders: Anatomic and physiologic relationships. *Neurogastroenterology & Motility*, *18*(2), 91–103.
[23] Chang, L. (2011). The role of stress on physiologic responses and clinical symptoms in irritable bowel syndrome. *Gastroenterology*, *140*(3), 761–765.
[24] Lechin, F. & van der Dijs, B. (2009). Central nervous system plus autonomic nervous system disorders responsible for gastrointestinal and pancreatobiliary diseases. *Digestive Diseases and Sciences*, *54*(3), 458–470.
[25] Bei, B., Seeman, T., Carroll, J., & Wiley, J. (2017). Sleep and physiological dysregulation: A closer look at sleep intraindividual variability. *Sleep*, *40*(9), zsx109.
[26] de Zambotti, M., Trinder, J., Silvani, A., Colrain, I., & Baker, F. (2018). Dynamic coupling between the central and autonomic nervous systems during sleep: A review. *Neuroscience & Biobehavioral Reviews*, *90*, 84–103.
[27] Bear, M., Connors, B., & Paradiso, M. (2015). *Neuroscience: Exploring the Brain* (4th ed.). Burlington, MA: Jones & Bartlett Learning.
[28] American Electroencephalographic Society Guidelines for Standard Electrode Position Nomenclature. (1991). *Journal of Clinical Neurophysiology*, *8*(2), 200–202.
[29] Nunez, P. & Srinivasan, R. (2006). *Electric Fields of the Brain: The Neurophysics of EEG* (2nd ed.). Oxford: Oxford University Press.

[30] Murakami, S. & Okada, Y. (2006). Contributions of principle neocortical neurons to magnetoencephalography and electroencephalography signals. *Journal of Neurophysiology*, *575*(3), 925–936.

[31] Baura, G. (2021). Chapter 1—Diagnosis and therapy. In *Medical Device Technologies: A Systems Based Overview Using Engineering Standards* (2nd ed., pp. 5–43). London: Academic Press.

[32] Selvaraj, N., Jaryal, A., Santhosh, J., Deepak, K., & Anand, S. (2008). Assessment of heart rate variability derived from finger-tip photoplethysmography as compared to electrocardiography. *Journal of Medical Engineering & Technology*, *32*(6), 479–484.

[33] Volynsky, M., Mamontov, O., Sidorov, I., & Kamshilin, A. (2016). Pulse wave transit time measured by imaging photoplethysmography in upper extremities. *Journal of Physics Conference Series*, *737*, 012053.

[34] Polanía, R., Nitsche, M., Korman, C., Batsikadze, G., & Paulus, W. (2012). The importance of timing in segregated theta phase-coupling for cognitive performance. *Current Biology*, *22*(14), 1314–1318.

[35] Badhwar, A., Tam, A., Dansereau, C., Orban, P., Hoffstaedter, F., & Bellec, P. (2017). Resting-state network dysfunction in Alzheimer's disease: A systematic review and meta-analysis. *Alzheimers & Dementia (Amsterdam).*, *18*(8), 73–85.

[36] Dauwels, J., Vialatte, F., & Cichocki, A. (2010). Diagnosis of Alzheimer's disease from EEG signals: Where are we standing? *Current Alzheimer Research*, *7*(6), 487–505. https://doi.org/10.2174/156720510792231720.

[37] Vitiello, B., Veith, R., Molchan, S., *et al.* (1993). Autonomic dysfunction in patients with dementia of the Alzheimer type. *Biological Psychiatry*, *34*(7), 428–433. https://doi.org/10.1016/0006-3223(93)90233-4.

[38] Wang, S.-J., Liao, K.-K., Fuh, J.-L., *et al.* (1994). Cardiovascular autonomic functions in Alzheimer's disease. *Age and Ageing*, *23*(5), 400–404. https://doi.org/10.1093/ageing/23.5.400.

[39] Issac, T., Chandra, S., Gupta, N., Rukmani, M., Deepika, S., & Sathyaprabha, T. (2017). Autonomic dysfunction: A comparative study of patients with Alzheimer's and frontotemporal dementia—A pilot study. *Journal of Neurosciences in Rural Practice*, *8*(1), 84–88. https://doi.org/10.4103/0976-3147.193545.

[40] Allan, L., Kerr, S., Ballard, C., *et al.* (2005). Autonomic function assessed by heart rate variability is normal in Alzheimer's disease and vascular dementia. *Dement Geriatr Cogn Disord*, *19*, 140–144. https://doi.org/10.1159/000082885.

[41] Czigler, B., Csikós, D., Hidasi, Z., *et al.* (2008). Quantitative EEG in early Alzheimer's disease patients—Power spectrum and complexity features. *International Journal of Psychophysiology: Official Journal of the International Organization of Psychophysiology*, *68*(1), 75–80. https://doi.org/10.1016/j.ijpsycho.2007.11.002.

[42] Ardissino, M., Nicolaou, N., & Vizcaychipi, M. (2019). Non-invasive real-time autonomic function characterization during surgery via continuous

Poincaré quantification of heart rate variability. *Journal of Clinical Monitoring and Computing*, *33*, 627–635. https://doi.org/10.1007/s10877-018-0206-4.

[43] Tasbihgou, S., Vogels, M., & Absalom, A. (2018). Accidental awareness during general anaesthesia – A narrative review. *Anaesthesia*, *73*(1), 112–122.

[44] Palanca, B., Mashour, G., & Avidan, M. (2009). Processed electroencephalogram in depth of anesthesia monitoring. *Current Opinion in Anaesthesiology*, *22*(5), 553–559.

[45] Kuhlmann, L., Freestone, D., Manton, J., *et al.* (2016). Neural mass model-based tracking of anesthetic brain states. *NeuroImage*, *133*, 438–456.

[46] Sheeba, J., Stefanovska, A., & McClintock, P. (2008). Neuronal synchrony during anesthesia: A thalamocortical model. *Biophysical Journal*, *95*, 2722–2727.

[47] Ching, S., Cimenser, A., Purdon, P., & Kopell, N. (2010). Thalamocortical model for a propofol-induced α-rhythm associated with loss of consciousness. *PNAS*, *107*(52), 22665–22670.

[48] Olofsen, E., Sleigh, J., & Dahan, A. (2008). Permutation entropy of the electroencephalogram: A measure of anaesthetic drug effect. *British Journal of Anaesthesia*, *101*(6), 810–821.

[49] Eagleman, S., & Drover, D. (2018). Calculations of consciousness: Electroencephalography analyses to determine anesthetic depth. *Current Opinion in Anaesthesiology*, *31*(4), 431–438.

[50] Litvan, H., Jensen, E., Revuelta, M., *et al.* (2002). Comparison of auditory evoked potentials and the A-line ARX index for monitoring the hypnotic level during sevoflurane and propofol induction. *Acta Anaesthesiologica Scandinavica*, *46*(3), 245–251.

[51] Sinha, P., & Koshy, T. (2007). Monitoring devices for measuring the depth of anaesthesia – An overview. *Indian Journal of Anaesthesia*, *51*(5), 365–381.

[52] Barr, G., Jakobsson, J., Owall, A., & Anderson, R. (1999). Nitrous oxide does not alter bispectral index: Study with nitrous oxide as sole agent and as an adjunct to i.v. anaesthesia. *British Journal of Anaesthesia*, *82*(6), 827–830.

[53] Anderson, R. & Jakobsson, J. (2004). Entropy of EEG during anaesthetic induction: A comparative study with propofol or nitrous oxide as sole agent. *British Journal of Anaesthesia*, *92*(2), 167–170.

[54] Liu, N., Chazot, T., Huybrechts, I., Law-Koune, J.-D., Barvais, L., & Fischler, M. (2005). The influence of a muscle relaxant bolus on bispectral and Datex-Ohmeda values during propofol-remifentanil induced loss of consciousness. *Anesthesia & Analgesia*, *101*, 1713–1718.

[55] Gajraj, R., Doi, M., Mantzaridis, H., & Kenny, G. (1999). Comparison of bispectral EEG analysis and auditory evoked potentials for monitoring depth of anaesthesia. *British Journal of Anaesthesia*, *82*(5), 672–678.

[56] Heir, T., & Steen, P. (1996). Assessment of anaesthesia depth. *Acta Anaesthesiologica Scandinavica*, *40*(9), 1087–1100.

[57] Rampil, I. (1998). A primer for EEG signal processing in anesthesia. *Anesthesiology*, *89*(4), 980–1002.

[58] Varela, F., Lachaux, J.-P., Rodriguez, E., & Martinerie, J. (2001). The brainweb: Phase synchronization and large-scale integration. *Nature Reviews Neuroscience*, *2*, 229–239.

[59] Kumagai, Y., Arvaneh, M., & Tanaka, T. (2017). Familiarity affects entrainment of EEG in music listening. *Frontiers in Human Neuroscience*, *11*, 384.

[60] Bradt, J., Dileo, C., & Shim, M. (2013). Music interventions for preoperative anxiety. *Cochrane Database of System Review*, *6*(6), CD006908.

[61] Chlan, L. (1998). Effectiveness of a music therapy intervention on relaxation and anxiety for patients receiving ventilatory assistance. *Heart Lung*, *27*, 169–176.

[62] Yung, P., Chui-Kam, S., French, P., & Chan, M. (2002). A controlled trial of music and pre-operative anxiety in Chinese men undergoing transurethral resection of the prostate. *Journal of Advanced Nursing*, *39*, 352–359. https://doi.org/10.1046/j.1365-2648.2002.02296.x.

[63] Andrillon, T., Pressnitzer, D., Léger, D., & Kouider, S. (2017). Formation and suppression of acoustic memories during human sleep. *Nature Communications*, *8*, 179.

[64] Binns-Turner, P., Wilson, L., Pryor, E., Boyd, G., & Prickett, C. (2011). Perioperative music and its effects on anxiety, hemodynamics, and pain in women undergoing mastectomy. *AANA Journal*, *79*(4), S21–S27.

[65] Bernatzky, G., Presch, M., Anderson, M., & Panksepp, J. (2011). Emotional foundations of music as a non-pharmacological pain management tool in modern medicine. *Neuroscience & Biobehavioral Reviews*, *35*(9), 1989–1999.

[66] Schaffer, F. & Ginsberg, J. (2017). An overview of heart rate variability metrics and norms. *Front Public Health*, *5*, 258. https://doi.org/10.3389/fpubh.2017.00258.

[67] Kleiger, R., Miller, J., Bigger Jr, J., & Moss, A. (1987). Decreased heart rate variability and its association with increased mortality after acute myocardial infarction. *American Journal of Cardiology*, *59*, 256–262. https://doi.org/10.1016/0002-9149(87)90795-8.

[68] Penttilä, J., Helminen, A., Jartti, T., *et al.* (2001). Time domain, geometrical and frequency domain analysis of cardiac vagal outflow: Effects of various respiratory patterns. *Clinical Physiology*, *21*(3), 365–376.

[69] Brennan, M., Palaniswami, M., & Kamen, P. (2001). Do existing measures of Poincare plot geometry reflect nonlinear features of heart rate variability? *IEEE Transactions on Biomedical Engineering*, *48*(11), 1342–1347.

[70] Piskorski, J., & Guzik, P. (2007). Geometry of the Poincaré plot of RR intervals and its asymmetry in healthy adults. *Physiological Measurement*, *28*(3), 287–300. https://doi.org/10.1088/0967-3334/28/3/005.

[71] Nicolaou, N. & Nasuto, S. (2007). Automatic artefact removal from event-related potentials via clustering. *Journal of VLSI Signal Processing*, *48*(1–2), 173–183.

[72] Rudrauf, D., Douiri, A., Kovach, C., *et al.* (2006). Frequency flows and the time-frequency dynamics of multivariate phase synchronization in brain signals. *Neuroimage*, *31*(1), 209–227.
[73] Le Van Quyen, M. & Bragin, A. (2007). Analysis of dynamic brain oscillations: Methodological advances. *TRENDS in Neurosciences*, *30*, 365–373.
[74] Koenig, T., Lehman, D., Saito, N., Kuginuki, T., Kinoshita, T., & Koukkou, M. (2001). Decreased functional connectivity of EEG theta-frequency activity in first-episode, neuroleptic-naive patients with schizophrenia: Preliminary results. *Schizophrenia Research*, *50*, 55–60.
[75] Nicolaou, N. & Georgiou, J. (2014). Global field synchrony during general anaesthesia. *British Journal of Anaesthesia*, *112*(3), 529–539. https://doi.org/10.1093/bja/aet350.
[76] Lehman, D. & Michel, C. (1989). Intracerebral dipole sources of EEG FFT power maps. *Brain Topography*, *2*, 155–164.
[77] Wiener, N. (1956). The theory of prediction. In E. Beckenbach (Ed.), *Modern Mathematics for Engineers*. New York, NY: McGraw-Hill.
[78] Granger, C. (1969). Investigating causal relations by econometric models and cross-spectral methods. *Econometrica*, *37*, 424–438.
[79] Granger, C. (1980). Testing for causality: A personal viewpoint. *Journal of Economic Dynamics and Control*, *2*, 329–352.
[80] Ding, M., Chen, Y., & Bressler, S. (2006). Granger causality: Basic theory and application to neuroscience. In B. Schelter, M. Winterhalder, & J. Timmer (Eds.), *Handbook of Time Series Analysis* (pp. 437–460). Wiley-VCH Verlag.
[81] Geweke, J. (1984). Measures of conditional linear dependence and feedback between time series. *Journal of the American Statistical Association*, *79*, 907–915.
[82] Bressler, S., & Seth, A. (2011). Wiener-Granger causality: A well established methodology. *Neuroimage*, *58*, 323–329.
[83] da Silva, F. (2005). EEG analysis: Theory and practice. In E. Niedermeyer & F. da Silva (Eds.), *Electroencephalography: Basic Principles, Clinical Applications, and Related Fields* (pp. 1199–1232). Philadelphia, PA: Lippincott Williams & Wilkins.
[84] Nicolaou, N. & Constandinou, T. (2016). A nonlinear causality estimator based on non-parametric multiplicative regression. *Frontiers in Neuroinformatics*, *10*, 19.
[85] Bandt, C. & Pompe, B. (2002). Permutation entropy: A natural complexity measure for time series. *Physical Review Letters*, *88*, 174102.
[86] Pincus, S., Gladstone, I., & Ehrenkranz, R. (1991). A regularity statistic for medical data analysis. *Journal of Clinical Monitoring*, *7*, 335–345.
[87] Zhou, Z.-H. (2021). *Machine learning* (1st ed.). New York, NY: Springer.
[88] Joshi, V., & Nanavati, N. (2021). A review of EEG signal analysis for diagnosis of neurological disorders using machine learning. *Journal of Biomedical Photonics & Engineering*, *7*(4), 040201-1-040201–040217.
[89] Rasheed, K., Qayyum, A., Qadir, J., *et al.* (2021). Machine learning for predicting epileptic seizures using EEG signals: A review. *IEEE Reviews*

in Biomedical Engineering, *14*, 139–155. https://doi.org/10.1109/RBME.2020.3008792.

[90] Sekkal, R., Bereksi-Reguig, F., Ruiz-Fernandez, D., Dib, N., & Sekkal, S. (2022). Automatic sleep stage classification: From classical machine learning methods to deep learning. *Biomedical Signal Processing and Control*, *77*, Article: 103751.

[91] Doma, V., & Pirouz, M. (2020). A comparative analysis of machine learning methods for emotion recognition using EEG and peripheral physiological signals. *Journal of Big Data*, *7*, Article number: 18.

[92] Gu, X., Cao, Z., Jolfaei, A., *et al.* (2021). EEG-based brain-computer interfaces (BCIs): A survey of recent studies on signal sensing technologies and computational intelligence approaches and their applications. *IEEE/ACM Transactions on Computational Biology and Bioinformatics*, *18*(5), 1645–1666.

[93] Ieracitano, C., Mammone, N., Hussain, A., & Morabito, F. (2020). A novel multi-modal machine learning based approach for automatic classification of EEG recordings in dementia. *Neural Networks*, *123*, 176–190.

[94] Gao, S., Calhoun, V., & Sui, J. (2018). Machine learning in major depression: From classification to treatment outcome prediction. *CNS Neuroscience & Therapeutics*, *24*(11 (Special Issue on Imaging Connectomics in Depression)), 1037–1052.

[95] Liu, Y., Pu, C., Xia, S., Deng, D., Wang, X., & Li, M. (2022). Machine learning approaches for diagnosing depression using EEG: A review. *Translational Neuroscience*, *13*(1), 224–235.

[96] Rahman, M., Usman, O., Muniyandi, R., Sahran, S., Mohamed, S., & Razak, R. (2020). A review of machine learning methods of feature selection and classification for autism spectrum disorder. *Brain Science*, *10*, 949.

[97] Ruksakulpiwat, S., Thongking, W., Zhou, W., *et al.* (2021). Machine learning-based patient classification system for adults with stroke: A systematic review. *Chronic Illness, 19*, 26–39. https://doi.org/10.1177/ 17423953211067435.

[98] Wu, S.-J., Nicolaou, N., & Bogdan, M. (2020). Consciousness detection in a complete locked-in syndrome patient through multiscale approach analysis. *Entropy*, *22*(12), 1411. https://doi.org/10.3390/e22121411.

[99] Gemein, L., Schirrmeister, R., Chrabaszcz, P., *et al.* (2020). Machine-learning-based diagnostics of EEG pathology. *NeuroImage*, *220*, Article number: 117021.

[100] Hosseini, M., Hosseini, A., & Ahi, K. (2021). A review on machine learning for EEG signal processing in bioengineering. *IEEE Reviews in Biomedical Engineering*, *14*, 204–218. https://doi.org/10.1109/RBME.2020.2969915.

[101] Chang, A., Cadaret, L., & Liu, K. (2020). Machine learning in electrocardiography and echocardiography: Technological advances in clinical cardiology. *Current Cardiology Reports*, *22*, Article number: 161.

[102] Lyon, A., Mincholé, A., Martínez, J., Laguna, P., & Rodriguez, B. (2018). Computational techniques for ECG analysis and interpretation in light of their contribution to medical advances. *Journal of the Royal Society Interface*, *15*, 20170821.

[103] Rizwan, A., Zoha, A., Mabrouk, I., *et al.* (2021). A review on the state of the art in atrial fibrillation detection enabled by machine learning. *IEEE Reviews in Biomedical Engineering, 14*, 219–239. https://doi.org/10.1109/RBME.2020.2976507.
[104] Sahoo, S., Dash, M., Behera, S., & Sabut, S. (2020). Machine learning approach to detect cardiac arrhythmias in ECG signals: A survey. *Innovation and Research in BioMedical Engineering, 41*(4), 185–194.
[105] Salari, N., Hosseinian-Far, A., Mohammadi, M., *et al.* (2022). Detection of sleep apnea using machine learning algorithms based on ECG signals: A comprehensive systematic review. *Expert Systems with Applications, 187*, 115950.
[106] Wasimuddin, M., Elleithy, K., Abuzneid, A.-S., Faezipour, M., & Abuzaghleh, O. (2020). Stages-based ECG signal analysis from traditional signal processing to machine learning approaches: A survey. *IEEE Access, 8*, 177782–177803. https://doi.org/10.1109/ACCESS.2020.3026968.

Chapter 6

Brain networking and early diagnosis of Alzheimer's disease with machine learning

Jingyang Yu[1] *and Marcela Vizcaychipi*[1]

Brain study has advanced significantly in the past decade and it is benefiting from the field of machine learning (ML). Indeed, ML is drawing brain researchers' attention for multiple reasons: first, the amount of data relating to a person's health is growing rapidly and it is becoming increasingly difficult for humans (including experts) to analyse all of those data. Therefore, ML becomes an indispensable technique to automatically extract pertinent information from the data, so that time-critical decisions such as medical diagnoses and tailored treatments can be made by clinicians with speed and confidence.

Second, medical ML is potentially more economical than human expertise. For some rural regions in developing countries, medical ML can improve the professional level and efficiency of local clinical practices at a comparatively low-cost [1]. Unlike resources that cannot be flexibly deployed (such as doctors), ML models are transferable since they are essentially software. Therefore, clinics in backward regions and/or less accessible regions can have access to these tools, particularly if they are deployed as web applications. Finally, ML can be trained for monitoring the progression of debilitating brain diseases such as Alzheimer's disease (AD), so that early treatment can be provided to improve health outcomes.

In this chapter, we discuss the basic structure of the brain and techniques to explore the brain's activities. Building on the knowledge of brain structure and techniques measuring brain activities, we will then move into reviewing the current state of the art in the application of ML to the early diagnosis of AD. Finally, we will conclude the chapter with future directions in this area of research.

6.1 Background

6.1.1 A brief history of brain study

The brain is arguably the most important part of the human body, as it commands most of our daily activities such as speaking, listening, visualisation and acting. Also, the

[1]Department of surgery and cancer, Imperial College London, UK

brain enables us to experience emotions and feelings. But what are the mechanisms behind all of these marvellous brain functions? Humans have been trying to unveil the mystery inside our skull for thousands of years: From Aristotle's (384–322 BC) coolant system theory to Vesalius's (1514–1564) work of the disserted brain, to Gall (1758–1828) and Spurzheim's (1776–1832) functional specialisation in brain's regions, to modern-day cognitive neuroscience, a lot of knowledge and theories about both the structures and functions of brains has been established and proposed. We have seen various ideas and theories explaining how the brain works.

6.1.2 Modern understanding of the brain

Generally, a brain can be divided into three main parts based on their positions: forebrain, midbrain, and hinderbrain. Each of them contains different subunits and each is responsible for different functions, as shown in Figure 6.1.

The forebrain contains the cerebrum (or the cerebral cortex) and diencephalon (or interbrain). The cerebrum, the largest part of the brain, is located in the uppermost region of the central nervous system and is associated with higher cognitive functions. Diencephalon is associated with sensory and motor impulses.

The midbrain is located beneath the diencephalon but above the hinderbrain. It is in the central position of the brain's 3D structure. The functions of the midbrain include helping to relay information for vision and hearing, helping to control body movements and containing a large number of dopamine-producing neurons. Moreover, the midbrain is part of the brain stem in the overall brain structure.

The hinderbrain is located at the bottom of the whole brain structure but its functions should not be underestimated. It is associated with multiple functions including coordinating muscle movements, maintaining posture and balance, especially for dexterity and smooth execution of movement, to organise life activities such as biting, swallowing breathing, heart rate, body temperature, wake and sleep cycles, and digestion.

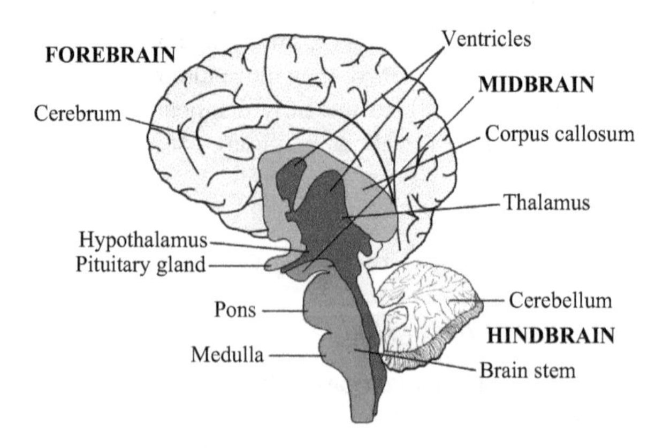

Figure 6.1 Brain anatomy [2]

6.2 Laboratory model of brain connectivity

The brain is like a dedicated machine where all parts of its units work collaboratively. But before discussing brain connectivity, we need to ask ourselves a question: what are we trying to "connect" where? In other words, at what scale are we going to look at brain connectivity? Neural cells? Neurons? Certain brain areas? Function modules? They are all correct answers. Models of brain connectivity vary on different levels of scale. At the microscale, neuron activities and their synaptic connections are studied. And at the mesoscale, people focus on the networks connecting neuron populations. As for the largest macroscale, brain regions formed by networks and neurons are studied to investigate regional interactions. In conclusion, cells, networks and regions are three different levels for modelling a brain.

Generally, models' construction is based on the existing understanding of brain connectivity from three different but related perspectives.

- Anatomical connectivity (AC), also called structural connectivity, forms the connectome through synaptic contacts between neighbouring neurons or fibre tracks connecting neuron pools in spatially distant brain regions. The whole set of such fibre tracks in the brain is called white matter. On short time scales (sec, min), anatomical connections are quite persistent and stable, while for longer time spans substantial changes may be observed. Any connection in an anatomical model biologically exists, or we say a real connection.
- Functional connectivity (FC) is defined as the temporal dependency of neuronal activation patterns of anatomically separated brain regions. It reflects statistical dependencies between distinct and distant regions of information processing neuronal populations. Hence, it is a statistical concept which relies on such statistical measures as correlation, covariance, spectral coherence, or phase locking. Statistical dependencies are highly time-dependent and fluctuate on multiple time scales ranging from milliseconds to seconds. Notably, a connection in functional connectivity does not have to refer directly and structural links. Therefore, functional connectivity inevitably carries uncertainty of a clear physical pathway behind it.
- Effective connectivity (EC) describes the influence one neuronal system exerts upon another, thus reflecting causal interactions between activated brain areas. It combines structural and effective connectivity into a wiring diagram which reflects directional effects within a neuronal network. The causal effect can be inferred from network perturbations or time series analysis (TSA).

The difference between a functional model and an effective model is that a functional model does not consider the direction. For example, suppose we find that brain region A and brain region B are connected. In a functional model, we consider the strength of the link between region A and region B. However, since direction does matter in an effective model, the relationship or link between regions A and B, and the link between B and A are distinct and studied separately.

From the discussion above, it is clear that creating a generic model for brain activities is extremely challenging because of the scale and complexity of the problem. Indeed, there are around 86 billion neurons in a human brain working together [3].

Computationally, it is very difficult for even an exascale computer to store and process information from a model containing these neurons. If we consider the possible permutations of the connections between these neurons, this will make the problem even more intractable. In addition, invasive methods are required to acquire measurements in order to model the extracellular interactions and molecular-scale processes of the brain, which is difficult if not impossible ethically in most cases. Consequently, the absence of this data often leads to the making of inaccurate assumptions as well as oversimplification, rendering them inaccurate and not particularly useful.

Rather than develop a generic model of the brain, many current approaches have focused on the detection of specific brain diseases, which are more tractable problems. In addition, current clinical practice for diagnosis involves examining data collected non-invasively from patients, in conjunction with patient history. For the sake of brevity, we will focus on the sensors, data, and methods that are relevant to Alzheimer's disease (AD).

6.3 Problem definition

Dementia refers to a group of symptoms associated with an ongoing decline of brain functions [4]. AD is the most common cause of dementia, contributing approximately 60–70% of all cases [5]. Specifically, symptoms include a gradual loss of memory, as well as cognitive functions such as orientation, language, and emotion [6]. Predictably, these symptoms often result in a severe degradation in the quality of life. Some argue that the number of people affected by dementia is estimated to double in the next two decades [7]. However, as one of the crucial parts of the treatment, accurate early diagnosis is difficult because the symptoms are often mistaken for those exhibited in normal ageing [8]. Therefore, finding an accurate solution for early diagnosis is of utmost importance.

AD is characterised by structural and functional degradation in brain structures, which takes place over 30–50 years [9]. Similarly, normal ageing results in Mild Cognitive Impairment (MCI), but MCI does not always regress into AD. More specifically, current research classifies an AD diagnosis result into one of three classes, as follows.

1. *Control Normal* (*CN*): healthy with no symptoms of AD.
2. *Mild Cognitive Impairment* (*MCI*): Mild structural changes in brain structure and function.
3. *AD*: Severe atrophy in brain structures, including medial temporal lobes, posterior cingulate, lateral and parietal temporal cortex [10].

Many methods have been devised successfully to distinguish between the CN and AD groups. However, the results are much poorer when they be applied to the classification between MCI and AD [9]. Consequently, much of the current research efforts are focused on the latter problem. Having described and defined the problem, we will discuss the relevant devices or sensors.

6.4 Devices used in AD diagnosis

The sensorial system of the brain is a part of the nervous system for processing sensory information. Sensory information comes from various sources including vision, hearing, touch, taste, smell, and balance. In another way, sensorial systems transfer signals in different forms like electromagnetic waves, sound waves, molecules into the bioelectrical pulses in the cortex, the surface of the cerebrum. These transformed bioelectrical signals can be monitored and measured through various devices. Three devices are commonly used in the diagnosis of AD, including the magnetic resonance imaging (fMRI), computerised tomography (CT), and electroencephalography (EEG).

- fMRI is a functional neuroimaging procedure measuring brain function by detecting changes in blood flow associated with neural activity, by measuring the radio waves in the magnetic fields [13]. For a subject who is doing a controlled cognitive task, the activated regions in the neural system require energy in the form of adenosine triphosphate (ATP). The increased demand for ATP results in an increase in the demand for oxygen and correspondingly, local cerebral blood flow (CBF). These changes in oxygenation concentration or BOLD (Blood Oxygen Level Dependent) contrast can be detected using an fMRI machine. Therefore, the fMRI can be used to track changes in brain physiology resulting from cognitive tasks. An example of a fMRI scanner is shown in Figure 6.2 [11]. Note that CBF and BOLD are used for different purposes. While CBF is mostly used for modelling the neurobiological mechanisms of activation or calibration of vasoreactivity, the BOLD contrast from resting state to stimulated state is used for mapping brain functions [14].
- CT is also a popular and useful imaging technology used in clinical practice. CT scanners apply X-rays to the human body from different angles. These X-rays penetrate the body and is captured in several images. Subsequently, these images

Figure 6.2 A fMRI scanner [11]

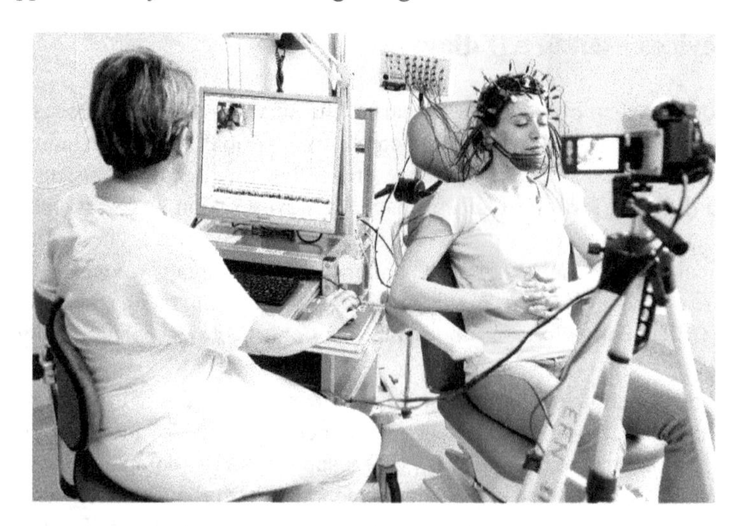

Figure 6.3 An EEG device [12]

are spliced together to form several cross sections of the body which provides useful information about different parts of the body.

- Electroencephalography (EEG) measures the electrical activity generated by the various cortical layers of the human brain. This is a powerful tool to measure sensorial activities because it can locate the changes in the electric field and amplify the electric signals in the grey matter that the cortex locates. The electric field changes during sensorial activities can be observed and studied. Since the communication between neurons takes place by means of electrical signals or impulses, these impulses can be picked up by means of the scalp electrodes that are positioned on the subject, amplified and displayed. Symptoms of AD can be found by analysing EEG graphs while the individual is subjected to specific stimuli. An important advantage of using EEG signals is that changes can be tracked with a very high time resolution (up to 1 ms depending on the sampling rate). Figure 6.3 [12] illustrates an example of an EEG acquisition device.

In this chapter, we will focus on MRI-based and CT imaging technologies to make the discussion more concrete.

6.5 Data types

From the devices discussed previously, different types of data can be obtained from the patient. These data sources usually afford different perspectives and allow the clinician to build a more holistic picture of the patient condition, which will in turn facilitate a more accurate diagnosis. In Figure 6.4 [15], we see the MRI and CT scans of the same human brain. Whilst the MRI scan shows the soft tissues at high resolution, hard tissue like the skull bones cannot be seen clearly. Instead, this information can

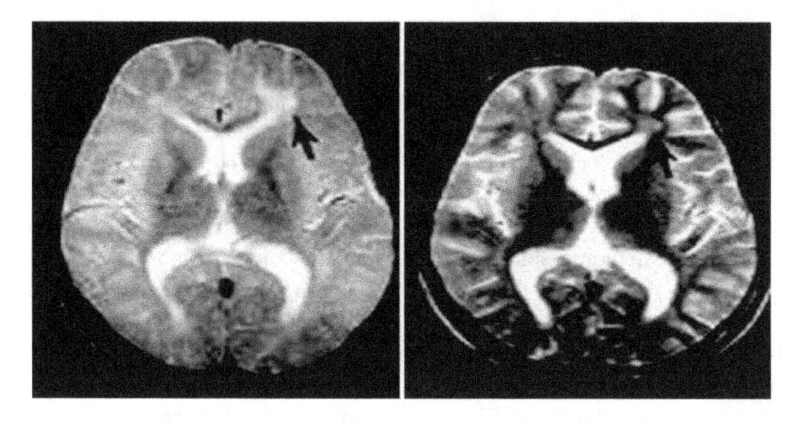

Figure 6.4 Left: MRI scan. Right: CT scan [15].

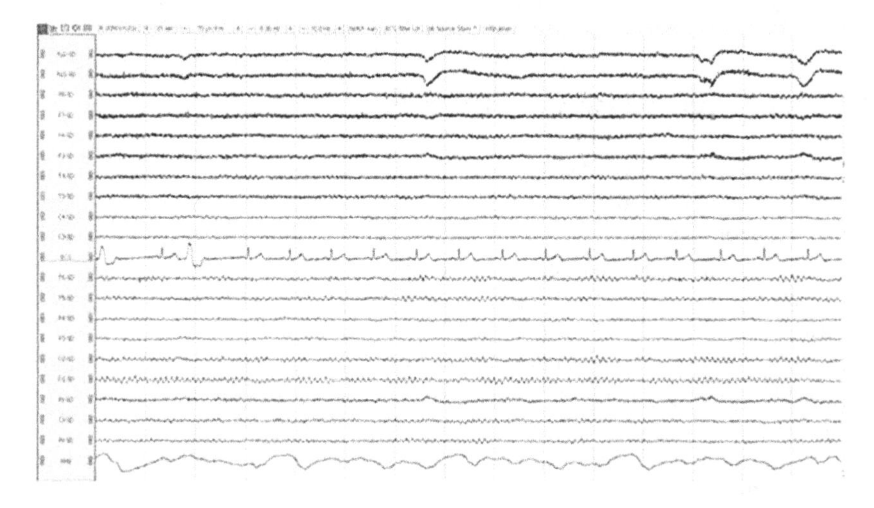

Figure 6.5 21-Channels resting state EEG waveform [16]

be obtained from the CT scan. This is valuable as it helps the clinician to localise any abnormal soft tissue, relative to known hard tissue landmarks in the CT scan. Therefore, similar to a clinician making a diagnosis, it is clear that automatic data analysis should consider different sources of data in order to attain a high level of diagnostic accuracy. We discuss some of these methods in the subsequent sections.

EEG waveforms are acquired as amplified electrical signals between scalp electrodes. Each pair of electrodes relates to brain activations at a different region and is known as a channel. Typical setups involve 21 channels, as shown in Figure 6.5 [16]. Abnormalities can be detected by analysing the EEG charts – Figure 6.5 shows the variation of potential with time. A useful method of analysis is to determine the power of the signal in specific frequency bands. This can be achieved by converting the time-based waveforms to the frequency domain, or equivalently the variation of the waveform with frequency rather than time (power spectral density). The resulting spectrum is then divided into different bands and the power of each band is obtained by calculating the area under the curve. Conventional bands with diagnostic content are delta (0.5–4 Hz), theta (4–8 Hz), alpha-1 (8–10 Hz), alpha-2 (10–13 Hz), beta (13–30 Hz), and gamma (30–48 Hz). Differences are observed between the power of these bands for CN and AD groups [16]. However, there are many confounding factors to consider, including signal artefacts and substantial variability even between normal individuals. We will discuss these factors and the methods to eliminate or reduce their extent in the next section.

To develop methods and algorithms for data analysis, it is essential for researchers and/or engineers to have access to substantial datasets acquired in a controlled manner, and accurately labelled (e.g. An EEG trace is annotated as belonging to an AD diagnosed subject) by expert clinicians. At the time of writing, EEG databases of sufficient quality are not readily available. On the other hand, there are several online databases for MRI and/or CT scans, including ADNI [17], AIBL [18], and OASIS [19].

6.6 Data preprocessing of MRI data

During an MRI scan, several snapshots of a patient are made. Each snapshot involves a cross-section of the patient's body, creating a two-dimensional image. These images are then spliced together to form a multi-dimensional map. The resulting map can be broken down into cubes, and each cube is known as a voxel [20]. The voxel is the smallest graphical unit and is assigned a set of unique three-dimensional coordinates in space, as illustrated in Figure 6.6 [20]. In a greyscale image, each voxel will be assigned a value indicating its intensity (or the brightness of the voxel). The analogue of a voxel is the pixel, which has two-dimensional coordinates to delineate its position in space.

Since the original snapshots happen at different times, unwanted noise or artefacts are often added. Apart from being a nuisance, this may result in the wrong diagnosis. As such, various data pre-processing techniques are frequently applied to the MRIs to attenuate these artefacts. These artefacts originate from different sources and involve different treatment techniques. Apart from denoising input data, image fusion is often carried out to combine different types of images (e.g. MRI and CT scans) into a single image. Doing so allows different types of information to be fed to the diagnostic modules that occur downstream of pre-processing operations, while allowing the dimensionality of the problem to be constrained within tractable bounds. We will discuss two popular preprocessing techniques, including Median filtering and Deconvolution, followed by Image fusion in the subsequent sections.

6.6.1 Median filters

Average weighted filters are simple operators, used for reducing high frequency noise in many applications, including image processing. However, they should be

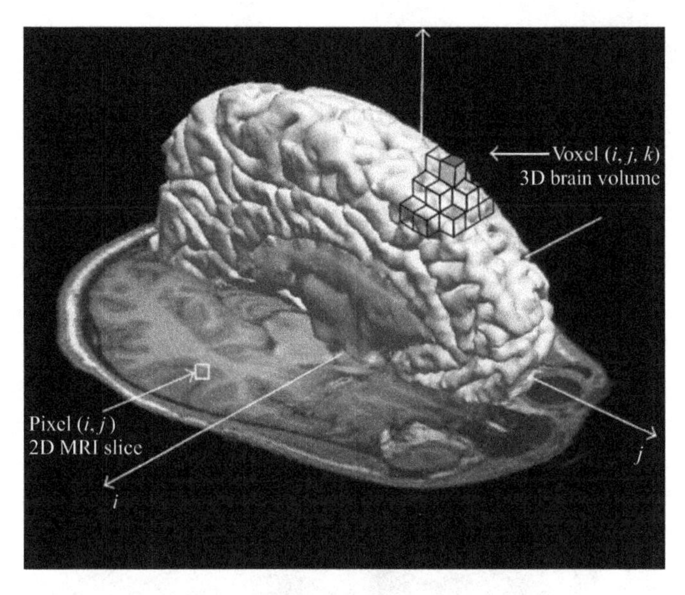

Figure 6.6 A multi-dimensional map of the brain is illustrated. Note how it is broken into individual voxels, such that unique three-dimensional coordinates can be assigned to each voxel [20].

judiciously used as they can potentially distort the very image that they are intended to improve. Typically, these filters are used in conjunction with another class of non-linear filters, known as Median filters. Unlike average weighted filters, median filters are less sensitive to outliers so filtered images are less susceptible to blurring and are likely to have higher contrast. We illustrate how a conventional median filter works in Figure 6.7. A filter kernel (3×3 pixels windows) is moved from the top left-hand corner of the image to the bottom right-hand corner, specifically from the left to the right and from the top to the bottom of the image (overlaps between the kernels are frequently allowed). At each position, the median operator, I_{med}, is applied within the kernel and the central pixel (marked 'X') is replaced by the result of the median operation, which consists of two steps.

1. Sort the pixels in the kernel according to their intensities, producing a sorted array $\{I_0, I_1, \ldots, I_{N-1}\}$, where $I_0 \leq I_1 \leq \ldots \leq I_{N-1}$ ($N = 9$ in Figure 6.7).
2. Pick the pixel with mid-level intensity i.e. $I_{med} = I_{\frac{N-1}{2}}$, N is odd, or $I_{med} = \frac{1}{2}(I_{\frac{N}{2}} + I_{\frac{N}{2}-1})$, N is even.

Consider the red kernel which lies at the boundary of two black and white clusters in Figure 6.7. The moving average filter will replace the central pixel with the mean of all members in the kernel, which is a very light grey colour. Adjacent kernels will produce light grey values as well – this will create a blurring and distorting effect at the edges of the image which is undesirable. On the other hand, the median

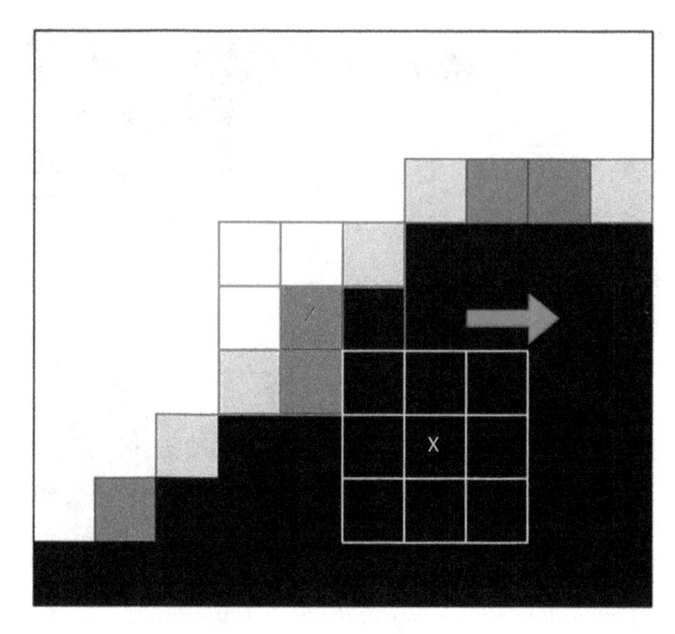

Figure 6.7 An image in grey scale. The red and green sub-blocks represent filter kernels and the pixels marked as 'X's represent the pixel that will be replaced with the output of the median operator. Overlaps between windows are permitted.

filter will replace the central pixel with a much darker pixel as it is less affected by the white and black pixels in the kernel. Therefore, image contrast is preserved by means of a median filter which is desirable. Note that we discussed median filters in the context of pixels, which exist in two-dimensional coordinate space with the pixel intensity as the third dimension, so as to illustrate the concept in a clear and simple manner. The approach can be trivially extended to voxels in three-dimensional coordinate space.

While a conventional median filter is a simple improvement over the moving average filter at the image boundaries, more sophisticated variants can be constructed to improve its performance. We demonstrate this using an example described in this work [21]. Broadly, an image compromises of different clusters, which are populated by voxels. The process of clustering is to determine the membership of each voxel. In a discrete clustering scheme, each voxel is permitted to belong to only one cluster whereas it is permitted to belong to multiple clusters in a fuzzy scheme. The extent to which a voxel belongs to a cluster is dependent on its degree of membership. For example, $u_j(x)$ refers to the degree of membership that Voxel x belong to Cluster j, where $0 \leq u_j(x) \leq 1$. The following scheme describes how fuzzy membership can be incorporated into median filters in order to further improve its performance. We assume that the membership of each voxel to all known clusters is known apriori. For the sake of clarity, we make certain modifications to the original algorithm described in [21].

1. Given a voxel x, we determine its most likely cluster by taking the maximum over memberships to all clusters and returning the corresponding cluster i.e. $C = \arg\max_j u_j(x)$.
2. Within the three-dimensional kernel K surrounding x, we reduce the neighbourhood by including voxels with at least the same degree of membership to C, as compared with x. This restricted neighbourhood is known as Ω.
3. The intensities of all members of Ω are sorted to obtain $\{I_0, I_1, \ldots, I_{N-1}\}$, where $I_0 \leq I_1 \leq \ldots \leq I_{M-1}$, and M refers to the number of members within Ω.
4. Rather than selecting the voxel with mid-level intensity, the membership to Cluster C of each corresponding element in the sorted intensity array is created, $\{u_0, u_1, \ldots, u_{M-1}\}$. Note that the subscript of u no longer refers to a particular cluster. Rather, it refers to the corresponding voxel index in the sorted array.
5. A weight array is then computed by accumulating the membership array. The index of the selected voxel, med_{index} within Ω, is then obtained by selecting the mid-point of the accumulated memberships, as shown in (6.1):

$$med_{index} = \min(\{i| \sum_{j=1}^{i} u_j > \frac{1}{2} \sum_{j=1}^{M} u_j, i \in [1, M]\}) \tag{6.1}$$

Indeed, selecting the voxel based on a combination of intensities and membership values, as part of the median operation, will further reduce the effects of outliers and preserve the sharpness of the image cluster boundaries, resulting in an overall improvement of image quality.

6.6.2 Physiological noise removal by means of deconvolution

Three main sources of noise in MRIs are as follows:

- *Bulkhead movement*: Patient movement during the scan will result in motion artefacts in the images.
- *Breathing*: This results in rhythmic changes in oxygen and carbon dioxide concentration in the blood and cerebrospinal fluid, which in turn results in an undesirable variance of the BOLD contrast of the MRI images.
- *Cardiac pulsatility*: The pulsing of the heart is another source of variance in the BOLD contrast. It is challenging to handle this artefact as there is a known and variable time lag (between 30 and 42 s [22]) between the cardiac cycle and its corresponding effect on BOLD.

While it is possible to largely remove bulkhead movement artefacts by asking patients to use a bite bar [22], it is impossible to stop the patient's heart or even for the patient to stop breathing for the duration of the scan. Therefore, it is necessary to attenuate the breathing and cardiac artefacts using an algorithm [22] that is able to accommodate the time lags in the respiratory and heart rate cycles. The first step involves the calibration of a linear model, described using (6.2), using heart rate and

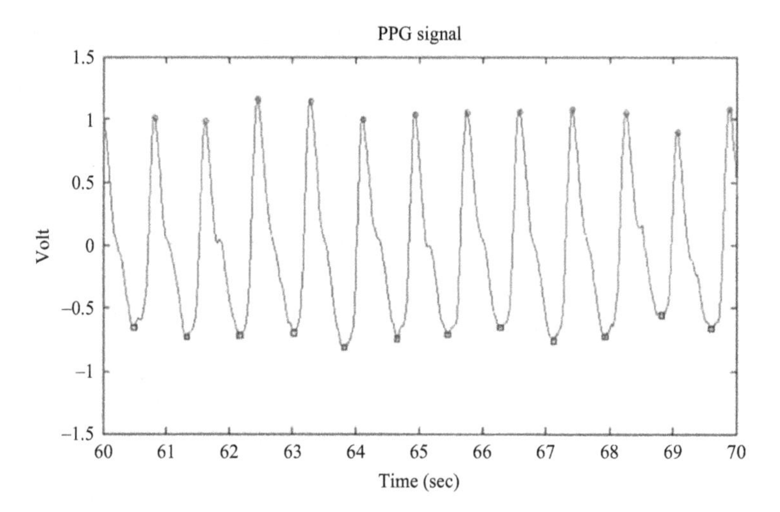

Figure 6.8 Photoplethysmogram. The peak to peak interval represents the time between heart beats or systoles [24].

respiratory data acquired while an MRI scan is carried out. The cardiac and respiratory rhythms can be obtained via pulse oximetry and a pneumatic belt respectively [23]. More specifically, the pulse oximeter produces a photoplethysmogram that is shown in Figure 6.8 [24]. Several peaks are seen in the waveform, each of which corresponds to individual heartbeats or systoles. By averaging the beat-to-beat intervals, inverting the average and scaling the result by 60, the averaged heart rate or *HR* can be obtained in beats per minute. Similarly, the respiration information can be obtained from the respiration volume signal. However, rather than computing the time intervals between breathing cycles, the average standard deviations of the raw signals are used instead:

$$y = X_r h_r + X_h h_h + \varepsilon \tag{6.2}$$

$$y = Xh + \varepsilon, \text{ where } X = [X_r, X_h]^T, h = [h_r, h_h]^T \tag{6.3}$$

$$h = \arg\max_h p(h|y) = (X^T X + \mu) X^T y \tag{6.4}$$

In (6.2), y refers to the intensity of a particular voxel. X_r and X_h refers to the time-shifted input averaged respiration and cardiac data streams respectively (formatted as matrices). The unknown variables, h_r and h_h, refer to the impulse responses of the respiratory and cardiac sub-systems that relate these artefacts to the voxel output. We can further simplify the model using (6.3) by combining the inputs and impulse responses as individual matrices X and h. Finally, the unknown impulse responses can be found using the least square approach as shown in (6.4). This process is known as deconvolution. Note that an additional parameter μ is added in (6.4) to account for the covariance between impulse responses. Further information about this approach can be found in [22].

Once calibration is completed, the impulse responses for individual voxels will be known and they can be applied to subsequent MRIs in a forward transform or convolution. In this way, the variable time lags between physiological sources and the resulting artefacts can be considered during data preprocessing to produce MRIs of better quality.

6.6.3 Image fusion

Different sources of information provide complementary information about the physiological state of the patient and can potentially allow the clinician to make a more accurate diagnosis. Examples include MRI, CT, and PET scans, which separately contain information about cellular metabolism and positional information of any abnormalities within the brain. Indeed, several studies were conducted [10] to compare the diagnostic efficacy between single modal and multi-modal techniques. Performance gains in terms of accuracy (up to about 9%), sensitivity (up to about 10%), and specificity (up to about 3%) were reported [10].

A computationally efficient way of handling different sources is to fuse those images together as a preprocessing step before further analysis takes place. Image fusion is an established technique and it can be broadly categorised into spatial and frequency domain techniques. We shall provide an example for each category in the subsequent sections.

6.6.3.1 A spatial technique for image fusion

As mentioned previously, the scans provide two-dimensional images, each of which can be broken down in pixels with two-dimensional coordinates. At the minimum, a pixel in a greyscale image has one channel of information, which indicate it's intensity. However, a pixel in a colour image has three channels of information, each of which represents the primary colours – Red, Green, and Blue. By varying the intensity of each of these components, any other colour may be obtained. Spatial techniques refer to the processing of these channels within the two-dimensional coordinate space. In this context, the objective is to fuse two different but aligned images together. We will further assume that an MRI (greyscale image) image and a PET (colour image) image of the same aspect ratios are the inputs and we wish to combine them together into a single image.

Colours can be represented using their primary components – Red (R), Green (G), and Blue (B) or the RGB colour space. However, it is by no means the only way to represent them. In fact, the processing method that we describe converts the images into a different colour space known as the Hue (H), Saturation (S), and Intensity (I), or the HSI colour space [25]. The intuitive explanations for each of these terms are provided below.

- *Hue*: this represents the pure colour component e.g. how red or yellow an observed colour appears.
- *Saturation*: the extent to which the pure colour component is corrupted by white light.

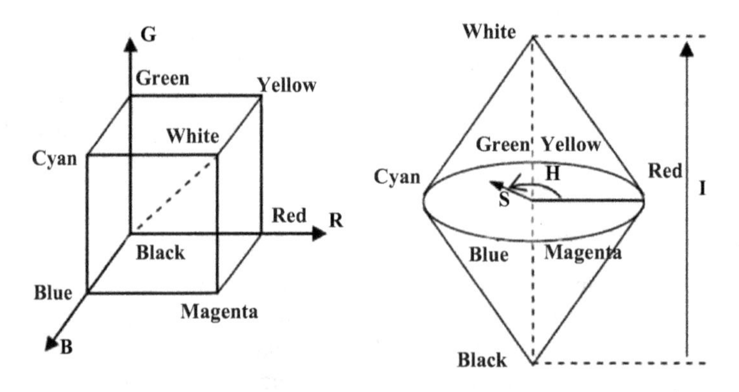

Figure 6.9 Left: RGB colour space. Right: HSI colour space [26].

- *Intensity*: the brightness of the colour. The two extremes are white and black in this case.

In comparison with the RGB space, the HSI space is more intuitive as it is similar to the manner in which the human eye perceives colour. Its relationship with the RGB colour space is illustrated in Figure 6.9. A particular colour can be represented as a data point in the RGB space d as it has unique values for the red, green and blue dimensions. Independently, a vector w can be created between the origin and the point where the red, green, and blue components are at their maximum values respectively (dotted line in the image on the left of Figure 6.9). Similarly, another vector v can be created between the origin and the Point d. Essentially, the angle between v and the red dimension (**R**-axis in the RGB space) is the *Hue* i.e. 0, 120, and 240 degrees refer to the colours Red, Green, and Blue respectively. In addition, Point d forms a plane that is orthogonal to the vector w, as shown in the image to the right of Figure 6.9. *Saturation* is then the length of Vector S in the HSI space shown in Figure 6.9. Finally, *Intensity* can be worked out as the normalised vector magnitude between the origin and Point d in the RGB space.

From the discussion above, obtaining the Hue, Saturation, and Intensity for each colour in the RGB space involves linear algebraic operations to determine the angle between vectors, projections of vectors on the HSI colour plane and their corresponding magnitudes. The conversion function from the RGB space to the HSI space can be succinctly specified via (6.5) [26]. The Hue and Saturation are determined using (6.6) and (6.7) respectively (the intensity can be trivially obtained from the R, G, and B values so it is omitted here):

$$\begin{bmatrix} 1 \\ V_1 \\ V_2 \end{bmatrix} = \begin{bmatrix} 1/3 & 1/3 & 1/3 \\ -\sqrt{(2)}/6 & -\sqrt{(2)}/6 & 2\sqrt{(2)}/6 \\ 1/\sqrt{(2)} & -1/\sqrt{(2)} & 0 \end{bmatrix} \begin{bmatrix} R \\ G \\ B \end{bmatrix} \tag{6.5}$$

$$H = \tan^{-1}\frac{V_1}{V_2} \tag{6.6}$$

$$S = \sqrt{(V_1^2 + V_2^2)} \tag{6.7}$$

HSI conversion allows us to move from the RGB to the HSI space. Since it is a reversible transform, moving back from the HSI to the RGB space is possible via (6.8) [26]. This is an important property that is exploited in the image fusion that we shall describe below.

$$\begin{bmatrix} R \\ G \\ B \end{bmatrix} = \begin{bmatrix} 1 & -1/\sqrt{(2)} & 1/\sqrt{(2)} \\ 1 & -1/\sqrt{(2)} & -1\sqrt{(2)} \\ 1 & \sqrt{(2)} & 0 \end{bmatrix} \begin{bmatrix} I \\ V_1 \\ V_2 \end{bmatrix} \tag{6.8}$$

Apart from HSI conversion, the Hilbert Transform (HT) is another important technique that is frequently used in image processing and this includes Image fusion. The primary purpose of HT is to identify regions within an image where there are step changes in pixel intensities, such as edges and corners. In the subsequent transform, these regions will be manifested as easily identifiable extrema whereas other regions have intensities far lower than those extrema. In addition, the HT has a directional feature as well. To illustrate this, we can imagine processing a two-dimensional image pixel by pixel, from the top to the bottom and from the left to the right-hand side of the image. The image is populated with objects residing in a background with very different pixel intensities. Therefore, at the object boundaries or edges, when there is a transition from the background to the object, there will be a step change in pixel intensity. Similarly, a transition from the object to the background will lead to step change in pixel intensity, but in the opposite direction. This is reflected as extrema of opposite signs in the corresponding HT of the pixel intensities. We will illustrate these two important properties mathematically. First, we define pixel intensity function $r(p)$, where p refers to the position within an image frame. We can model r as a random process of an unknown statistical distribution and zero mean. Therefore, at the object boundaries, there will be a change in the sign signal r. Second, we can then define a function $x(p)$ which detects changes in sign in Signal r or equivalently, the detection of object boundaries in (6.9). More precisely, the signal x is defined in (6.10), which is further broken into separate functions involving the object boundaries i.e. Function x_i will assume a value of 0 by default, except when there is a step change of pixel intensity from pixels p_i to p_{i+1}:

$$x(p) = sgn(r(p)) \tag{6.9}$$

$$= \sum_{i=0}^{N} k_i x_i(p) \tag{6.10}$$

$$x_i(p) = \begin{cases} 1, p_i < i < p_{i+1} \\ 0, \text{otherwise} \end{cases} \tag{6.11}$$

We carry on to define the Hilbert Transform function, which is essentially the convolution of Signal x with function $\frac{1}{\pi p}$, as seen in (6.11). By substituting (6.10) and

(6.11) into (6.12) and integrating the resulting argument, we can show that the HT will result in a sum of logarithms or equivalently (6.12).

$$H(x(p)) = \frac{1}{\pi}\int_{-\infty}^{\infty}\frac{x(\tau)}{\tau - p}d\tau \tag{6.12}$$

$$= -k_0 \log|p - p_0| + \sum_{i=1}^{N}(k_{i-1} - k_i)\log|p - p_i| + k_N \log|p - p_{N+1}| \tag{6.13}$$

Equation (6.13) demonstrates that singularities (∞ or $-\infty$) will occur at $\{p = p_i | 0 \leq i \leq N + 1\}$ or at the edges. In addition, it will either be positive or negative infinity depending on whether there is a step transition from low to high intensities or from low to high intensities in the original image indicating that it has a directional property. Thus far, we discussed the problem in the context of large step changes in the signal. In real images, these step changes are likely to be much smaller and the corresponding Hilbert Transform will result in large finite positive or negative values rather than infinity. Further, using HT is found to be more robust to image noise compared with other techniques [27]. For this reason, it is a popular image processing technique [10].

Having described HSI colour space conversion and the Hibert Transform, we will discuss an image fusion method that makes use of both techniques. The block diagram explaining this method is illustrated in Figure 6.10. The Hilbert Transform is first applied to a Greyscale MRI image to obtain a set of coefficients $MRI_{coefficient}$ (note that this is a two-dimensional analogue of (6.12), and it is implemented by means of Fast Fourier Transforms). On a separate stream, the PET scan image is converted from the RGB colour space to the HSI colour space using the linear transformation described in (6.5). The HT is then applied to the intensity map to obtain a set of coefficients $PET_{coefficient}$. Subsequently, a new intensity map is created by applying the Rule (6.14) to each of the pixels in the input intensity maps to produce H_F^K. Intuitively, the maximum operator is chosen to allow dominant features such as clear edges to be visualised in the final image. The Inverse Hilbert Transform is then applied to the fused coefficients H_F^K to obtain a fused intensity map. Finally, the Inverse Hilbert Transform is applied to the fused intensity map, in combination with the original Hue and Saturation maps to obtain the final fused image, which exists in the RGB space. In this way, the resolution preserving qualities of the HT and IHS are exploited in the process of fusing two types of images together:

$$H_F^K(i,j) = \text{Max}\{H_{MRI}^K(i,j), H_{PET}^K(i,j)\} \tag{6.14}$$

An example where the image fusion method is applied is shown in Figure 6.11. As we expect, different levels of image details, concerning the tissues and bones, are seen in the MRI and PET images. The fused image in Figure 6.11 represents the best of both worlds. Note that the image size remains the same in the fused image, so the computational burden for downstream operations should remain the same as processing only one image type even though there is increased clarity in

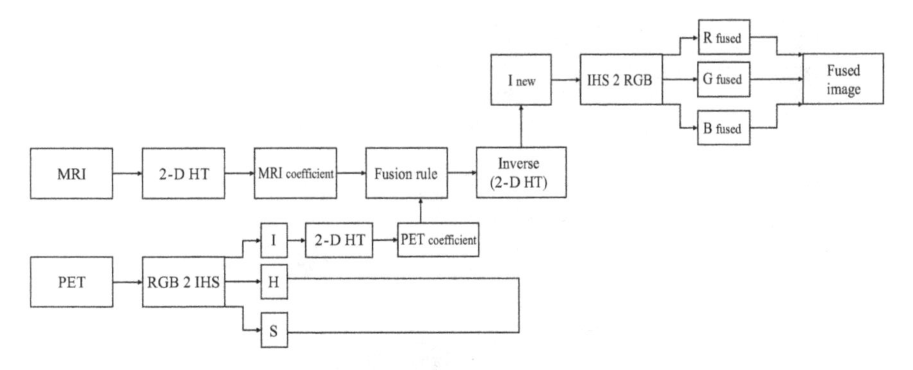

Figure 6.10 HSI colour space image fusion [26]

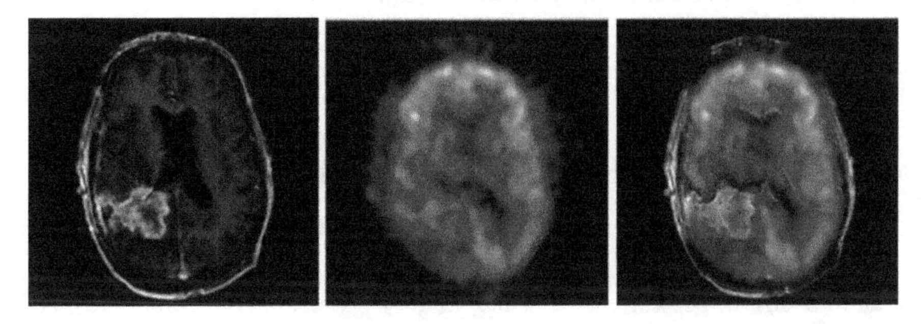

Figure 6.11 Left: Brain MRI image with tumour. Middle: brain PET image. Right: fused image [26].

the fused image. However, there are two caveats in this case – first, the increased computational overhead from this additional preprocessing step needs to be taken into account. Second, undesirable artefacts may be introduced as a result of image fusion, which may in turn affect the final diagnosis [10].

6.6.3.2 A frequency domain technique for image fusion

Another technique that can be used to fuse two aligned images together is based on a spatial-to-frequency transformation method known as Discrete Cosine Transform (DCT). To understand DCT, we look at a typical image in Figure 6.12. If we extract a small block from the image and examine its colour variation, we notice that there is a large degree of similarity in colour as we move from one point (or pixel) in the image to another point (an adjacent pixel) in its neighbourhood. In other words, colour varies slowly with respect to space – this is known as spatial correlation. Exceptions occur in the form of edges and corners, such as image blocks that contain tree branches along with bits of the blue sky. However, if we divide the image into small sub-blocks, we will find that blocks with limited colour variations will occur at a much higher frequency as compared to those with abrupt colour changes. In addition, an alternative to representing colour information with respect to space is to do so with

Figure 6.12 A slice of sky from the image to the left

respect to two-dimensional frequency space instead. By looking at this alternative representation (which is afforded using frequency domain techniques like DCT), we can better quantify the extent of colour variation or equivalently, the amount of detail within that image, and exploit this property for image fusion.

Before DCT can be applied, the image needs to be divided into sub-blocks of a uniform size (a popular choice is an 8 by 8 pixels block). The two-dimensional DCT, shown in (6.15), is then applied to each sub-block to produce another block of an equal size. The resulting block represents the variation of colour with respect to frequency. Specifically, the 'pixel' or DCT coefficient at the top left of the new block represents the colour variation at the frequency of zero Hz (this is equivalent to the mean pixel value in the sub-block). The frequency will increase as we move from the left to the right, and from the top to the bottom of the sub-block. For many of the blocks, the high absolute value DCT coefficients will be concentrated at the low-frequency bins within the block whereas the high-frequency coefficients would have very small absolute values. In this way, DCT can achieve energy compaction (a removal of spatial redundancy) within the signal. In fact, the extent to which energy compaction can be achieved directly relates to the amount of details within that image region. Therefore, by examining the absolute amplitudes of DCT coefficients for both blocks as well as how they vary with frequency, we are able to determine which image region has more content. In addition, the DCT is a reversible transform and it allows us to move from the frequency to the spatial domain and this is shown in (6.16):

$$F(u,v) = \frac{2}{N}C(u)C(v)\sum_{y=0}^{N-1}\sum_{x=0}^{N-1} f(x,y)cos[\frac{(2x+1)u\pi}{2N}]cos[\frac{(2y+1)u\pi}{2N}]$$

$$C(u) = \begin{cases} \frac{1}{\sqrt{(2)}}, u = 0 \\ 1, u \neq 0 \end{cases} \tag{6.15}$$

$$f(x,y) = \frac{2}{N}\sum_{v=0}^{N-1}\sum_{u=0}^{N-1} X(u,v)cos[\frac{(2x+1)u\pi}{2N}]cos[\frac{(2y+1)u\pi}{2N}] \tag{6.16}$$

$$X(u) = C(u)C(v)F(u,v) \tag{6.17}$$

We can fuse two aligned images, I_0 and I_1, based on the method described in [28]. Broadly, this is done by performing DCT on both images and selecting sets of coefficients (among I_0 and I_1) that has a larger number of coefficients with higher absolute values, and selecting the corresponding fused image block based on this criteria. In this way, we ensure that the technique preserves image detail in the final fused image. The specific steps to do so are shown below [28]:

1. Divide both images into blocks of uniform size and perform DCT on each block to obtain separate sets of DCT coefficients.
2. For each paired set of DCT coefficients, corresponding to Images I_0 and I_1, maintain counters C_0 and C_1 for each set of coefficients and initialise both to zero.
3. If the absolute coefficient value at block coordinates (u, v) is larger for Image I_0, as compared with I_1, increment Counter C_0. If the opposite is true, increment Counter C_1 instead.
4. After going through all the coefficients within the block, set the sub-block of the fused image with the one from I_0 if $C_0 >= C_1$. Otherwise, set the fused image sub-block as the one from I_1.
5. Keep repeating the above operations until all image blocks have been considered.

There are several advantages to using the above method. First, it is more computationally efficient as no floating-point calculations are required after DCT. Second, in comparison with traditional methods like FFT, it is more memory efficient as it produces only real-valued coefficients rather than imaginary ones so its storage requirements are lower. Also, comparisons between real numbers are quicker as opposed to comparisons between imaginary numbers. Third, unlike other techniques that compute the fused DCT coefficients by aggregating those from the input images, this method selects one set of coefficients or the other in its entirety and therefore avoids the need to perform IDCT, which results in substantial savings. Finally, there is scope for integrating it into conventional compression algorithms as I/DCT are standard modules within JPEG and MPEG algorithms. Doing this can potentially bring about further savings in terms of storage and/or transmission bandwidth [28].

6.7 Machine learning for early AD diagnosis

In this section, we will discuss the application of machine learning (ML) to the problem of the early diagnosis of AD. In general, we can consider a Supervised ML algorithm as a black box where the inputs, I, and outputs, O, are known. The goal of ML is to determine the contents of this black box (Function $f(I)$), so as to predict the value of O via a process known as 'Training'. The prediction efficacy of mutable Function *f* improves as it is fed with more input data. Indeed, successful AI created by ML techniques has been known to approach and even exceed human performance [29]. The generality of ML makes it a powerful technique, one that can be applied across vastly different application domains.

We look at a popular technique that has been widely used to solve this problem – support vector machines (SVMs) [30] as well as a more recent deep learning [31] approach to the problem.

6.7.1 SVMs

Let us suppose that we made a set of observations, each of which can be characterised based on its features. Therefore, each observation will occupy a point in feature space as shown in Figure 6.13. Assuming that the observations can be categorised into two classes, we will obtain red and blue clusters separately. The primary objective of ML is to determine a function or *hyperplane* that separates these two clusters. Subsequently, the type of observation (or which class the data point belongs to) can be determined based on its relative location from the hyperplane.

In the SVM approach, the hyperplane is described using (6.18). In this example, Vector x is a two-dimensional plane that spans the feature space (note that this is not restricted to two and is dependent on the number of features pertaining to our problem). The solid black line in the figure is the hyperplane and it is bounded by two parallel lines at equal distance from it that are known as the margins. The support vectors are actual feature points that lie on those margins. Given this context, the SVM technique determines the best combination for the weight vector w and bias vector

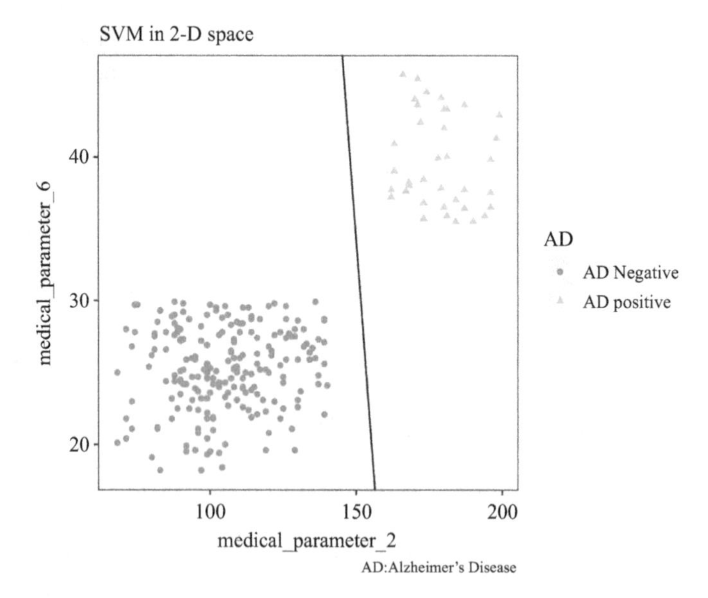

Figure 6.13 A linear hyperplane constructed for the separation of two clusters

b that will maximise *both* the distance between the margins and correspondingly, the distance between the clusters. Function f produces a discrete result – 'Positive' or 'Negative' and the sign determines the class membership of each data point. The solution that is obtained is a globally optimal one (not trapped in local minima) and it is an important reason why the SVM is such an effective technique.

$$f(x) = sign(w^T x + b) \tag{6.18}$$

Real-world datasets are usually more complex so much so that the original SVM may not yield results that are good enough. Several augmentation methods may be used to improve the classification results – two such methods are shown below:

1. *Soft margin*: this relaxes constraints in the linear optimisation problem and allows a small portion of the data to be misclassified [32].
2. *Kernel trick*: this involves the modification of (6.18) into a non-linear function to produce a hyperplane with a higher degree of freedom and potentially obtain better classification results.

6.7.2 Deep learning

Similar to the SVM, the objective of deep learning approaches is to construct a hyperplane that is able to carry out a classification or cluster separation in multi-dimensional feature space. However, instead of formulating the hyperplane using (6.18), it makes use of a topology known as Artificial Neural Network (ANN), as shown in Figure 6.14. The network is organised into three different types of layers known as input, hidden, and output layers. The input and output layers usually consist of only one column of nodes, whereas the hidden layer can contain multiple columns.

Each node consists of a weighted sum of inputs (including Biases $\{b_0, b_1\}$), which is fed into an activation function. The activation function at the output node, O, normalises the inputs from the preceding stage to a real number between 0 and 1 – this can be interpreted as a prediction of the conditional probability that the current observation falls in a certain class, given the input features, $\{I_0, I_1\}$. Essentially, the goal of ANN training is to determine the set of weights and biases that will reduce the errors between the predicted conditional probabilities and the reference values, which is achieved using backpropagation. Refer to Chapter 1 for an explanation of backpropagation.

From the above discussion, there is an implied need to select and compute the features before the location of observation in the feature space can be ascertained. This selection process is a manual and frequently difficult task [33]. Instead, feature engineering can be avoided altogether by using structures like Convolutional Neural Networks (CNNs). Without going into too many details, this method involves the use of 3-layered kernel filters with mutable coefficients that are directly applied to the input images, as shown in Figure 6.15. These filter coefficients are adjusted as part of the back-propagation process, and it will gradually be able to filter out the redundant aspects of the input image and obtain the features that are required in the subsequent stages of the neural network. Typically, the use of CNN networks usually contains many parameters that need to be determined during training. Consequently,

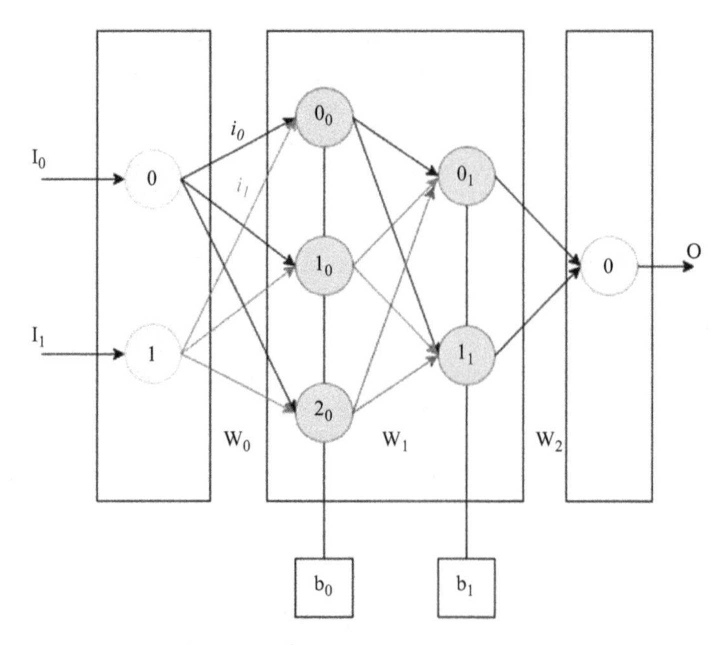

Figure 6.14 Artificial neural network (ANN). Yellow, red, and blue nodes correspond to neurons in the input, hidden, and output layers respectively [Chapter 1].

it is essential to provide enough training input data in order to obtain a network that is sufficiently effective. For further information regarding CNNs, refer to Chapter 1.

6.7.3 SVM techniques

The brain is made up of grey matter (GM), white matter (WM), and cerebrospinal fluid (CSF), and these regions can be uniquely distinguished in an MRI [34]. Differences in the distribution of these components are useful features that are used in conjunction with machine learning to classify between the three different groups – CN, MCI, and AD [34,35]. After cleaning up the three-dimensional MRI images with image pre-processing techniques, the feature extraction process is then performed by dividing the brain into Regions of Interest (ROIs), such that one or more features can be computed from each ROI.

An example feature involves applying histogram modelling to the pixel intensities in each ROI. Due to the co-existence of GM, WM, and CSF, a multi-modal distribution will be produced since each component takes on different intensities, as shown in Figure 6.16(a). The Gaussian Mixture Modelling technique [36] can then be applied to decompose the composite distribution into its individual components (see Figure 6.16(b)), such that they can be described using (6.19), where μ_1, μ_2 and μ_3 refer to the mean values for the distributions for CSF, GM, and WM, and α_1, α_2, and α_3 refer to the relative contributions or weights of each component to the overall

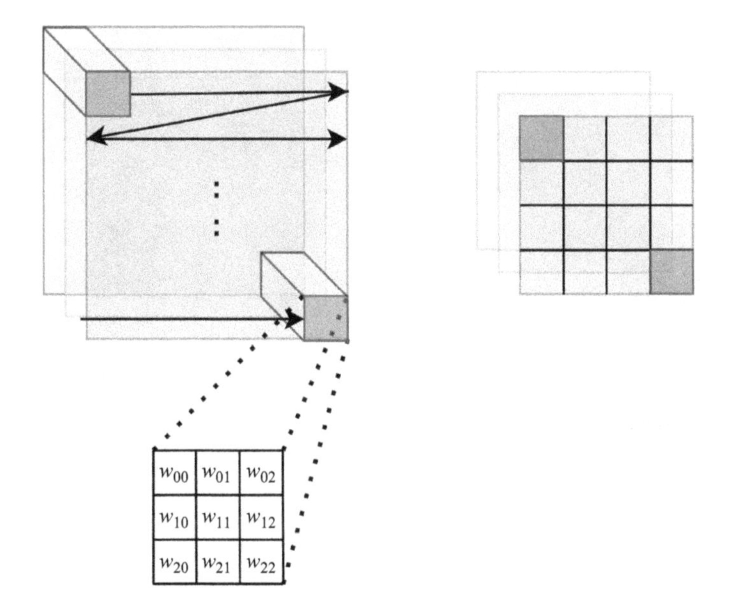

Figure 6.15 A CNN for a colour image. This consists of three channels – Red, Green, and Blue. The filter kernel is three-dimensional matrix that slides across the image from left to right, top to bottom. At each location, a sum of products is computed with the 3-by-3 kernel (shaded in blue) to arrive at a corresponding sum on the right (coloured in blue), forming a 3-dimensional structure on the right [citation: Chapter 1].

distribution. Physiologically, there is a progressive decline in the proportion of grey matter relative to white matter in AD patients [34]. Therefore, the obvious features that should be effective are the mean intensities for grey and white matter, μ_2 and μ_3. However, these values vary widely across individuals. In contrast, α_2 was found to be a more effective and invariant feature, as it measures the amount of white matter as a proportion of GM and CSF (rather than the absolute level). Another example feature is the 'cortical thickness', which is defined as the closest distance between the grey/white boundary and the grey/CSF boundary in the MRI. Since these distances are obtained from spatial intensity gradient maps, they are less affected by variations in intensity and voxel resolutions of the MRI images [35]:

$$N(\mu, \sigma^2) = \alpha_1 N(\mu_1, \sigma_1^2) + \alpha_2 N(\mu_2, \sigma_2^2) + \alpha_3 N(\mu_3, \sigma_3^2) \quad (6.19)$$

$$\text{where } \alpha_1 + \alpha_2 + \alpha_3 = 1$$

Clusters residing in the original feature space with dimensions defined by α_2 and/or the cortical thickness can be separated via hyperplanes that are computed from SVMs. In this case, a non-linear hyperplane is required to achieve good separability between the CN and AD clusters. As discussed in previously, the kernel trick can be

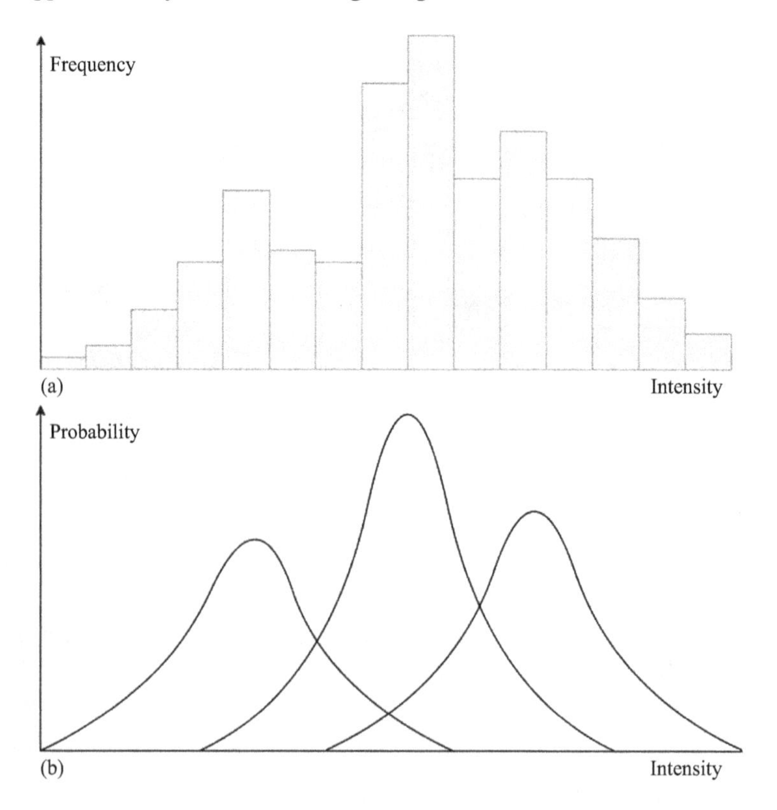

Figure 6.16 (a) Stylised histogram of intensities in the MRI of a control normal subject and the modelled envelope of the multi-modal distribution. (b) Separation of composite distribution via the Gaussian Mixture Model technique.

employed to do so. Specifically, the prediction function used to predict the class of a given feature, x, is described using (6.18). A cost function is formulated in order to obtain the optimum values for w and b, as shown in (6.20) – soft margins are assumed in this case to relax the constraints and prevent overfitting. y_i refers to the actual labels in the dataset and are assigned -1 or 1 depending on whether they belong to the CN or AD group. Further, the optimum values for w and b are obtained via (6.21) for this linear SVM classification problem:

$$\text{maximise}\, f(c_1 \dots c_n) = \sum_{i=1}^{n} c_i - \frac{1}{2}\sum_{i=1}^{n}\sum_{j=1}^{n} y_i c_i (x_i^T x_j) y_j c_j) \tag{6.20}$$

$$\text{where} \sum_{i=1}^{n} c_i y_i = 0,\ 0 \le c_i \le \frac{1}{2n\lambda} \forall i$$

$$w = \sum_{i=1}^{n} c_i y_i x_i, \text{ and } b = w^T x_i - y_i \tag{6.21}$$

In (6.20), the inner product between x_i and x_j is effectively a measure of the distance between two points in the feature space. The Kernel trick involves replacing this linear operation with a non-linear function as shown in (6.22) – this is known as the Radial basis function kernel [37]. With this transformation, the originally linear relationship between distance and the cost function has been converted to an exponentially decreasing value with Euclidean distance. The advantage of performing the kernel trick is that the optimisation framework described in (6.20) and (6.21) still applies, while achieving a non-linear hyperplane that can afford better separability between clusters:

$$\text{maximise} f(c_1 ... c_n) = \sum_{i=1}^{n} c_i - \frac{1}{2} \sum_{i=1}^{n} \sum_{j=1}^{n} y_i c_i (k(x_i, x_j)) y_j c_j) \tag{6.22}$$

$$k(x_i, x_j) = e^{\frac{1}{2\sigma^2 \|x_i - x_j\|^2}} \tag{6.23}$$

Training and testing schemes are important ingredients in creating an effective ML classifier. A bootstrap method is employed in [34] to train and test the classifier. The steps taken are as follows. Using this technique, high mean specificity, sensitivity, and accuracy scores of 96.9%, 91.5%, and 94.5% were achieved [34].

1. 25% of the initial dataset are drawn at random and without replacement from each class (CN and AD).
2. 75% of the remaining data are then used for training under the non-linear SVM framework.
3. The above steps are repeated 5,000 times, which is deemed a suitable number of iterations for the SVM parameters to converge.

6.7.4 Deep learning techniques

It is common practice to reuse deep learning architectures across different applications. Indeed, reusability is a particular virtue of machine learning that makes it such a powerful technique. An example is the *LeNet5* (Figure 6.17) [38] which was originally designed for automatic text recognition. This network has since been re-purposed for 'reading' fMRI images to classify them into *CN* and *AD* categories [39].

There are various types of layers in [38], which are used to remove redundancies within the original image and 'compress' it into pertinent information, used to determine which class or category the input image falls input. It is important that these layers are carefully designed so as not to remove any essential information used for the decision-making process. Brief explanations for various layers are given below. For a more in-depth discussion, refer to Chapter 1.

1. *Convolution layer*: This comprises two-dimensional filter kernels that are applied to the original image. A stack of six filters is applied to the same image in Layer C1 to obtain feature maps. The operating characteristics of convolution filter kernels are described in Section 6.7.2. The network can either be 'expanded' or 'narrowed' based on the number of filters that are applied at that stage so that different characteristics of the image can be captured by each filter, weighted

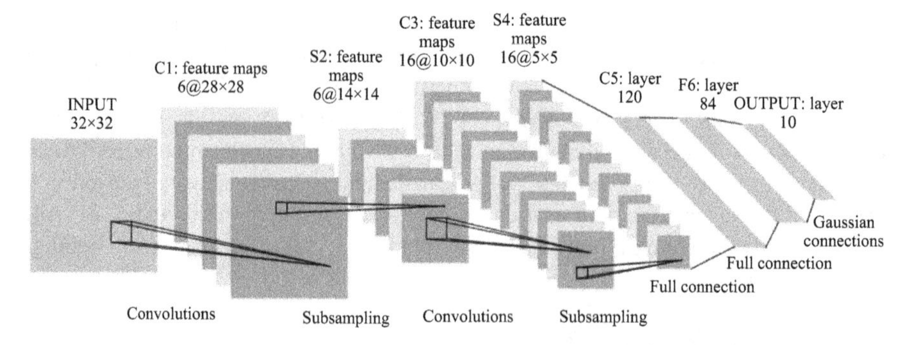

Figure 6.17 The architecture LeNet5, originally used for text recognition [38]

and aggregated in a meaningful way further downstream of the network. At the extreme, a stack of 1 by 1 filters can be applied (Layer C5) – this is useful as a preceding stage to the fully connected layer as it reduces the network dimension to one.

2. *Subsampling layer*: This operation is also known as average pooling where aggregating kernels are applied to each feature map. The average value of the image covered by each kernel is then computed and assigned to a new feature map. In Layer S2, non-overlapping 2 by 2 kernels are used without overlap such that the resulting feature maps are halved in size.
3. *Fully connected layer*: Convolution layers consist of filter kernels that are only applied to a small portion of the feature map or image at any one time. In contrast, fully connected layers connect every output to every possible input of the subsequent layers, so that all information from the preceding stages can be weighted and considered (Figure 6.14). While this allows for more informed decision-making, there is a vastly increased number of connections, and by implication the number of network parameters that need to be trained. For this reason, fully connected layers are usually positioned at the last few stages of the network where a substantial amount of information aggregation has been carried out by preceding stages.
4. *Output layer*: This consists of one node for each corresponding class, and the node that computes the highest value will be the 'winning class'. A Soft Max function (comprises of aggregated exponentials) (Chapter 1) is typically used instead to compute these output values.

fMRI comprises of more than two dimensions whereas *LeNet5* takes in two-dimensional images as images. In [39], a simple approach is taken to overcome this problem. Slices of the original MRIs were taken and labelled by an expert. Subsequently, the dataset is used to train and test the network using five-fold cross-validation. An accuracy of 96.9% was achieved when applied to *CN* and *AD* datasets. Unfortunately, no other information such as the sensitivity, specificity, and F1 scores were made available by the authors.

Slicing multi-dimensional data and considering each slice may well be a computationally efficient approach to solving the problem. However, this would raise the problem of information loss. Therefore, instead of converting the MRIs into two dimensions before feeding them to the network, three-dimensional convolution is directly applied to the three-dimensional voxel grid to produce three-dimensional feature maps [40]. The large volume of data that needs to be processed necessitates very deep networks and the authors encountered the problem of diminishing gradients. Specifically, network training involves error back-propagation where a set of weights W_{k-1} is adjusted in order to minimise error $E(W)$, to obtain the new weights W_k. The abridged equation describing error back-propagation is shown in (6.24). For very deep networks, the partial derivative $\partial E(W)/\partial W$ tends to zero and these null values propagate upstream of the network and tend to persist, leading to very slow convergence of the network parameters:

$$W_{k-1} = W_{k-1} - \eta \frac{\partial E(W)}{\partial W} \tag{6.24}$$

To overcome this problem, some techniques such as batch normalisation or *ResNets* [41] can be used. The authors of [40] make use of a novel layer which combines known methods with some novel techniques, as illustrated in Figure 6.18. At the heart of their method is the dense block (Figure 6.18(a)). This involves connections between the current layer and all preceding layers, as shown in (6.25), where x_i refers to the three-dimensional feature volumes that are computed at each stage. This is followed by batch normalisation (refer to Chapter 1 for more information), a rectified linear unit (ReLU), and three-dimensional convolution. Gradients are prevented from becoming too low by including a weighted sum of outputs from previous stages and as well as through batch normalisation. Including the ReLU increases the rate of convergence and thus speeds up the training process [40]. Further, they augment their architecture by using an ensemble of networks and aggregating the results of multiple networks (of similar architectures) to increase the overall effectiveness [40]:

$$x_l = H_l([x_0, x_1, \ldots, x_{l-1}]) \tag{6.25}$$

By considering the problem beyond two dimensions, the authors were able to obtain good results for the more difficult problem of classifying between three classes – *CN*, *MCI*, and *AD* (as opposed to differentiating between *CN* and *AD*). The accuracy and F1-score obtained are 0.98 and 0.96 respectively.

Which method is better for *AD*, *MCI*, and *CN* classification – SVM or deep learning? By definition, SVM is a globally optimum technique whereas deep learning is more likely to get trapped at the local minima. On the other hand, the fact that no feature engineering is required for deep learning is an important advantage. Despite best intentions, experts might be unable to spot pertinent features and the utility of information from experts is reduced when data is transformed into a higher-dimensional space. Therefore, it may be argued that automatic feature recognition may be a better strategy to create a more effective classifier, particularly as more data becomes available. This is especially true as fMRI images are combined with other data types such as EEG, genetic makeup, and demographics [9].

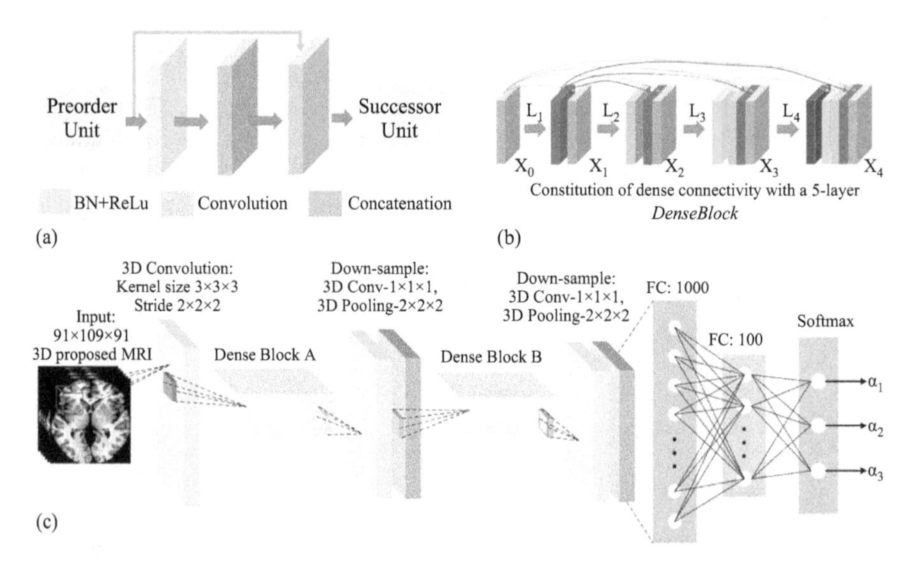

Figure 6.18 (a) Dense unit, (b) dense block, and (c) overall architecture [40]

We have considered the technical aspects of both methods so far. The technical effectiveness of Computer-Aided Diagnostic tools is only one (albeit important) aspect out of many that manufacturers should consider. For example, they have to apply for regulatory approval for their tool in the country where they are intended to sell their product. Often, they should provide clear explanations of how their product operates and how it would not affect patient safety. In contrast to SVM, deep learning is frequently seen as a 'black box' where internal operations are opaque [9]. Hence, it is more difficult to explain and rationalise that the product is safe for use. For these reasons, it is unlikely that non-deep learning approaches will be superseded in the near future.

6.8 Conclusion

The brain is a complex multi-dimensional structure with multiple biological activities and cellular functions that orchestrate body functions and sensorial activities. Over the years, multiple attempts have been made to understand and interpret brain activities. Postmortem analysis of anatomical structures to functional images and interpretation of electrical activities led to the use of more advanced techniques, to facilitate early diagnosis and targeted treatments.

A review of various ML-based techniques was carried out in this chapter, including pre-processing and predictive models that are applied to MRI scans. It is clear that substantial progress has been made in improving the efficacy of these models used for the early diagnosis of Alzheimer's disease, particularly in distinguishing between MCI, normal, and AD cases. However, these models are extremely compute intensive. We believe that an important area of research is the development of real-time versions of these algorithms to enable rapid and accurate predictions, so that the results can be quickly made known to the physicians, who can then affect early management and treatment schemes for their patients.

To conclude, the early diagnosis of Alzheimer's disease is but one of a vast array of applications relating to the brain. We envision that with the increasing amount and availability of computing resources, effective models can be developed to detect brain anomalies in general. The mystery of brain connectivity can finally be decoded and understood in this process. Clearly, we are still a long way from achieving this vision but the rate of progress in recent years, particularly with the aid of ML, have been encouraging and definitely a cause for optimism.

References

[1] Guo J and Li B. The application of medical artificial intelligence technology in rural areas of developing countries. *Health Equity*. 2018;2(1):174–181. Available from: https://www.ncbi.nlm.nih.gov/pmc/articles/PMC6110188/.

[2] Northern Brain Injury Association. *Brain Structure and Function: Brain Injury*. British Columbia: Northern Brain Injury Association; 2017. Available from: https://www.nbia.ca/brain-structure-function/.

[3] Makin S. The four biggest challenges in brain simulation. *Nature*. 2019;571(7766):S9–S9. Available from: https://www.nature.com/articles/d41586-019-02209-z.

[4] NHS. Overview – Alzheimer's Disease. NHS; 2021. Available from: https://www.nhs.uk/conditions/alzheimers-disease/.

[5] World Health Organisation. *Dementia*; 2021. Available from: https://www.who. int/news-room/fact-sheets/detail/dementia.

[6] Duong S, Patel T, and Chang F. Dementia. *Canadian Pharmacists Journal/Revue des Pharmaciens du Canada*. 2017;150(2):118–129.

[7] Khan A and Usman M. IEEE, editors. *Early diagnosis of Alzheimer's disease using machine learning techniques: A Review Paper*, 2015. 7th International Joint Conference on Knowledge Discovery, Knowledge Engineering and Knowledge Management (IC3K), Lisbon, Portugal, 2015, pp. 380–387, Available from: https://ieeexplore.ieee.org/document/7526944.

[8] Diogo VS, Ferreira HA, and Prata D. Alzheimer's Disease Neuroimaging Initiative. Early diagnosis of Alzheimer's disease using machine learning: a multi-diagnostic, generalizable approach. *Alzheimer's Research & Therapy*. 2022;14(1):107.

[9] Tanveer M, Richhariya B, and Khan RU. Machine learning techniques for the diagnosis of Alzheimer's disease. *ACM Transactions on Multimedia Computing, Communications, and Applications*. 2020;16(1s):1–35.

[10] Lazli L, Boukadoum M, and Mohamed OA. A survey on computer-aided diagnosis of brain disorders through MRI based on machine learning and data mining methodologies with an emphasis on Alzheimer disease diagnosis and the contribution of the multimodal fusion. *Applied Sciences*. 2020;10(5):1894.

[11] Dartmouth. State-of-the-Art fMRI Brain Scanner Arrives at Dartmouth; 2016. Available from: https://pbs.dartmouth.edu/news/2016/08/state-art-fmri-brain-scanner-arrives-dartmouth.

[12] NHS. Electroencephalogram (EEG); 2019. Available from: https://www.nhs.uk/conditions/electroencephalogram/.

[13] Belliveau JW, Rosen BR, and Kantor HL. Functional cerebral imaging by susceptibility-contrast NMR. *Magnetic Resonance in Medicine*. 1990;14(3): 538–546.

[14] Glover GH. Overview of functional magnetic resonance imaging. *Neurosurgery Clinics of North America*. 2011;22(2):133–139.

[15] Agarwal J and Bedi SS. Implementation of hybrid image fusion technique for feature enhancement in medical diagnosis. *Human-Centric Computing and Information Sciences*. 2015;5(1):1–17.

[16] van der Zande JJ, Gouw AA, van Steenoven I, *et al.* EEG characteristics of dementia with Lewy bodies, Alzheimer's disease and mixed pathology. *Frontiers in Aging Neuroscience*. 2018;10:190. Available from: https://www.frontiersin.org/articles/10.3389/fnagi.2018.00190/full.

[17] ADNI. Available from: https://adni.loni.usc.edu/.

[18] AIBL. Available from: https://aibl.csiro.au/.

[19] OASIS. Available from: https://www.oasis-brains.org/.

[20] Despotović I, Goossens B, and Philips W. MRI segmentation of the human brain: challenges, methods, and applications. *Computational and Mathematical Methods in Medicine*. 2015;2015:1–23. Available from: https://www.hindawi.com/journals/cmmm/2015/450341/.

[21] Xue JH, Pizurica A, Philips W, *et al.* An integrated method of adaptive enhancement for unsupervised segmentation of MRI brain images. *Pattern Recognition Letters*. 2003;24(15):2549–2560.

[22] Chang C, Cunningham JP, and Glover GH. Influence of heart rate on the BOLD signal: the cardiac response function. *NeuroImage*. 2009;44(3):857–869.

[23] Deckers RH, van Gelderen P, Ries M, *et al.* An adaptive filter for suppression of cardiac and respiratory noise in MRI time series data. *Neuroimage*. 2006;33(4):1072–1081.

[24] Akdemir Akar S, Kara S, Latifoğlu F, *et al.* Spectral analysis of photoplethysmographic signals: the importance of preprocessing. *Biomedical Signal Processing and Control*. 2013;8(1):16–22. Available from: https://www.sciencedirect.com/science/article/abs/pii/S1746809412000468.

[25] Zhang C, Xiao X, and Li X. White blood cell segmentation by color-space-based K-means clustering. *Sensors*. 2014;14(9):16128–16147.

[26] Haddadpour M, Daneshvar S, and Seyedarabi H. PET and MRI image fusion based on combination of 2-D Hilbert transform and IHS method. *Biomedical Journal*. 2017;40(4):219–225.

[27] Livadas GM and Constantinides AG. Image edge detection and segmentation based on the Hilbert transform. In: *ICASSP-88, International Conference on Acoustics, Speech, and Signal Processing.*

[28] Phamila AV and Amutha R. Low complexity multifocus image fusion in discrete cosine transform domain. *Optica Applicata*. 2013;43(4):693–706. Available from: 10.5277/oa130406

[29] Wiredelta. 5 Things AI Does Better Than Humans in 2019. WIREDELTA; 2019. Available from: https://wiredelta.com/5-things-ai-does-better-than-humans-2019/.

[30] Vapnik V and Izmailov R. Knowledge transfer in SVM and neural networks. *Annals of Mathematics and Artificial Intelligence*. 2017;81(1):3–19.

[31] LeCun Y, Bengio Y, and Hinton G. Deep learning. *Nature*. 2015;521(7553):436–444.

[32] Manning CD, Raghavan P, Schutze H, *et al. Introduction to Information Retrieval*. Cambridge: Cambridge University Press; 2009.

[33] Verdonk C, Verdonk F, and Dreyfus G. How machine learning could be used in clinical practice during an epidemic. *Critical Care*. 2020;24(1):265. doi: 10.1186/s13054-020-02962-y.

[34] Magnin B, Mesrob L, and Kinkingnéhun S. Support vector machine-based classification of Alzheimer's disease from whole-brain anatomical MRI. *Neuroradiology*. 2008;51(2):73–83.

[35] Oliveira PPdM, Nitrini R, Busatto G, *et al*. Use of SVM methods with surface-based cortical and volumetric subcortical measurements to detect Alzheimer's disease. *Journal of Alzheimer's Disease*. 2010;19(4):1263–1272.

[36] Bilmes JA. A gentle tutorial of the EM algorithm and its application to parameter estimation for Gaussian mixture and hidden Markov models. *International Computer Science Institute*. 1998;4(510):126.

[37] Buhmann MD. Radial basis functions. *Acta Numerica*. 2000;9:1–38.

[38] Tra V, Kim J, Khan SA, *et al*. Bearing fault diagnosis under variable speed using convolutional neural networks and the stochastic diagonal levenberg-marquardt algorithm. *Sensors*. 2017;17(12):2834.

[39] Sarraf S and Tofighi G. Deep learning-based pipeline to recognize Alzheimer's disease using fMRI data. In: *2016 Future Technologies Conference (FTC)*; December 2016.

[40] Wang H, Shen Y, Wang S, *et al*. Ensemble of 3D densely connected convolutional network for diagnosis of mild cognitive impairment and Alzheimer's disease. *Neurocomputing*. 2019;333:145–156.

[41] He K, Zhang X, Ren S, *et al*. Deep residual learning for image recognition. In: *2016 IEEE Conference on Computer Vision and Pattern Recognition (CVPR)*, June 2016. p. 770–778.

Chapter 7

From classic machine learning to deep learning advances in atrial fibrillation detection

Miguel Hernandez Silveira[1] *and Su-Shin Ang*[1]

Cardiac arrhythmias are disorders in the electrical activity of the heart, mainly characterized by fast or slow heartbeats and often accompanied by an irregular rhythmic pattern. Some arrhythmias also exhibit morphological changes in the electrocardiogram (ECG) because of damage of heart cells, resulting in abnormal conduction of the impulses through different pathways. While some arrhythmias are benign, others are life-threatening or may lead to severe health complications.

Automatic arrhythmia detection is not something novel; and in fact, most of medical bedside monitors and cardiac telemetry systems these days have embedded capabilities for detecting a wide range of these. Algorithms for arrhythmia detection have been developed using different kinds and even combinations of input data – e.g. photoplethysmography (PPG), seismocardiograms (ECG), ballistocardiograms (BCG), and electrocardiograms (ECG). The latter is the oldest, most popular, and perhaps the most reliable data source – as it directly represents the electro-chemical conductive properties of the heart.

Unfortunately, ECG signals can be easily corrupted by motion artefacts resulting from locomotion or other forms of physical activity. Furthermore, physical activity results in increased heart rate, introducing more confounding factors into the problem – for example, some forms of tachycardia are characterized by high heart rates as those seen during exercise. Even worse, high heart rates shorten the distance between heartbeats – which makes the detection of irregular rhythms more difficult. Moreover, some arrhythmias share similar characteristics among them or with noisy signals, adding more complications into the equation. Thus, despite the advances in commercial wireless wearable/portable systems and apps for arrhythmia detection (e.g., KardiaMobile https://www.alivecor.com), the task of arrhythmia detection can still be challenging.

While all arrhythmias are worth of research, atrial fibrillation (AF) is of particular interest. The rationale for this is twofold: first, this is the most common arrhythmia with the highest incidence and prevalence worldwide. According to the British Heart Foundation [1] 110,102 patients were admitted in hospitals in the UK due to AF as the main diagnosis. In addition, data from the Framingham Heart Study [2] showed

[1]MF Technology Ltd, UK.

a three-fold increase in the prevalence of AF over the last 50 years; whereas the Global Burden of Disease report [3] estimated a worldwide prevalence of more than 40 millions of patients in 2016 [4]. Second, AF is a life-threatening arrhythmia and a lead cause of stroke, which in turn often results in dead or permanent disabilities. Therefore, the central focus of this chapter is on presenting a comprehensive review of advances in algorithmic development for detection of AF from ECG signals in the last three decades; as well as providing the reader with our own experience on the development of different methods to detect this condition based on aspects such as data availability and labeling.

7.1 Physiology essentials

7.1.1 The healthy heart

Understanding the mechanisms associated with atrial fibrillation requires familiarization with basic anatomical and physiological aspects of the healthy heart. Together with the arteries and veins, the heart is responsible for circulation of blood within the human body. A healthy heart acts like a pump that works in a synchronized manner owing to a sequence of electrical organized impulses responsible for its sequential mechanic actions during every beat. In simple terms, these electrical pulses are first originated within a structure rich on nerve conductive cells (sino-atrial or sinus node, SA). Subsequently such pulses are propagated from this natural pacemaker across the atria from left to right; first to the atrio-ventricular node (AV), wherein the pulse stay for a brief period until the ventricle conductive cells are ready (repolarized) for subsequent excitation. Finally, the ventricles receive and propagate the impulses through their conductive arrangement of cells (e.g., Bundle of His and Purkinje fibres) resulting in a ventricular contraction. As a result of this orchestrating electrical activity, the atrium first contracts pumping the blood to the ventricles, which in turn contract once they are ready for pumping more blood towards different parts of the body. Finally, the heart muscle relaxes (the potential difference of the membrane of the conductive cells returns to equilibrium because of the action of the sodium-potassium pump) ready for a new sequence of pulses/contractions [5].

In the ECG, the electro-ionic activity of the normal heart is represented by the summation of different cells potentials taking place at different parts of the organ. This results in a few signatures waveforms or complexes representing the polarization, depolarization, and repolarization of different nerve cells across the organ. As shown in the top graph of Figure 7.1, the normal ECG exhibits a steady and regular rhythm. Deviation from these parameters may indicate the presence of an abnormal rhythm or heart condition.

7.1.2 Atrial fibrillation

AF is an arrhythmia characterized not only by the presence of an irregular rhythm but also by partial distortion of the ECG signal (normal P wave replaced by fluttering wavelets (coarse or fine) reflecting aberrant conduction of nerve impulses across the

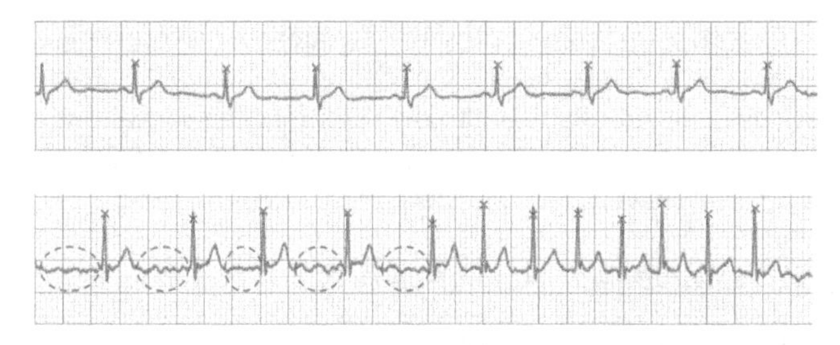

Figure 7.1 Normal ECG (A) characterized by regular heartbeats (equidistant R-peaks marked as red crosses) and presence of the signal components (P-QRS-T components) representing the polarization and depolarization of atrium and ventricles. AF (B) instead is characterized by irregular rhythm emerging as fluctuations in length between RR intervals (unequal distances between consecutive R-peaks), as well as fluttering activity and absence of P-waves (encircled in red) are evident in AF.

atrium (Figure 7.1 – bottom graph)). Another important feature of AF is the rate of the heartbeats. Although AF is often accompanied by ventricular rates above 100 beats per minute (bpm); the range can vary between 50 and 250 bpm, depending on the degree of AV conduction, patient age, and medications (such as beta-blockers). AF can also be paroxysmal (intermittent seizures of short duration that last for about a week), persistent (lasts for longer periods of time, often months and can be stopped with a clinical intervention), or permanent (chronic, cannot be stopped with any form of treatment).

7.2 Detection of AF

There exist a variety of methods to detect/classify AF. While some of these focus on abnormal changes in ECG morphology (absence of P waves and atrial fluttering activity as seen in Figure 7.1), others focus on rhythm irregularities (successive differences between R peaks and RR intervals) or both.

Morphology and hybrid methods (based on ECG morphology and rhythm) have been developed by adopting digital signal processing (DSP) techniques (such as digital filter banks, Fourier spectral analysis, Wavelet transforms, template matching and QRS cancellation) and machine learning approaches – such as Hidden Markov models (HMMs), artificial neural networks (ANNs), genetic programming (GP), support vector machines (SVM), among others. Such techniques allow for the determination of presence/absence of P waves, abnormal atrial activity, and inter-beat irregularity. Furthermore, recent developments incorporate the use of more complex algorithms including ensemble methods (random forests and extreme gradient boosting) as well as deep neural networks (convolutional networks, sequential models and autoencoders). As a result, hundreds of works have been published during the last few decades; and therefore, it is important to emphasize that the intention of this chapter is to provide the reader with a panoramic light-weight view of the evolution of AF detection over the years; not only by reviewing some of the key work forming part of our own research journey; but also with examples based on our own experience, that can be useful to enlighten those who wanted to get immersed in this exciting field.

7.2.1 AF detection based on beat-to-beat irregularities

These pattern recognition methods focus on the analysis irregularities of consecutive inter-beat intervals. The motivation for RR interval (RRI) methods is that locating and identifying morphological characteristics of the ECG (like P waves) can be difficult due to low amplitudes and poor signal-to-noise ratios (SNRs). This problem becomes more complicated in the presence of high heart rates – e.g., where fluttering waveforms blend in with other parts of the ECG complex, and therefore they can be misclassified as true P or T waves (as seen in the second half of the bottom ECG trace of Figure 7.1).

Figure 7.2 shows the different components forming part of the architecture of this type of AF detection algorithms. It combines elements of digital signal processing (DSP) with classic ECG beats detection and machine learning techniques.

In a nutshell, this pipeline can be described as follows:

- The ECG signal is first processed with an arrangement of filters to remove contaminants and enhance the events of importance (R-peaks for RRI-based algorithms). Normally, these are band pass filters with cut-off frequencies to clean the signal first (e.g., 0.5–40 Hz), and then other filters such as high-pass and differentiators are applied to enhance the heartbeats whilst minimizing near-DC contaminants (such as baseline wandering due to motion or respiratory modulation). Once the R peaks are salient events, the resultant signal is rectified and smoothed.
- Next, QRS detection is applied to the conditioned signal to find the location and amplitude of the fiducial marks corresponding to R-peaks [6–8].

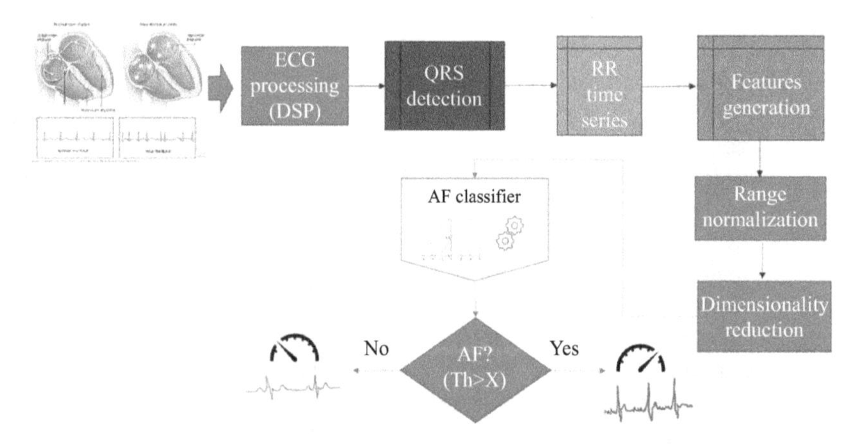

Figure 7.2 AF detection algorithm framework – the big picture (courtesy of MF Technology Ltd, UK)

- Subsequently, RR intervals are calculated – and in some cases – correction for ectopic and missing beats is applied.
- Candidate features are generated, and a subset is selected. Feature engineering is a crucial step of algorithm development. Initially features are chosen based merely in subject matter knowledge, and then visualization techniques (e.g., using different kinds of plots such as histograms, scatter plots, box-whisker plots, etc.); and/or feature selection algorithms – e.g., Recursive Feature Elimination Relief and mRMR [9–11] – can be applied to reduce the feature space to an optimal size.
- Once the feature space has been reduced, the need for normalization must be assessed before moving to the classification stage. Certain models ranging from a single classification tree to ensembles of trees (e.g., random forest, extra trees, and gradient boosting machines) are good feature selectors in nature, and do not require normalization of the feature space during their greedy optimization process. In contrast, other algorithms such as neural networks and SVMs require normalization of their inputs to ensure a right balance in their weights span, and thus good optimizer performance and convergence. For models with probabilistic outputs such as linear regression and neural networks with sigmoid and/or SoftMax activations it is customary to apply range (Minmax) normalization to ensure all inputs lie within 0 and 1. This operation is simple, and can be expressed as follows:

$$f(i)_{normalised} = \frac{f(i) - \min(f_{train})}{max(f_{train}) - min(f_{train})}, \tag{7.1}$$

  where $f(i)$ is the ith sample of the incoming feature to be normalized, and f_{train} the corresponding feature column data from the training set.
  Alternatively, zero mean and unit variance normalization (so-called standard scaling or z-scores) can also be applied to feature spaces targeted to certain classifiers for which Gaussian assumptions are made. Standard scaling is expressed as:

$$f(i)_{normalised} = \frac{f(i) - \mu(f_{train})}{\sigma^2(f_{train})}, \tag{7.2}$$

  where μ and σ^2 correspond to the mean and variance of the feature undergoing normalization. Note that in both cases $f(i)$ can be any sample of the training or testing set.
- If the resultant feature space is still highly dimensional, alternative feature extraction techniques can be applied to further reduce the dimensionality. Principal component analysis (PCA), Kernel PCA, locally linear embeddings (LLE), manifold methods (Isomap, T-distributed stochastic neighbor embedding), and multidimensional scaling are among some of the popular techniques.
- The last stage is AF detection. Here is where classification models perform this ultimate task.

Several RRIs-based AF detection algorithms use statistical features and uncertainty to construct threshold-based heuristic rules computed from RR time series of different lengths.

Tateno and Glass developed a statistical model applied to the RR intervals and their successive differences (ΔRR). Standard density histograms of both parameters were first generated and then used as templates for AF detection by applying the Kolmogorov–Smirnov (KS) statistical test to unseen data. Thus, cases where no statistically significant differences between standard and unseen data histograms were found were labeled as episodes of AF. Likewise, the coefficient of variation (CV) of both RR and ΔRR intervals were used as a feature to detect AF in combination with a predefined range. Data from both MIT-BIH arrhythmia and MIT-BIH atrial fibrillation databases was divided into segments of different lengths (from 20 to 200 beats) and then used for development and evaluation of the algorithms. The results of this study showed that the use of KS test applied to the successive difference of RR intervals yielded sensitivity and specificity values of 94.4% and 97.2% respectively.

Dash *et al.* [12] developed an algorithm using both – heart rate variability and complexity metrics as features. These constituents are namely the turning point ratio (TPR), the root mean squared of the successive differences (RMSSD), and Shannon entropy. While the first is a metric used to quantify randomness in a time series, the latter two correspond to time domain heart rate variability (HRV) and complexity features. As we will see later in this chapter, RMSSD and other HRV metrics have been adopted by other authors for the development of AF detection algorithms. RMSSD is expected to be higher in the presence of AF. Once ectopic beats (spurious peaks occurring during normal sinus rhythm intervals) are statistically filtered from the RRIs time series, this metric is calculated as follows:

$$RMSSD = \sqrt{\frac{1}{(L-1)}\sum_{j=1}^{L-1}[x(j+1)-x(j)]^2} \tag{7.3}$$

where L is the segment length, and x a given segment of RR intervals. The authors emphasized that during the optimization phase of this algorithm the selected threshold was compared against the mean value of the RR interval being processed; to compensate for potential extrasystole beats and other outliers. Therefore, this threshold was selected for the ratio of RMSSD divided by the mean RR as this value remains constant across different RR segments. In contrast, Shannon entropy was used as a feature to determine the uncertainty of rhythm regularity; so that this complexity value is expected to be lower for normal rhythms than for AF. Finally, cut-off values for all these parameters were estimated for best possible sensitivity and specificity by applying ROC analysis. The algorithm was trained and tested with a subset of data from the MIT-BIH arrhythmia and AF databases. Evaluation results revealed a high classification performance (Se = 94.4% and Sp = 95.1%) for the AF database, but lower (Se = 91.2% and Sp = 87.2%) for the arrhythmia database (record series 200, involving different types of abnormal rhythms). It is also important to bear in mind the computation of Shannon's entropy requires processing power and memory for the generation of the probability density functions and logarithmic calculations per ECG input segment.

Moody and Mark [13] reported on the development of a more complex probabilistic method where RR intervals sequences where modeled as Markov's processes.

The authors stated that this technique is equivalent to estimating the arithmetic mean of a set of probabilistic scores associated with the relative likelihood of observing RR intervals in AF versus making the same observations outside of AF. The model was developed using a subset of 12 records of 30 min each from the MIT-BIH Arrhythmia database. Thus, compilation of statistics of transitions between states, and subsequent generation of probability matrices for different rhythms were possible. The model was tested with a subset of 26 records also from the MIT-AF database. Each record comprised dual channel ECGs of 10 h each, for a total of 319 episodes of atrial fibrillation together with RR intervals delineation, beats annotations and time stamps for the whole record. One baseline and three additional variants of models with different processing stages were trained and evaluated. Receiver operating characteristic (ROC) analysis showed that the model using the first-order filtering and interpolation as pre-processing stage yielded the best compromise between sensitivity (Se = 96.09%) and positive predictivity ($p^+ = 86.79\%$) on the learning set; and Se = 93.58% and $p^+ = 85.92\%$ in the testing set. Although the authors concluded that the addition of these filtering steps to the time series of generated scores reduce false predictions, it is evident that the generalization capability of the model still requires improvement.

For our example, we developed two simple models using a subset of MIT-BIH Arrhythmia and MIT-AF databases. ECG segments of 30 s – including normal sinus rhythms and excluding irregular arrhythmias other than AF – were selected. The rationale for this is that the use case scenario for these algorithms was a clinical setting where patients with no history of heart disease but with AF symptoms will be screened using a stationary single-lead ECG recording device and in resting conditions Likewise previous approaches described here, we extracted a subset of data from both MIT-BIH Arrhythmia and AF databases for development purposes. A total of 2,900 epochs for training and 726 segments for testing conformed the dataset. The proportion of AF and NSR episodes in both partitions was reasonably balanced.

After processing the ECG signal, more than 20 pre-selected features – mainly short-term HRV metrics [14,15], statistical measures and SNR calculations – were generated from the resultant RRIs segments. These features were first normalized using standard scaling (see (7.2)). Subsequently, the dimensionality of the feature space was reduced from 20 to 2 dimensions using a feature extraction method. This step was essential since:

- Complex models and large size datasets are computationally expensive in terms of processing and storage costs. Consequently, high-dimensional classification algorithms are often difficult to implement in wearable devices due to memory, processing and power consumption constrains.
- High-dimensional spaces are likely to contain redundant and irrelevant features that can add confounding information to the model, leading to increased misclassification errors and limited ability to generalize on unseen data (overfitting).

Thus, principal component analysis (PCA) was selected for this task. The latter is an unsupervised algorithm that applies a linear transformation to find orthogonal projections of the original data in a lower dimensional space. In simple terms, it searches for the directions of largest variance of the high-dimensional dataset and projects it into a new subspace with less dimensions than the original one. By applying linear algebra operations, the algorithm generates a OxN dimensional transformation matrix W that is used to map the original input vectors into a new subspace with less dimensions than the original. Thus, for each observation of the original feature space, we have:

$$z = xW, W \in \mathbb{R}^n \tag{7.4}$$

where $x = [x_1, x_2, \ldots, x_o]$ is the original o-dimensional input vector, and $z = [z_1, z_2, \ldots, z_n]$ is the new n-dimensional vector projected in the new lower-dimensional space. Note that O >> N.

For a new n-dimensional feature space, building the matrix W involve the following steps:

- Center the data by subtracting the mean from the dataset.
- Calculate the covariance matrix.
- Decompose the covariance matrix into its eigenvectors and eigenvalues.
- Select the n-eigenvectors associated with the largest n-eigenvalues (components with the largest variance) corresponding to the n-dimensions of the new feature space.
- Construct the transformation matrix W using these n-eigenvectors.
- Apply the transformation to each observation of the original dataset (see equation (7.4)) to obtain (or map a single input vector to) the new n-dimensional feature space.

It should be borne in mind that the number of components forming the new dimension should be large enough such that it represents a reasonable percent of the variance – i.e., the number of selected components must contain sufficient information to allow reconstruction of the data back to its original form when applying the inverse transformation. In our example, selection of the two principal components meets the best trade-off between reconstruction ability and illustration purposes with a reasonable variance threshold of 85%. For a real-case scenario, it is suggested to increase the number of selected components to 5, to yield a variance threshold of at least 95%. Figure 7.3 shows the resultant feature space after the application of PCA for both training and testing data splits.

Following dimensionality reduction of the feature space, the data is ready to be processed by classification algorithms. As mentioned above, we implemented two methods. The first uses a Gaussian Naïve Bayes classifier, which assumes that the input data is drawn from a normal distribution. Thus, classification can be derived from the Bayes's rule as follows:

$$P(C_k|x) = \frac{P(x|C_k)P(C_k)}{P(x)} \tag{7.5}$$

where C_k is the k-class, and x the input vector of features. In this binary class problem ($k = 2$), the decision of whether a particular RRI segment corresponds to an AF seizure is made by the ratio of posterior probabilities for each label:

$$\frac{P(C_1|x)}{P(C_2|x)} = \frac{P(x|C_1)P(C_1)}{P(x|C_2)P(C_2)} \tag{7.6}$$

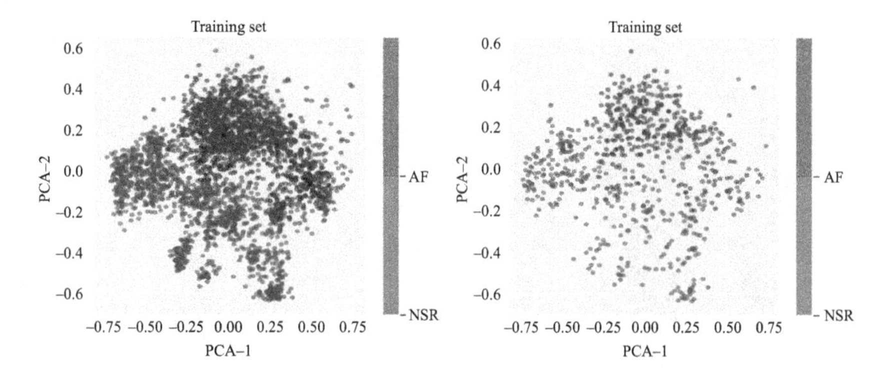

Figure 7.3 Bidimensional feature space following transformation of training and testing data using PCA (courtesy of MF Technology Ltd, UK)

The training process of this technique leads to the estimation of a generative model that is assumed to be associated with the random process that generates the original input data. It should be borne in mind that finding the best model representing the data is not a trivial task – hence, the naïve assumption that the data follows a Gaussian distribution (and this explains the use of the naïve classifier term).

Figure 7.4 shows the probabilistic decision function of the gaussian classifier acting on the testing set, and its ROC curve. This shows that the probability of AF given the observed data increases as the datapoints move inwards to the center of the ellipse (red area). Also note that most of the AF points lie inside the 2D Gaussian distribution of the generative model. Conversely, the majority of the non-AF segments lie outside the classifier elliptical region. This demonstrated good separability between AF and NSR data, which in turn is further confirmed by the area under the curve (AUC = 0.97), true positive rate (TPR = 0.94) and false positive rate (FPR = 0.14). Note that most misclassifications occur at the boundary regions between the two classes – where the probability is about 0.5. This suggests that implementing a heuristic where the classification decision is made with a rule conditioned to some probability thresholds with a margin of tolerance (Tol) can reduce classification errors. For example, an input data vector x would correspond to AF if $P(C_{AF} = 1|x) > 0.5 + Tol$. On the contrary, an input vector would correspond to NSR if $P(C_{NSR} = 1|x) < 0.5 - Tol$. Lastly, a region of uncertainty where the class of the input sample is undetermined can also be defined. For this example, this “rejection region” can be defined as probabilities lying within the interval *[0.5 – Tol, 0.5 + Tol]*. Both probability thresholds and tolerances can be finely tuned by means of ROC analysis – i.e. specifically for a target TPR and FPR.

To improve the accuracy further, a second attempt was made to obtain a model closer to the actual distribution of the data. Therefore, we implemented another Bayesian Classifier, this time using Kernel Density Estimation (KDE) to find the probability density functions (pdfs) that best fit the data. When using KDE one can

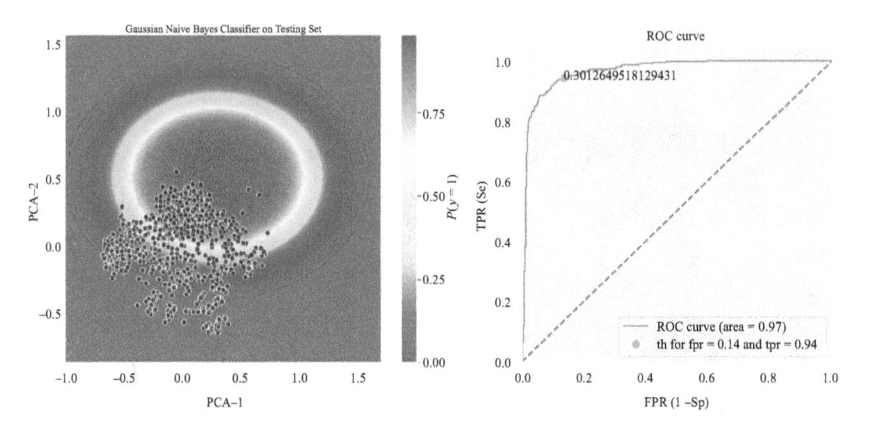

Figure 7.4 Gaussian Naïve Bayes Classifier. Decision function (left) – blue dots correspond to input vectors associated with NSR, whereas red dots are linked to AF samples. The ROC curve (right).

estimate the density function **p(x)** to find the probability of any sample corresponding to a specific class in the feature space. Thereby, the shape of the pdf is then estimated using the following expression:

$$\hat{p}(x) = \frac{1}{nh^d}\sum_{i=1}^{n} K\left(\frac{x - x_i}{h}\right) \tag{7.7}$$

where K is the kernel (positive and symmetric function which area under the curve equals to 1); d the number of dimensions of the feature space; $\boldsymbol{x}$ the average (or central) point at which KDE is performed for any $\boldsymbol{x}_i$ point; and h is the bandwidth that serves as a smoothing parameter that controls kernel's size at each point. Training KDE models focuses on implementing the right kernel and finding the optimal value for h based on minimization of a given cost function – such as the mean integrated square error – MISE. For this purpose, we implemented a machine learning hyperparameter tuning pipeline using grid search and cross-validation to select the best kernel and bandwidth value for the final implementation of the model. In this occasion, we used a Gaussian kernel such that (7.7) turned into the following expression:

$$\hat{p}(x) = \frac{1}{N}\sum_{i=1}^{n} \frac{1}{\left(2\pi h^2\right)^{\frac{d}{2}}} exp\left\{\frac{\|x - x_i\|}{2h^2}\right\} \tag{7.8}$$

In summary, the steps for developing a Bayesian classifier for AF detection using generative models from KDE are:

1. Stratify and divide the training dataset into partitions grouped by class.
2. Apply KDE to each partition to obtain the generative model corresponding to each class. This makes possible computation of the conditional probability $P(x|C_k)$, or likelihood of class k for any observation x.

3. Compute the prior probabilities $P(C_k)$ for group of samples corresponding to each class k.
4. Lastly, once the classifier is trained, a new prediction can be made for any unseen point x by calculating the posterior probability $P(C_k|x) \propto P(x|C_k)P(C_k)$ for each class k. Thus, the class with the highest probability is assigned to the current point x.

Figure 7.5 shows the decision function and ROC curve for this Bayesian KDE classifier. The results showed that only a slight improvement in sensitivity (TPR = 96%) was possible to obtain. The other ROC parameters maintained their values (AUC = 0.97 and FPR = 0.14).

The methods described so far provide valuable insights – not only about the use of different features for heuristic detection of AF but also the adoption of statistical and probabilistic methods (Bayesian inference) for recognition of irregular patterns in RRI time series.

Other classification techniques for AF detection adopt discriminative models to map the input feature vectors onto a target class, bypassing the estimation of probability distributions unlike generative models. In this category, support vector machines (SVMs) are one of the favorite choices; as their decision boundaries can be trained to obtain the maximum separation between classes regardless of the statistical distribution of the training data. Figure 7.6 shows a graphical representation of a linear SVM in a 2D feature space. The decision boundary (H) is formed by all the points where $H_{w,b} = 0$. The dashed lines form the margin that separate both classes (labelled as −1 and 1), and are parallel and equidistant with respect to the decision boundary ($d_+ = d_-$). Also note that w is orthogonal to the decision line, hence the slope of H is equivalent to the norm of this vector. It is also evident that the distance between the two dashed lines is twice the distance of any of them to the decision boundary. Taking all these factors into account, the width of the margin is given by the scalar projection

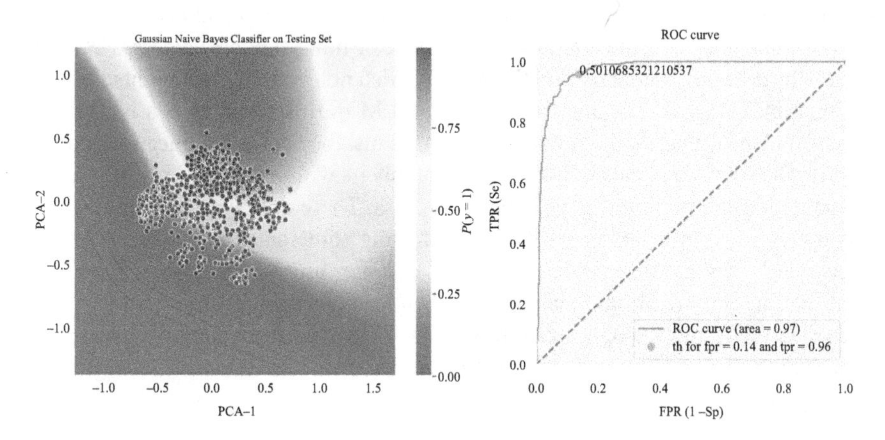

Figure 7.5 Bayesian KDE generative AF classifier decision function (left) and its ROC curve (right) – courtesy of MF Technology UK, Ltd

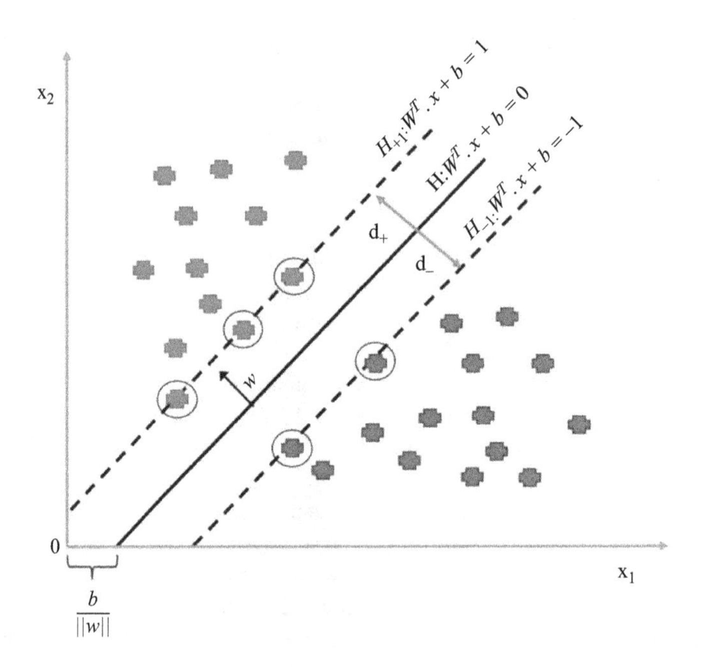

Figure 7.6 Maximum hard-margin classification with a linear SVM

of any two support vectors (one positive and one negative) on the unit normal vector, and it is expressed as $\frac{2}{\|w\|}$.

Since the classes are separated by the "street" formed by the two dashed lines the training process is focused on finding the values of w and b that maximize its width. Finding the solution to this problem is rather complex and involves the use of Lagrange multipliers and quadratic optimization solvers.

One of the main advantages of SVMs is that changes in the structure of the data (excluding the support vectors) do not affect the position and orientation of the decision boundary. On the contrary, moving one or more support vectors will impact both – the decision boundary and the margin. Also note that no data points lie within the margin in Figure 2.5, and therefore the SVM in this scenario is a hard margin classifier. Unfortunately, in most real-life problems, the data is not linearly separable. This implies that some data points will regrettably lie within the margin street, or even worse within the region of their antagonist class. There are two ways of addressing this issue – first by relaxing the margin (allowing some datapoints to lie within the street), or by trying to separate the data in a higher-dimensional space. The former (soft margin classification) is carried out by introducing slack variables ($\zeta_i \geq 0$) for each data point into the optimization process. These variables quantify how much a data point is allowed to violate the margin for. Based on this principle, the aim of the training process is to:

$$\underset{w,b,\zeta}{\text{minimise}} \frac{1}{2} w^T w + C \sum_{i=1}^{n} \zeta_i s.t. y_i \left(w^T x_i + b \right) \geq 1 - \zeta; fori = 1, \ldots, N. \tag{7.9}$$

where C is a hyperparameter that controls the strictness (width) of the margin.

The second option consists of finding linear separable boundaries in a higher dimensional space. This can be done using a non-linear basis function. However, such operations tend to be computationally expensive. Fortunately, there is a shortcut that enables this transformation without incurring into complex operations and high computational costs. Thus, under some special conditions (outside the scope of this book), this is carried out by using a kernel function whose dot product is equivalent to the required non-linear transformation. Applying such a trick – i.e., $K(x, x') = \varphi(x)^T \varphi(x')$ – yields to:

$$\underset{\mathrm{w,b},\zeta}{\mathrm{minimise}} \frac{1}{2} w^T w + C \sum_{i=1}^{n} \zeta_i \mathrm{s.t.} y_i \left(w^T \varphi(x_i) + b \right) \geq 1 - \zeta ; \mathrm{for\ i} = 1, \ldots, \mathrm{N}. \tag{7.10}$$

Mohebbi and Ghassemian [16] used the MIT-BIH database to develop a SVM classifier for arrhythmia detection. Following signal pre-processing the authors first extracted the RR intervals of the ECG and then calculated time-domain and non-linear HRV features. Next, linear discriminant analysis (LDA) was used to reduce the feature space from 9 to 4 dimensions that were passed to a SVM classifier as inputs for AF detection. The algorithm was evaluated with 835 normal and 322 AF segments of 32 RR intervals each; yielding Se = 99.07%, Sp = 99.29% when using a non-linear kernel combined with a C = 10. Another example on the use of SVM for AF detection was reported in [17]. The authors derived HRV features from Poincare plots using interbeat intervals obtained by means of a wavelet QRS detector. Three features – number of clusters, namely mean stepping increment of RRIs, and dispersion of datapoints around the diagonal line – were derived from the plots. The distribution of the number of clusters obtained with k-means was used to discriminate between AF and non-AF data. The two latter features were used as inputs of a Gaussian Kernel SVM classifier. Selected records from Computers in Cardiology 2001 and 2004 challenges (https://www.physionet.org/content/aftdb/1.0.0/, https://www.physionet.org/content/afpdb/1.0.0/) were selected and segmented in epochs of 1 min of duration.

Other discriminative approaches based on RR regularity features involved the use of artificial neural networks (ANNs). These models imitate the synaptic behavior of neural circuits in the brain and higher centers of the central nervous system. In a dense, fully connected network (Figure 7.7), the simplest processing unit is the neuron. For each neuron in a single layer synapsis occurs when the combination of all inputs with their respective weights (sum of products) passes through a differentiable function and then exceeds an activation threshold. The outputs of neurons in a single layer may serve as inputs to subsequent layers, or simply constitute the outputs of the network which is iteratively used to optimize the weights and gradients of the network in a reverse pass (i.e., backpropagation) until a cost function is minimized via gradient descent or other more sophisticated optimization method.

Ramirez and Hernandez [18] developed a classifier based on a fuzzy neural network for discrimination of AF from extra-systole and normal sinus rhythms. A

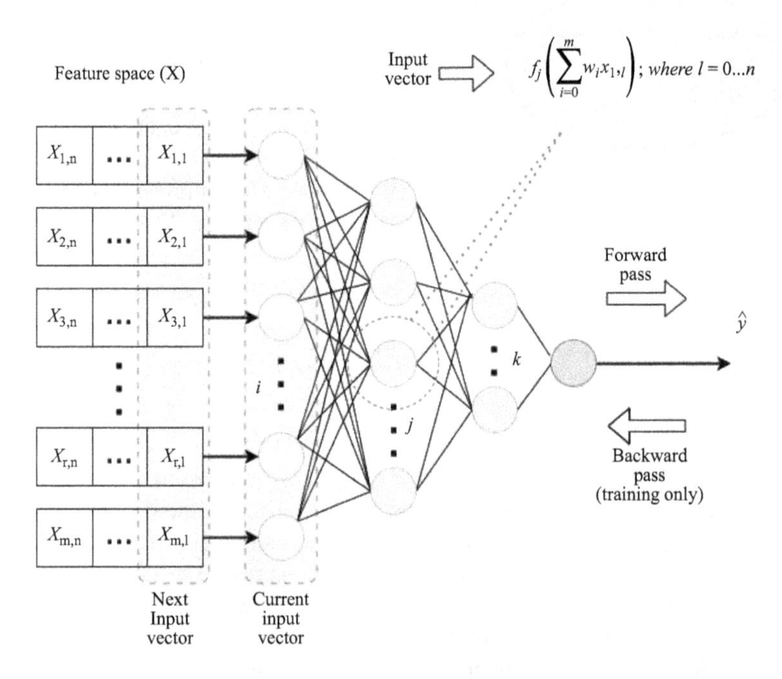

*Figure 7.7 Structure of an artificial neural network. The feature space of size **n** is formed by columns vectors containing **m** features that are passed as inputs to the first hidden layer of the network. Each processing unit of each layer processes all inputs by first estimating the sum of products of the weights by the current input vector, and then applying a differentiable transfer function (e.g., sigmoid or tanh) which output is propagated to the next layer and so on during the forward pass. In each iteration of the training phase, the weights can be set to random values which are updated in the forward pass after the gradients have been calculated in the backward pass. This process corresponds to the back propagation algorithm, which in turn runs in cyclical fashion until the error function is minimized.*

modification of the conventional learning algorithm allowed the feed-forward multilayer network to generate fuzzy outputs that could be compared against fuzzy target labels. Results from testing the algorithm with a subset of records from the MIT-BIH database showed values of Se = 91.4% and +P = 92.3% when detecting AF.

7.2.2 AF detection based on the ECG waveform morphology and hybrid methods

An early method to diagnose AF based on atrial activity was proposed by Slocum *et al.* [19]. The authors emphasized that approaches based on irregularity of ventricular activity are prone to fail when discriminating AF from other anomalous rhythms that also present with irregular ventricular beats. Based on this premise, an AF multi-lead detection algorithm reliant on fibrillatory activity in the frequency domain was developed with 73 12-lead ECGs. The method first cancels out the ventricular complexes to determine next whether the spectrum of atrial fibrillation waves is present in the remainder of the ECG signal. Thus, after applying the Fast Fourier Transform (FFT) and defining the spectral metric – i.e. ratio of the AF power (5–9 Hz) to the overall power of the modified spectrum (2–57 Hz) – the AF discrimination algorithm is executed. First it verifies if non-coupled P waves between leads are present, and then check whether AF is present – i.e., when the percent in one of two leads (II and V1) of power exceeds 32%. After testing the algorithm with 148 ECGs, the results showed poor sensitivity (Se = 68.3%) and moderate specificity (87.8%).

Parvaresh and Ayatollahi [20] developed and evaluated AF classifiers based on different machine learning methods (SVMs, LDA, and k-nearest neighbors). An autoregressive (AR) model was created and applied to extract features (coefficients) directly from 15-s segments of selected ECG records of the MIT-AF database. AR coefficients were computed using the Burg method, after applying a median filter to the signals to remove the baseline wandering trend. These coefficients were fed as inputs to the different classification models. The results from the evaluation revealed that LDA combined with an AR model of order 7 yielded the best performance (Se = 96.18%, Sp = 91.58% and +P = 91.09%) when tested with 640 segments of 30 s length.

Kara and Okandan [21] applied the wavelets transform to ECG episodes of one minute, and subsequently estimated the power spectral density of the resultant level 6 "db10" wavelets using the Welch method, aiming to extract features of AF and normal rhythms for subsequently classification using a multilayer back propagation neural network with logarithmic sigmoid activations. This fully connected network consisted of three layers – 22 inputs, 11 hidden, and 2 output neurons; and it was trained and evaluated with a database containing 80 and 72 records of AF and normal ECG, respectively. The duration of the ECG epochs was 1 min, sampled at 128 Hz. Although the results were overoptimistic (i.e., 100% of classification accuracy, due to the small sample size), this work presents an interesting framework to derive frequency domain features from the ECG signal by decomposing the signals at various frequency bands with different resolutions using the discrete wavelet transform combined with a non-parametric spectral density estimation method.

Couceiro *et al.* proposed a hybrid algorithm that combines P-wave detection, RR intervals, and atrial activity analysis to detect episodes of AF. The algorithm first applied filtering (using morphological operations like erosion and dilation) to ECG segments of at least 12 beats to remove baseline wandering and other types of noise. Following, detection of the ECG waves (P, QRS and T) is performed. P-wave absence was determined by the correlation between the detected P-waves and a template. QRS detection provided information of the location of R peaks, leading to calculation of the RR intervals and subsequently generation of features based on the approach proposed by [13]. The Kullback–Leibler divergence (KLD) between the distribution of the AF irregularity model and the RR model under analysis was used as a feature for the final classifier. Atrial activity was estimated by first cancelling QRS and T waves, and then using FFT and entropy to estimate an AF spectrum distribution template from training data as well as the spectrum of distribution of the segment under analysis. The KLD was also used as a similarity feature to detect AF. Finally, the features obtained in previous phases were fed into a dense multilayer neural network classifier (three layers and sigmoid activation functions) using the Levenberg–Marquardt algorithm as the optimizer in the training process. Twenty-three records from the MIT AF database (347160 ECG segments) were partitioned and used for the development and evaluation of the algorithm. High sensitivity and specificity values (Se = 93.80% and Sp = 96.09% were obtained for the testing partition.

The 2017 Computers in Cardiology Challenge provided a window for the development of algorithms focused on discriminating AF from noise, normal and other abnormal rhythms using wearable single-lead ECG recordings. A total of 8,528 records were given to the participants to develop and validate their algorithms; while 3,658 records were kept hidden for final scoring after submission to the organizers [22]. The data was sampled at 300 Hz and bandpass filtered (0.5–40 Hz). Four categories of labels were available: AF, normal, noisy, and other. Out of the 75 participating teams, we selected 2 out of the 4 algorithms that obtained the highest score (all ties) for this review.

The first of these works consisted of a two-layer binary cascaded classifier [23]. Noise removal was based on the generation and utilization of spectrograms to localize and remove the noise fragments along the entire recordings (i.e., those containing spectral information above 50 Hz). Following, baseline wandering was removed using high-pass filtering. Signals wherein 80% of the waveform were flagged by the spectrogram as corrupted were classified directly as noise, bypassing the execution of the two classification layers. Otherwise, the signals underwent the following process: first, feature extraction is performed by detecting the P-QRS-T complexes, leading to subsequent generation of 150 features – some morphological and some based on regularity of the RR intervals (HRV metrics in both frequency and time-domain, as well as non-linear features). Feature selection was applied to remove irrelevant and redundant features using methods such as maximal information coefficient (MIC) and minimum redundancy and maximum relevance (mRMR). The cascade comprised two levels of classification. In the first level, normal and abnormal rhythms were treated as one class, and AF and noisy signal as another one. Depending on the result of the first stage, the algorithm branched to one of the two classifiers in the second level – here is where classification results in one category of the four mentioned above. Regardless of the classification level, the method adopted by the authors was Adaptive Boosting (Adaboost). The latter belongs to the family of ensemble methods based on decision trees; whereby several weak learners (usually single trees with one split – i.e., stumps) are added and trained during subsequent iterations such that the weights assigned to each training sample are adjusted to calculate the error of each predictor that minimizes the classification error. The algorithm accuracy was evaluated using the f1-scores, resulting in values of 0.92, 0.82, and 0.75 for normal rhythms, AF, and other rhythms; and an average overall score of 0.83.

The second approach encompassed the generation of a combination of expert, statistical, signal processing and deep learning features that are ultimately feed into an ensemble classifier [24]. Data preprocessing consisted of splitting the records into five categories: long time-series data, short waves data using QRS detection, length of QRS complexes, center waves (most representative waveform obtained from short waves data found using dynamic time warping and spectral clustering and graph theory), and expanded data using a sliding window with overlap to increase the size and balance the dataset by creating many segments from each record. Of particular interest, deep neural network classifiers (formed by a combination of convolutional, recurrent, and dense layers) were turned into feature extractors by removing the output layer and using the values of last hidden layers as feature space. This is possible as deeper layers tend to remove irrelevant data and just keep the most representative information. Further details of how these deep models work is given later in this chapter. All the features converged into an extreme gradient boosting classifier – i.e., XGBoost. Likewise Adaboost, this model relies on weak learners as based classifiers and then iteratively train new classifiers using until achieving the desired performance – however, the difference is that gradient boosting fits new predictors based on residual errors resulting from gradient descent optimization. This is advantageous as it allows the choice of different cost functions (if they are differentiable) that can be optimally minimized. Furthermore, using extreme boosting implies the adoption of more advance techniques in terms of minimizing the cost function (as it uses second-order gradients via Newton's method) and in terms of preventing overfitting – more advance L1 and L2 regularization than conventional gradient boosting machines; resulting in better accuracy and generalization capabilities. Thus, after applying k-fold cross validation a f1-score of 0.84 was obtained for a model comprising five XGBoost classifiers with a maximum depth of 9 and 3,000 trees.

As described above, one of the novelties of the algorithm developed by Hong *et al.* was the use of deep learning models as feature extractors. However, there are also approaches wherein deep learning models are also used as AF classifiers. Before reviewing some of these works, it is necessary to go through some background knowledge. Deep learning networks (DNNs) are basically artificial neural networks formed by combinations of many hidden layers. However, one important difference is that the architecture of most shallow neural networks is based on dense and fully connected layers of artificial neurons (i.e., minimal computing unit of the formed by several inputs from previous stages and their associated weights) only. In contrast, DNNs implement convolutional (see Figure 7.8) and/or sequential and/or pooling (down-sampling) layers to produce features maps of simple representations of the input data in the first layers and more complex in the deeper layers. The rationale for using these types of layers is twofold: first, reduce the number of trainable parameters (weights) of the network – mainly to prevent overfitting. For example, in computer vision tasks, a conventional shallow network is suboptimal as a flattened image (i.e., converted from a 2D matrix to a 1D vector) combined with many processing units even with less hidden layers often results in larger numbers of weights. On the contrary, a convolutional neural network (CNN) applies convolutional kernels and pooling operations across its layers – resulting in a lesser number of parameters. In some

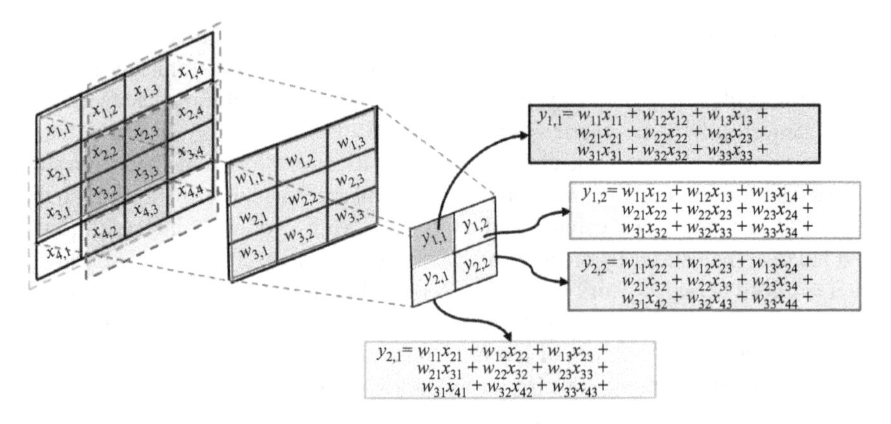

Figure 7.8 Illustration of a convolution operation between a 4×4 input matrix and a 3×3 kernel. Different colors illustrate each one of the convolutions as the kernel as the kernel is shifted in steps of 1 along the input matrix columns and rows. Note that in any case the output of the convolution process is the element-wise sum of products of the weights and input values.

cases, the last layer of the CNN is flattened and then fully connected to dense layers towards the output of the network (see Figure 7.9). Second, their main advantage is that the feature engineering step can be omitted due to the generation of feature maps of the raw inputs at different depths of the network.

One of the main enablers of DNNs (of course, besides the wide availability of high-performance computational resources) is that different solutions to the vanishing/exploding gradients problem now exist – i.e., weight initialization methods, use of different activation functions, and batch normalization layers. Like ANNs, backpropagation is typically used in combination with cross-entropy loss functions and learning rates to optimize the network and converge to an optimal solution. See Chollet [25], Lecun *et al.* [26] for further details of DNNs and implementations in Python.

The current state-of-the-art of using deep learning for arrhythmia classification revealed different architectures when it comes to AF detection. A comprehensive review on the topic can be found in Murat *et al.* [27]. Here we described some of the recent advances encompassing simple models based on one-dimensional CNNs [28,29], two-dimensional CNNs [30,31], recurrent neural networks RNNs (mainly long-short time networks – LSTMs) [32], and hybrid models combining CNNs and LSTMs [33,34].

Hsieh *et al.* [28] developed a deep learning model based on a series of one-dimensional convolutional blocks. Ten convolution layers with kernels of length 5 and rectified linear units (ReLU) as activation functions were stacked and combined with max pooling operations with kernels of size 2. Only in the first layer batch normalization (BN) was inserted between the convolution and pooling layers; such that the input layer is normalised to zero mean and unit variance, to enhance the

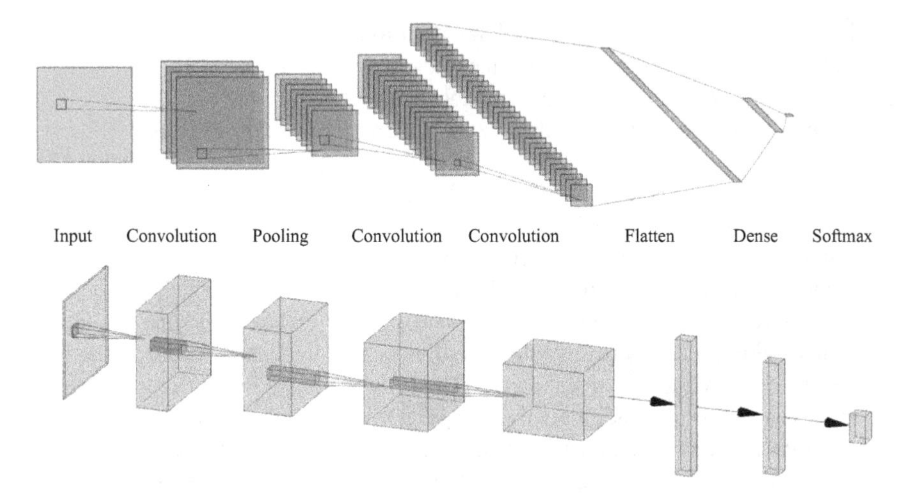

Figure 7.9 Architecture of CNN. The same network was drawn in two different styles – LeNet (top) and AlexNet (bottom) – using the online tool NN-SVG (https://alexlenail.me/NN-SVG/). Note that the convolution operations are performed across 2D tensors arranged in parallel in accordance with the number of filters assigned for each layer – 4, 8, 16, and 32 for this example.

stability and performance of the model. Likewise, dropout regularization layers (with a threshold $p = 0.5$) were inserted in the sixth, eighth, and ninth convolutional blocks – between the convolutions and max pooling layers. The number of feature maps started with 32 in the first layer and increased by a factor every two convolution blocks – i.e., 32, 32, 64, 64, 128, 128, 256, 256, 512, and 512). The last convolutional layer was flattened, and then fully connected to a dense layer of 128 hidden neurons which outputs were in turn connected to the last layer formed by 4 neurons with Softmax activations to generate probabilistic predictions for each class (NSR, AF, OTHER, and NOISE). The training process adopted adaptive moment estimation (Adam) optimizer, which in turn utilizes backpropagation to update the model weights until the cost function (in this case cross-entropy) is minimized. The initial learning rate was set to 0.001, and then adjusted by Adam as appropriate at different epochs/iterations during training. The data used for training and testing the network was the same used in the Physionet Challenge of 2017 – but following segmentation and data length normalization, the number of data samples increased from 8,528 to 10,151 single-lead ECGs. Selection of the best possible model was carried out applying 5-fold stratified cross validation and hyperparameter tuning (number of hidden layers, batch size, kernel size, and learning rate) via grid search. Thus, the best out of 300 models (the one with the highest average f1-score) was subsequently modified to generate several configurations with various numbers of max pooling, average pooling, and BN layers. The best of these configurations was then selected as the final model to be implemented. The authors reported that partitioning the data into 80% for training and 20% for testing yielded the best accuracy (average f1-score = 76.75) for model with slightly more than 3.2 million trainable parameters.

Further techniques include the use of recurrent networks. RNNs architectures (Figure 7.10A) are most suited for the task of processing time-series signals (such as ECGs and RR intervals segments), and widely used for classification and forecasting. The main reason behind this is that these sequential models preserve both the time information and the order of the data (i.e., spatiotemporal structure of the data). Unlike conventional dense ANNs and CNNs, these models have memory, which enables them to process real-time sequences of data as new points or beats become available – and thus making them more suitable for online applications. Nevertheless, the main issues that involve the use of standard RNNs is that they are prone to more unstable gradients (which is often addressed with layer normalization across the feature space dimensions and regularization), as well as their limited ability to recall short term events. Fortunately, the solution for the latter issue is to use networks such as LSTMs (Figure 7.10B) which are formed by specialized cells. Thus, each cell has the ability of discern what to keep stored in the long term (controlled by an input gate), what to discard over time (controlled by a forget gate) and what long-term information to pass to the next cell and output (output gate). The gating process is regulated by multiplication and additive operations that combine different inputs and outputs with fully connected dense layers and sigmoid or tangential hyperbolical activation functions. Based on sequential deep learning models, Macknickas and Macknickas [32] developed a model that learns from time series of time domain features generated from the ECG following QRS complex detection. Three LSTM layers were stacked

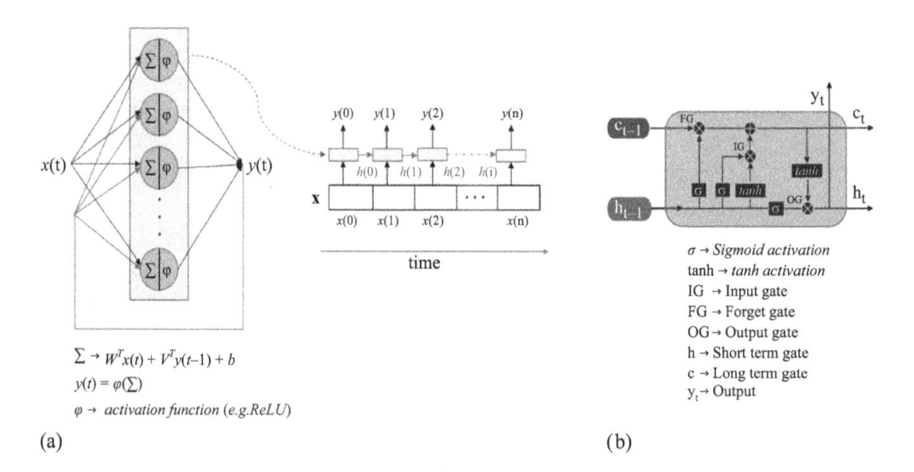

Figure 7.10 (A) Illustration of a simple RNN with its layers unfolding over time (right) W and V are the weight matrices for the input x(t) and previous state y(t–1). (B) Internal structure of a LSTM cell.

and the time-distributed output of the last one connected to a dense ANN. The model was trained with 80% of the 2017 Physionet Challenge Dataset, yielding an average accuracy of 0.78 when tested with the remainder 20% split.

An alternative to time-series deep models comprises the use of bidimensional CNNs, where each ECG signals is converted into an image (2D tensor). The main advantage of this approach is that signal processing aspects such as sampling rate and filtering become irrelevant; allowing the use of ECGs that stem from different datasets without having to pre-process the data. In this context, Jun *et al.* [30] developed an algorithm pipeline that first segments the ECG signals into individual beats, then convert these segments into images (128 × 128 × 1 – grayscale), followed by data augmentation using matrix operations (translation, shifting, rotations, and scaling) to increase the size and variance of the training and validation sets by a factor of ten. The classification model consisted in the adoption and modification of an existing off-the-shelf deep 2D convolutional network known as Very Deep Convolutional Network – VCGNet [35] that became popular after obtaining the second place in a prestigious image recognition competition. The original version contains 16 blocks, each comprising two convolutional layers (with 3×3 kernels) and one pooling layer. The output of the last block is connected to a dense ANN comprising two hidden and one output layer. The modified version contained only six blocks combined with BN and pooling operations at the outputs. The final block was then connected to a fully connected ANN that contained one dense layer in between BN and pooling layers respectively. The authors remarked the following important aspects: first, initial setting of the weights of each kernel using Glorot initialization [36] to ensure convergence is achieved during the training process. Second, the use of exponential linear units (ELUs) to ensure that no weights are initialized to zero values and thus avoid exclusion of some nodes from the learning process. Last, Adam optimization combined with an exponential decaying learning rate scheduler to ensure the best possible compromise between training speed and minimization of the cost function (cross-entropy). After applying 10-fold cross validation the best model contained 8 convolutional blocks connected to a dense network regularized (dropout) and normalized before reaching its Softmax probabilistic output. An average classification accuracy of (ROC area under the curve – AUC = 99.05%) was achieved after training and testing with more than 100,000 beats of normal and 8 different arrhythmias extracted from the MIT-BIH Arrhythmia database (https://www.physionet.org/content/mitdb/1.0.0/) and converted into grayscale images. Although the proposed DNN did not include AF detection, we strongly felt that it should be included in this review as it could be modified to detect this rhythm.

Using ECG signals digitized as images is the closest to method to real clinical assessments performed by cardiologists on electrocardiograms when looking for arrhythmias. However, for an AI model, running in a computer is always more difficult to discriminate between various morphologies corresponding to different or the same arrhythmia due to inter/intra patient variability and ECG leads and electrodes placement. Therefore, some researchers have attempted to facilitate characterization of different rhythms by converting ECG signals directly into their time–frequency representation. One way of doing this is by the application of the short-time Fourier transform (STFT). The essence of this method is that any periodic signal can be synthetized by adding multiple sine waveforms of different frequencies. Thus, the basis of the Fourier transform (FT) is to decompose a signal into its frequency components – i.e., those corresponding to each of the sine waveforms that constitute a given time invariant periodic signal. However, ECG signals – mainly those associated with arrhythmias – are stochastic in nature (their statistical properties such as the mean and standard deviation vary over time); hence, computing the FFT to the totality of an ECG signal may result in an inaccurate representation of the frequency content as the time information is lost. The STFT method divides a signal into time bins, and then calculate the FFT over each of them as follows:

$$STFT\{x[n]\} = X(m, \omega) = \sum_{n=-\infty}^{\infty} x(n)w(n-m)e^{-j\omega n} \tag{7.11}$$

where $\boldsymbol{x(n)}$ represents the ECG segment, $\boldsymbol{w(n)}$ is the window function (usually Hanning is chosen to avoid frequency leakage), $\boldsymbol{m}$ is the time index, and $\boldsymbol{\omega}$ the frequency.

Thereby, the STFT algorithm is often used to find the frequency and phase components of a whole ECG signal as it changes over time. In simple terms, its output is used to construct spectrograms where the horizontal and vertical axis correspond to the time and frequency values where 2D points of different colors represent the intensity at different times and frequencies. Consequently, spectrograms are images that characterised different patterns of heart rhythms and morphologies (Figure 7.11) in time and frequency domains; and thus, they can be fed directly as inputs to CNNs for arrhythmia classification. Spectrograms are calculated as the magnitude squared of the STFT.

Ross-Howe and Tizhoosh [31] developed and evaluated the performance of CNNs for AF detection using spectrograms as inputs to two classification approaches. First, an existing pre-trained Densely connected CNN (DenseNet) was used to generate features from the ECG spectrum. DenseNets are deep learning networks that connect each layer with their successor ones in a feedforward manner [37]. The main advantage of this architecture is that very deep CNNs can be implemented with a reduced number of parameters and without being affected by the vanishing gradient problem as a result of the shorter connections between layers. Thus, a DenseNet with 201 layers was first trained with more than 14 million images from a publicly available database, and then executed with the spectrograms as inputs to extract features from the last average pooling layer. The resultant feature space was fed into a linear SVM classifier that discriminates between AF and any other rhythms or noise. The model was trained and cross-validated (5-fold) using the MIT-AF database. Alternatively, a second classification model was tailored by the researchers. The latter consisted of a CNN architecture that takes 128×64 grayscale spectrograms as inputs, with two blocks with 5×5 and 2×2 convolutional and average pooling kernels and batch normalization at their outputs. The output of the last block was connected to a fully connected network with two dense layers, dropout regularization and a two-node Softmax output.

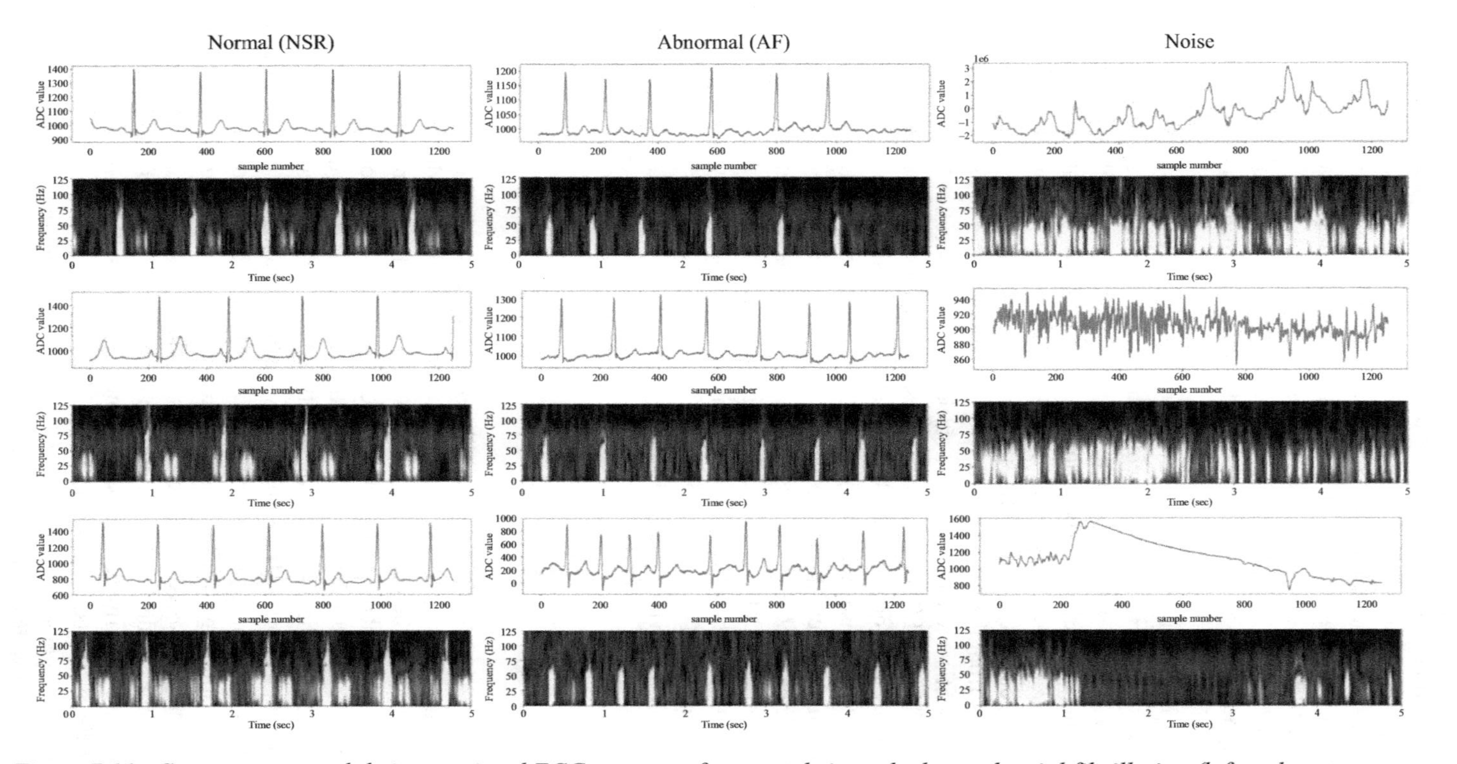

Figure 7.11 Spectrograms and their associated ECG segments for normal sinus rhythm and atrial fibrillation (left and center columns), and noisy segments (right column)

Adam optimization was used obtain the network parameters corresponding to the minimum possible mean squared error value, when training the model with 70% of the MIT-BIH AF dataset (about 407,577 instances – after segmenting the ECGs in 6 s epochs and data augmentation). Following testing, a classification accuracy of $92.18 \pm 0.48\%$ with Se = $88.38 \pm 0.02\%$ and Sp = 95.14 ± 0.01 was obtained for the first model (DenseNet+SVM); whereas slightly better results (accuracy = 93.16, Se = 98.33, and Sp = 89.74) was obtained for the tailor-made CNN AF detection model.

Hybrid approaches incorporate combinations of different types of deep neural network. Warrick and Homsi [34] developed an architecture that first processes ECG raw segments (30 s of duration) with a first CNN one-dimensional layer combined with a batch normalization, a max pooling layer and a ReLU activation. The authors stated that the rationale for adopting this configuration as the first layer was not only to reduce computational load at the upper layers but also to extract the most salient features from the ECG complexes. The output of the CNN block was then fed into a sequential model formed by a stack of three consecutive batch-normalised LSTM layers. The last LSTM layer was connected to a fully connected dense network with four nodes at its output to enable probabilistic classification of four different rhythms (AF, NSR, noise, and other) via Softmax activation. The reason behind the adoption of this 3-layer LSTM approach was to learn hidden sequential patterns inherent to different ECGs. Training and testing were carried out using the 2017 Physionet Challenge Database. Model parameters were optimized using the Root Mean Square Propagation (RMSProp) algorithm and cross-entropy as the objective function. An average f1-score of 0.83 was obtained after several 10-fold stratified cross-validation runs with different batch sizes and class weights to adjust for class imbalances.

7.3 Conclusions

AF is a serious arrhythmia affecting millions of people across the world. This has motivated many researchers in academia and industry to develop different methods to catch this condition before the onset of severe and life-threatening complications.

More than a critical review this chapter describes a journey – providing the reader with insights of a variety of methods for detection of this arrhythmia developed during the last few decades. The different choices outlined in this chapter can fulfill different use cases, target platforms (from wearable embedded systems to cloud computing), and datasets of different sizes used during development. To the best of our knowledge, we covered a wide range of techniques including ruled-based algorithms, statistical pattern recognition, classic and probabilistic machine learning, ensemble methods, and deep learning models. Furthermore, we expanded in some of the methods we used in the past for arrhythmia classification – i.e., Gaussian Naïve and Bayesian KDE classifiers, SVMs, ANNs, and DNNs. Thus, we hope this chapter serves as a guideline to enlighten enthusiasts to adopt, improve, and develop their own AF detection algorithms that best suit their needs.

It is important to emphasize that the totality of the approaches reviewed in this chapter were developed with labeled datasets, and therefore they belong to the supervised learning category. It is also arguable that the size of the datasets used for the development of some these algorithms is small, limiting their generalization capability to a wider population. In addition, arrhythmias such as AF are anomalies that sporadically occur in some patients, adding a significant class imbalance. In an ideal world, one solution to this issue could be the creation of a vast and balanced annotated arrhythmia dataset to extend the generalization capability of complex models. However, it is common knowledge that gathering and annotating such a dataset can be extremely costly and time consuming. Therefore, we are in favor of adopting semi-supervised learning approaches whereby partially annotated datasets are used in combination with unsupervised clustering algorithms and anomaly detection to create new labels and detect AF episodes. For example, deep CNN autoencoders (AEs) (Figure 7.12) can be designed to take input ECG data (in the form of time series, 2D digitized image, or spectrograms) to obtain low-dimensional representations (i.e. latent space or bottleneck) enclosing the most salient characteristics associated with a particular rhythm. For example, AE can be trained with NSR data only. The aim of

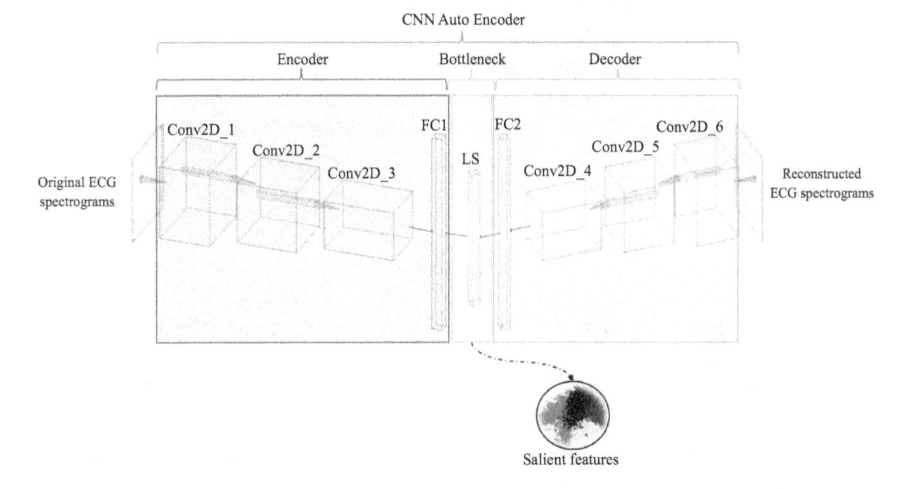

Figure 7.12 CNN autoencoder architecture. The encoder comprises three convolutional blocks and one fully connected layer (FC1) which output is connected to a final dense layer containing the latent space (LS) with the most salient features of the inputs. The decoder is a mirrored image of the encoder and takes LS as input to its first layer (FC2). The output of FC2 is then passed to the first convolutional module (Conv2D_4), and then propagated to the subsequent layers. Note that the convolutional blocks in the encoder perform down-sampling operations (max pooling layers) from one block to the next, whereas up-sampling layers are added to the decoder until obtaining the reconstructed ECG spectrums.

the training process is to minimize the reconstruction error by comparing the reconstructed versus the original inputs (often the mean squared error is adopted as this cost function). Thus, this method is unsupervised as the target outputs are the same inputs used to train the model. Since the AE is trained with only one class of data (in our case normal ECGs), it can be utilized as anomaly detector. An anomaly threshold can be set (e.g., the mean plus the standard deviation of the training reconstruction error). Thus, when running the AE with unseen input data, normal ECGs will hypothetically generate equal or lower errors below the threshold, whereas abnormal rhythms and noise will produce errors above the preset threshold. An example of the use of AEs as anomaly detection with ECG data can be found in the TensorFlow official website (https://www.tensorflow.org/tutorials/generative/autoencoder). It should be borne in mind that the latter is a toy example using limited data (only from one patient with congestive heart failure), and it is not expected to be able to detect AF with high accuracy. However, it is a good starting point for those who wish to follow this approach. Lastly, AEs are therefore very powerful as they perform dimensionality reduction, feature saliency, and anomaly detection tasks in a single model. If required, further dimensionality reduction can be performed on the latent space (bottleneck) using PCA, kernel PCA, feature agglomeration or manifold methods. The latent space or an even further reduced representation can also be utilized in combination with clustering techniques for stratification of different rhythms.

As a final remark, we hope you find this chapter useful either to enrich your general knowledge about arrhythmia detection, or for pre-selection of different algorithmic alternatives for arrhythmia detection; bearing in mind all possible scenarios and clinical applications that you might be willing to embark.

References

[1] British Heart Foundation. (2021). Heart and Circulatory Disease Statistics 2021. *Heart and Circulatory Disease Statistics 2021, June*, 157. https://www.bhf.org.uk/what-we-do/our-research/heart-statistics/heart-statistics-publications/cardiovascular-disease-statistics-2021%0A https://www.bhf.org.uk/what-we-do/our-research/heart-statistics/heart-statistics-publications/cardiovascular-disease-sta.

[2] Schnabel, R. B., Yin, X., Gona, P., *et al.* (2015). 50 year trends in atrial fibrillation prevalence, incidence, risk factors, and mortality in the Framingham Heart Study: a cohort study. *The Lancet, 386*, 154–162.

[3] Benjamin EJ, Muntner P, Alonso A, *et al.* (2019). American Heart Association Council on Epidemiology and Prevention Statistics Committee and Stroke Statistics Subcommittee. Heart Disease and Stroke Statistics – 2019 Update: A Report From the American Heart Association. *Circulation, 139*, e56–e528.

[4] Kornej, J., Börschel, C. S., Benjamin, E. J., & Schnabel, R. B. (2020). Epidemiology of atrial fibrillation in the 21st century: novel methods and new insights. *Circulation Research, 127*(1), 4–20. https://doi.org/10.1161/CIRCRESAHA.120.316340.

[5] Khandpur, R. (2004). *Biomedical Instrumentation: Technology and Applications.* New York, NY: McGraw Hill.

[6] Hamilton, P. S. & Tompkins, W. J. (1986). Quantitative investigation of QRS detection rules using the MIT/BIH arrhythmia database. *IEEE Transactions on Biomedical Engineering, BME-33*(12), 1157–1165. https://doi.org/10.1109/TBME.1986.325695.

[7] Köhler, B. U., Hennig, C., & Orglmeister, R. (2002). The principles of software QRS detection. *IEEE Engineering in Medicine and Biology Magazine, 21*(1), 42–57. https://doi.org/10.1109/51.993193.

[8] Pan, J. & Tompkins, W. J. (1985). A real-time QRS detection algorithm. *IEEE Transactions on Biomedical Engineering, BME-32*(3), 230–236. https://doi.org/10.1109/TBME.1985.325532.

[9] Kira, K. & Rendell, A. L. (1997). A practical approach to feature selection (ReliefF).pdf. In: *Proceedings of ICML* (pp. 249–256). http://citeseerx.ist.psu.edu/viewdoc/summary?doi=10.1.1.49.7856.

[10] Miao, J. & Niu, L. (2016). A survey on feature selection. *Procedia Computer Science, 91*(Itqm), 919–926. https://doi.org/10.1016/j.procs.2016.07.111.

[11] Peng, H., Long, F., & Ding, C. (2005). Feature selection based on mutual information: criteria of max-dependency, max-relevance, and min-redundancy. *IEEE Transactions on Pattern Analysis and Machine Intelligence, 27*(8), 1226–1238. https://doi.org/10.1109/TPAMI.2005.159.

[12] Dash, S., Chon, K. H., Lu, S., & Raeder, E. A. (2009). Automatic real time detection of atrial fibrillation. *Annals of Biomedical Engineering, 37*(9), 1701–1709. https://doi.org/10.1007/s10439-009-9740-z

[13] Moody, G. B. & Mark, R. G. (1983). A new method for detecting atrial fibrillation using R-R intervals. *Computers in Cardiology, 10*, 227–230.

[14] Malik, M. (1996). Heart rate variability – standards of measurement, physiological interpretation and clinical use. *European Heart Journal, 17*, 354–381.

[15] Vanderlei, L. C. M., Pastre, C. M., Hoshi, R. A., de Carvalho, T. D., & de Godoy, M. F. (2009). Basic notions of heart rate variability and its clinical applicability. *Brazilian Journal of Cardiovascular Surgery, 24*(2), 205–217. https://doi.org/10.1590/s0102-76382009000200018.

[16] Mohebbi, M. & Ghassemian, H. (2008). Detection of atrial fibrillation episodes using SVM. In *Proceedings of the 30th Annual International Conference of the IEEE Engineering in Medicine and Biology Society, EMBS'08 – "Personalized Healthcare through Technology,"* pp. 177–180. https://doi.org/10.1109/iembs.2008.4649119.

[17] Park, J., Lee, S., & Jeon, M. (2009). Atrial fibrillation detection by heart rate variability in Poincare plot. *BioMedical Engineering Online, 8*, 1–12. https://doi.org/10.1186/1475-925X-8-38.

[18] Ramirez, C. A. & Hernandez-Silveira, M. (2000). Multithread implementation of a fuzzy neural method for arrhythmia classification. *Computers in Cardiology, 28*, 297–300.

[19] Slocum, J., Sahakian, A., & Swiryn, S. (1992). Diagnosis of atrial fibrillation from surface electrocardiograms based on computer-detected atrial activity.

Journal of Electrocardiology, *25*(1), 1–8. https://doi.org/10.1016/0022-0736(92)90123-H.

[20] Parvaresh, S. & Ayatollahi, A. (2011). Automatic atrial fibrillation detection using autoregressive modeling. In: *Proceedings of International Conference on Biomedical Engineering and Technology (ICBET 2011)*, *11*, 11–14.

[21] Kara, S. & Okandan, M. (2007). Atrial fibrillation classification with artificial neural networks. *Pattern Recognition*, *40*(11), 2967–2973. https://doi.org/10.1016/j.patcog.2007.03.008.

[22] Clifford, G. D., Liu, C., Moody, B., *et al.* (2017). AF classification from a short single lead ECG recording: the PhysioNet/computing in cardiology challenge 2017. *Computing in Cardiology*, *44*, 1–4. https://doi.org/10.22489/CinC.2017.065-469

[23] Datta, S., Puri, C., Mukherjee, A., *et al.* (2017). Identifying normal, AF and other abnormal ECG rhythms using a cascaded binary classifier. *Computing in Cardiology*, *44*, 1–4. https://doi.org/10.22489/CinC.2017. 173–154

[24] Hong, S., Wu, M., Zhou, Y., *et al.* (2017). ENCASE: an ensemble classifier for ECG classification using expert features and deep neural networks. *Computing in Cardiology*, *44*, 1–4. https://doi.org/10.22489/CinC.2017.178-245.

[25] Chollet, F. (2021). Deep learning with Python. In *Deep Learning with Python*, Second Edition. https://www.manning.com/books/deep-learning-with-python-second-edition.

[26] Lecun, Y., Bengio, Y., & Hinton, G. (2015). Deep learning. *Nature*, *521*(7553), 436–444. https://doi.org/10.1038/nature14539.

[27] Murat, F., Sadak, F., Yildirim, O., *et al.* (2021). Review of deep learning-based atrial fibrillation detection studies. *International Journal of Environmental Research and Public Health*, *18*(21), 11302. https://doi.org/10.3390/ijerph182111302.

[28] Hsieh, C. H., Li, Y. S., Hwang, B. J., & Hsiao, C. H. (2020). Detection of atrial fibrillation using 1D convolutional neural network. *Sensors (Switzerland)*, *20*(7), 2136. https://doi.org/10.3390/s20072136.

[29] Tutuko, B., Nurmaini, S., Tondas, A. E., *et al.* (2021). AFibNet: an implementation of atrial fibrillation detection with convolutional neural network. *BMC Medical Informatics and Decision Making*, *21*(1), 1–21. https://doi.org/10.1186/s12911-021-01571-1.

[30] Jun, T. J., Nguyen, H. M., Kang, D., Kim, D., Kim, D., & Kim, Y.-H. (2018). *ECG arrhythmia classification using a 2-D convolutional neural network.* http://arxiv.org/abs/1804.06812.

[31] Ross-Howe, S. & Tizhoosh, H. R. (2019). Atrial fibrillation detection using deep features and convolutional networks. In: *2019 IEEE EMBS International Conference on Biomedical and Health Informatics, BHI 2019 – Proceedings*, *Bhi*, 0–4. https://doi.org/10.1109/BHI.2019.8834583.

[32] Maknickas, V. & Maknickas, A. (2017). Atrial fibrillation classification using QRS complex features and LSTM. *Computing in Cardiology*, *44*, 1–4. https://doi.org/10.22489/CinC.2017.350-114.

[33] Pandey, S. K., Kumar, G., Shukla, S., Kumar, A., Singh, K. U., & Mahato, S. (2022). Automatic detection of atrial fibrillation from ECG signal using hybrid deep learning techniques. *Journal of Sensors, 2022*, 6732150. https://doi.org/10.1155/2022/6732150.

[34] Warrick, P. & Homsi, M. N. (2017). Cardiac arrhythmia detection from ECG combining convolutional and long short-term memory networks. *Computing in Cardiology*, *44*, 1–4. https://doi.org/10.22489/CinC.2017.161-460.

[35] Simonyan, K. & Zisserman, A. (2015). Very deep convolutional networks for large-scale image recognition. In: *3rd International Conference on Learning Representations, ICLR 2015 – Conference Track Proceedings*, pp. 1–14.

[36] Glorot, X. & Bengio, Y. (2010). Understanding the difficulty of training deep feedforward neural networks. *Journal of Machine Learning Research*, *9*, 249–256.

[37] Huang, G., Liu, Z., Van Der Maaten, L., & Weinberger, K. Q. (2017). Densely connected convolutional networks. In: *Proceedings – 30th IEEE Conference on Computer Vision and Pattern Recognition, CVPR 2017*, *2017-Janua*, pp. 2261–2269. https://doi.org/10.1109/CVPR.2017.243.

Chapter 8

Dictionary learning techniques for left ventricle (LV) analysis and fibrosis detection in cardiac magnetic resonance imaging (MRI)

Juan José Mantilla[1], José Luis Paredes[2], Jean-Jacques Bellanger[1], François Carré[1], Frédéric Schnell[1] and Mireille Garreau[1]

The characterization of cardiac function is of high clinical interest for early diagnosis and better patient follow-up in cardiovascular diseases. A large number of cardiac image analysis methods and more precisely in cine-magnetic resonance imaging (MRI) have been proposed to quantify both shape and motion parameters. However, the first major problem to address lies in the cardiac image segmentation that is most often needed to extract the myocardium before any other process such as motion tracking, or registration. Moreover, intelligent systems based on classification and learning techniques have emerged over the last years in medical imaging. In this chapter we focus in the use of sparse representation and dictionary learning (DL) in order to get insights about the diseased heart in the context of cardiovascular diseases (CVDs). Specifically, this work focuses on fibrosis detection in patients with hypertrophic cardiomyopathy (HCM).

8.1 Introduction

The goal in medical image analysis is the extraction of biomedical information from medical images. Three key aspects have contributed to this domain: (i) the construction of large biomedical datasets from different sensors, (ii) the design of different methods to better describe and analyze the data including feature extraction procedures, and (iii) the development of advanced learning and statistical computing algorithms to analyze the extracted features. Typical biomedical image datasets include those acquired from: magnetic resonance (MR), computed tomography (CT), ultrasound (US), nuclear medicine, and X-ray. Potential applications in medical imaging involves the enhancement of contrast and edges, noise reduction, feature detection

[1]Univ Rennes, CHU Rennes, Inserm, LTSI UMR 1099, France
[2]Center for Biomedical Engineering and Telemedicine (CIByTEL), Universidad de Los Andes, Venezuela

and segmentation techniques, registration, object detection, as well as classification based on segment attributes such as shape, appearance or spatio-temporal information. Different machine learning techniques have been applied for post processing, including among others, support vector machines, random forest, graphical models and compressed sensing/sparse dictionary learning methods.

Sparse signal representation has proven to be an extremely powerful tool for acquiring, representing, compressing and classifying high-dimensional signals [1–5]. There are many areas of science and technology which have greatly benefited from advances involving sparse representation. For example, image and signal processing have been influenced in numerous ways such as denoising, image compression, feature extraction and many more. The sparse representation problem is a parsimonious principle that a sample can be approximated by a sparse linear combination of basis vectors over a redundant dictionary. Three main aspects deals with this problem: optimization techniques for solving sparse approximation problems (an inverse problem that arises in the representation), the choice of a dictionary, and the applications of sparse representations.

The main goal in sparse signal representation is to find an overcomplete basis or dictionary $\boldsymbol{D}$ that is good for representing a given set of vectors as sparsely as possible. Two major approaches have been followed to answer this issue. The first is to use some standard overcomplete basis, such as Wavelets, Curvelets, Contourlets, steerable Wavelet filters, short-time-Fourier, and the DCT basis. The success of such dictionaries in applications depends on how suitable they are to sparsely describe the signals to analyze. The second approach is to obtain an overcomplete basis from a given set of vectors through training. These approaches are relevant for dictionary learning techniques. While choosing a prespecified standard basis is appealing due to its simplicity, the training-based approach intuitively appears to be a better option as it generates dictionaries that are well suited to the class of signals in the training set.

8.2 Basics of dictionary learning

Dictionary learning is a recent approach to dictionary design that has been strongly influenced by the latest advances in sparse representation theory and algorithms. This approach suggests the use of machine learning-based techniques to infer the dictionary from a set of examples. In this case, the dictionary is typically represented as an explicit matrix, and a training algorithm is employed to adapt the matrix coefficients to the examples. The most recent training methods are focusing on ℓ^0 and ℓ^1 sparsity measures, which lead to simple formulations and enable the use of recently developed efficient sparse coding techniques. An illustration of the dictionary learning problem is shown in Figure 8.1. Next, we briefly introduce the basic framework for DL.

Let $\boldsymbol{Y} = [\boldsymbol{y}_1, \boldsymbol{y}_2, \ldots, \boldsymbol{y}_N] \in \mathbb{R}^{n \times N}$ be a data matrix (a finite training set of signals) where each column is a n-dimensional input signal. Learning a reconstructive

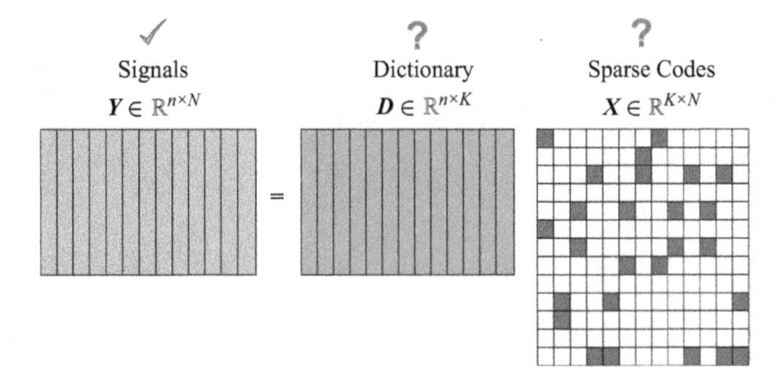

Figure 8.1 An illustration of the dictionary learning (DL) process

dictionary $\boldsymbol{D}$ with K items for sparse signal representation of $\boldsymbol{Y}$ can be accomplished by solving:

$$\langle \boldsymbol{D}, \boldsymbol{X} \rangle = \operatorname{argmin}_{\boldsymbol{D},\boldsymbol{X}} \|\boldsymbol{Y} - \boldsymbol{D}\boldsymbol{X}\|_F^2 \quad s.t. \quad \forall i, \|\boldsymbol{x}_i\|_0 \leq T, \tag{8.1}$$

where: $\boldsymbol{D} = [\boldsymbol{d}_1, \boldsymbol{d}_2, \boldsymbol{d}_3, \ldots, \boldsymbol{d}_K] \in \mathbb{R}^{n \times K}$ $(K > n,$ making the dictionary over-complete) is the dictionary to be learned from the data samples $\boldsymbol{Y}$. The N-column vector $\boldsymbol{X} = [\boldsymbol{x}_1, \boldsymbol{x}_2, \ldots, \boldsymbol{x}_N] \in \mathbb{R}^{K \times N}$ are the sparse codes coefficients of input signals $\boldsymbol{Y}$. T is a sparsity constraint parameter (each signal has fewer than nonzero T items in its decomposition). $\|\boldsymbol{Y} - \boldsymbol{D}\boldsymbol{X}\|_F$ denotes the reconstruction error and $\|\boldsymbol{x}\|_0$ denotes the ℓ^0-pseudo norm that counts the number of non-zero elements of $\boldsymbol{X}$. The Frobenius norm $\| \cdot \|_F$ defined as $\|\boldsymbol{A}\|_F = \sqrt{\sum_{ij} \boldsymbol{A}_{ij}^2}$ is an indication of the reconstruction error. The construction of $\boldsymbol{D}$ is achieved by minimizing the reconstruction error and satisfying simultaneously the sparsity constraint [1].

In [6] a classification of dictionary learning algorithms is presented in three main directions: (i) probabilistic learning methods; (ii) learning methods based on clustering or vector quantization; and (iii) methods for learning dictionaries with a particular construction.

8.2.1 Probabilistic methods

Probabilistic and non-probabilistic approaches have been adopted for the derivation of DL algorithms. Maximum-likelihood (ML) dictionary learning method for natural images was introduced in [7] under the sparse approximation assumption. In [8] another ML algorithm is presented, which uses the Laplacian prior to enforce sparsity. Given the training examples $\boldsymbol{Y} \in \mathbb{R}^{n \times N}$, to obtain the likelihood function $P(\boldsymbol{Y}|\boldsymbol{D})$ and seek the dictionary $\boldsymbol{D}$ that maximizes it, two assumptions are necessary: the first one is that the measurements are drawn independently, therefore:

$$P(\boldsymbol{Y}|\boldsymbol{D}) = \prod_{i=1}^{N} P(\boldsymbol{y}_i|\boldsymbol{D}) \tag{8.2}$$

The second one is critical and refers to the coefficient vector $\boldsymbol{x}$ which is considered as a random vector with prior distribution $P(\boldsymbol{x})$. The components of the likelihood function are computed using the relation:

$$P(\boldsymbol{y}_i|\boldsymbol{D}) = \int P(\boldsymbol{y}_i, \boldsymbol{x}|\boldsymbol{D})dx = \int P(\boldsymbol{y}_i|\boldsymbol{x}, \boldsymbol{D}) \cdot P(\boldsymbol{x})d\boldsymbol{x} \tag{8.3}$$

Formally, the goal of ML learning method is to maximize the likelihood that signals have efficient sparse representations in a redundant dictionary given by the matrix $\boldsymbol{D}$. This can be accomplished by finding the overcomplete dictionary $\hat{\boldsymbol{D}}$ such that

$$\hat{\boldsymbol{D}} = \text{argmax}_{\boldsymbol{D}} \left[log \int_{\boldsymbol{X}} P(\boldsymbol{Y}|\boldsymbol{X}, \boldsymbol{D}) \cdot P(\boldsymbol{X})d\boldsymbol{X} \right] \tag{8.4}$$

Here, all the examples $\boldsymbol{y}_i$ are concatenated as columns to construct the data matrix $\boldsymbol{Y}$. Likewise, the representations coefficient vectors $\boldsymbol{x}_i$ are gathered together to build the matrix $\boldsymbol{X}$. The optimization problem in (8.4) can be reduced to an energy minimization problem where it is possible to compute an estimation $\hat{\boldsymbol{X}}$:

$$\begin{aligned} (\hat{\boldsymbol{D}}, \hat{X}) &= \text{argmin}_{\boldsymbol{D},\boldsymbol{X}} \left[-log[P(\boldsymbol{Y}|\boldsymbol{X}, \boldsymbol{D}) \cdot P(\boldsymbol{X})] \right] \\ &= \text{argmin}_{\boldsymbol{D},\boldsymbol{X}} \left[\|\boldsymbol{Y} - \boldsymbol{D}\boldsymbol{X}\|_2^2 + \lambda \|\boldsymbol{X}\|_1 \right] \end{aligned} \tag{8.5}$$

This problem can be solved by a heuristic, which involves iterating between 2 steps. In the first step (sparse approximation step), $\boldsymbol{D}$ is kept constant and the energy function is minimized with respect to a set of coefficient vectors $\boldsymbol{x}_i$. It can be solved, for example, by convex optimization for each vector $\boldsymbol{y}_i$. The second step (dictionary update step) keeps the sparse codes coefficients $\boldsymbol{x}_i$ constant, while performing, for example, the gradient descent on $\boldsymbol{D}$ to minimize the average energy.

The probabilistic inference approach in overcomplete dictionary learning has subsequently been adopted by other researchers. For simplicity, the two-step optimization structure has been preserved in most of these works, and the modifications usually appeared in either the sparse approximation step, or the dictionary update step, or in both, for example, the method of optimal directions (MOD) [9]. It uses the Orthogonal Matching pursuit algorithm (OMP) [10,11] or FOCUSS* optimization [12] algorithm in the sparse coding stage and introduces an analytic solution of the quadratic problem in the dictionary update step given by $\boldsymbol{D} = \boldsymbol{Y}\boldsymbol{X}^+$, with $\boldsymbol{X}^+$ denoting the Moore–Penrose pseudo-inverse.

The same researchers that conceived the MOD method also suggested a maximum a posteriori probability (MAP) setting for the training of dictionaries. Instead of maximizing the likelihood $P(\boldsymbol{Y}|\boldsymbol{D})$, the MAP method maximizes the posterior probability $P(\boldsymbol{D}|\boldsymbol{Y})$.

*FOCUSS stands for FOcal Underdetermined System Solver: an algorithm designed to obtain sub-optimally sparse solutions to the $m \times n$, underdetermined linear inverse problem: $\boldsymbol{Ax} = \boldsymbol{y}$.

8.2.2 Clustering-based methods

Approaches of this type are based on vector quantization (VQ)[†] achieved by K-means clustering [6]. In clustering, a set of descriptive vectors $\{\boldsymbol{d}_k\}_{k=1}^{K}$ of the searched dictionary is learned, and each sample is represented by one and only one of those vectors (the one closest to it, based on the ℓ^2 distance measure). In contrast, in sparse representations, each example is represented as a linear combination of several vectors $\{\boldsymbol{d}_k\}_{k=1}^{K}$. A generalization of the K-means algorithm for dictionary learning, called the K-SVD algorithm, has been proposed by [1]. The K-SVD algorithm takes its name from the singular-value-decomposition (SVD) process that forms the core of the atom update step, and which is repeated K times, as the number of atoms.

Dictionaries learned with K-SVD have been initially used in synthetic signals to test whether the algorithm recovers the original dictionary that generate the data. Then the K-SVD algorithm has been applied on natural image data with two different main goals: filling in missing pixels (inpainting) and compression. K-SVD has been widely used in other signal processing tasks like denoising [2], image restoration [13], and signal separation [14].

8.2.3 Parametric training methods

A parametric dictionary is constructed typically driven by prior on the structure of the data or to the target usage of the learned dictionary. The advantages of parametric dictionaries reside in the short description of the atoms. Parametric dictionary learning tries to find better parameters for atoms based on some criteria yielding to better and more adaptive representations of signals. It also gains the benefits of dictionary design approaches which are the simplicity and better matching to the structure of a special class of signals. An important advantage of parametric dictionary learning is that only the parameters of an atom (which is as few as five parameters in typical applications) should be stored instead of all the samples of the atom. So, it is very well suited to the applications with large matrix dimensions [15]. Some examples of parametric dictionary structures are related to: translation-invariant dictionaries, multiscale dictionaries, and sparse dictionaries. A widely description can be found in [16].

As we have seen, most important methods for DL include the maximum-likelihood method, the method of optimal direction MOD and the K-SVD algorithm. Among these, the K-SVD algorithm has been the most popularly used technique for dictionary learning. It provides a good trade-off between sparsity and convergence [17].

8.3 DL in medical imaging – fibrosis detection in cardiac MRI

Confirming what has been observed for natural images, we would like to show how in medical imaging, sparse representation and dictionaries, directly learned from a

[†]Vector quantization (VQ) is a classical quantization technique from signal processing which allows the modelling of probability density functions by the distribution of prototype vectors. It was originally used for data compression.

set of training images, can better capture the distribution of the data and later, can be used in statistical inference tasks such as detection. Sparse representation and dictionary learning are closely related to each other in the framework of compress sensing theory. Some successful applications in medical imaging have been recently explored for sparse representation and DL approaches. They concerned, among others, image reconstruction, image denoising, image fusion, image segmentation, multimodal images analysis, and classification. Image modalities involve MRI, CT, ultrasound, and microscopy images. In cardiac medical images, works are focused on segmentation on epicardial and endocardial contours of LV in echocardiography images. In cardiac MRI, to our knowledge, there are no works based on sparse representation and DL for detection tasks.

In the next section, we address the problem of fibrosis detection in patients with hypertrophic cardiomyopathy (HCM) by using a sparse-based clustering approach and DL. HCM, as a genetic cardiovascular disease, is characterized by the abnormal thickening of left ventricular myocardium. Myocardial fibrosis commonly presented in HCM can be detected in LGE cardiac magnetic resonance imaging (MRI). In this chapter, we present the use of a DL-based clustering technique for the detection of fibrosis in LGE-MRI. The addressed issue in this part that concerns HCM and fibrosis is depicted in Section 8.4. Furthermore, a description of the LGE-MRI cardiac imaging modality is presented in Section 8.5. A brief state-of-the-art for the assessment of myocardial fibrosis in LGE-MRI is presented in Section 8.6. Next, in Section 8.7, we present our proposed approach that represents a novel approach based on clustering and DL techniques for the detection of fibrosis in cardiac LGE-SAX MRI.

8.4 HCM and fibrosis

Cardiomyopathy is a type of progressive heart muscle disease in which the heart is abnormally enlarged, thickened, and/or stiffened. As a result, the heart muscle's ability to pump blood is weakened, often causing heart failure (HF). The disease can also cause abnormal heart rhythms. The World Health Organization/International Society and Federation of Cardiology (WHO/ISFC) classification of 1996 associated the term cardiomyopathies to all heart muscle diseases that lead to functional disturbances of the heart. A classification is described with four main phenotypes, which can be assessed by invasive and noninvasive imaging methods: dilated cardiomyopathy (DCM), HCM, restrictive cardiomyopathy (RCM), and unclassified forms. The right ventricular cardiomyopathy, previously right ventricular dysplasia (ARVCM), was then added to this classification [18].

This work aims at characterizing the hypertrophic cardiomyopathy. HCM is the most common genetic cardiovascular disease [19]. The estimated prevalence in the general adult population with phenotypic evidence of HCM is 1 per 500 [20]. Men are more often affected than women and black patients more so than white patients. In young adults, HCM is the most common cause of sudden cardiac death [21,22].

Cardiac muscles hypertrophy in HCM is described as "concentric" or "eccentric." Concentric heart hypertrophy means increased heart muscle bulk and wall thickness, and it is best assessed on cardiac MRI by looking the heart thickness in the short axis view. Eccentric heart hypertrophy means general increase in the heart muscles with preservation of the normal cardiac wall thickness (isometric) [23].

Sometimes, the thickened heart muscle does not block blood flow out of the left ventricle. This is called non-obstructive hypertrophic cardiomyopathy. The entire ventricle may thicken, or the thickening may happen only at the basal part of the heart. Hypertrophic nonobstructive cardiomyopathy may be found as an apical, a midventricular isolated septal form, or as hypertrophy of the papillary muscles [18,21].

In both types of HCM (obstructive and non-obstructive), the thickened muscle makes the left ventricle cavity smaller, so it holds less blood. The walls of the ventricle also may stiffen. As a result, the ventricle is less able to relax and fill with blood.

HCM is a disease with an extremely variable prognosis. Some patients will die from sudden death [24], others will develop atrial fibrillation [25] with accidents due to arterial embolism, others will suffer from heart failure in most cases with a preserved left ventricular ejection fraction, and some will be life-long asymptomatic.

Some people who have HCM have no signs or symptoms. The disease does not affect their live. Others have severe symptoms and complications. For example, they may have shortness of breath, serious arrhythmias, or an inability to exercise [26].

8.4.1 Myocardial fibrosis in HCM

Myocardial fibrosis is a condition that involves the impairment of the heart's muscle cells called cardiomyocytes. When fibrosis occurs in response to injury, the term "scarring" is used. It belongs to a class of diseases collectively known as fibrosis, which denotes hardening or scarring of tissue. This is a condition that not only affects the heart but also other organs such as the lungs and the liver. Myocardial fibrosis is also referred to by the more general term of cardiac fibrosis.

Cardiomyocytes, which come from originating cells called myoblasts, are instrumental in controlling the heart rate by producing electrical impulses. Each cardiomyocyte cell is organized as a collection of cylindrical filaments called myofibrils. These are the cell units that enable the heart to contract. Normally, cardiomyocytes form lines of cells in the heart [27].

In myocardial fibrosis, cardiomyocytes are replaced by tissue that is unable to contract. This happens when fibroblasts, which produce collagen to enable wound healing, provide excessive amounts of the protein. This results in a case of abnormal scarring, or fibrosis.

Myocardial fibrosis leads to both systolic and diastolic dysfunctions. On the one hand, fibrotic segments will deform less than normal ones (in systolic phases). On the other hand, fibrosis will lead to a decrease in compliance thus to difficulties to fill the left cardiac chambers (in diastolic phases). Other studies proved that the extent of regions with LGE observed in MRI was also correlated with adverse cardiac events such as sudden cardiac death, fatal arrhythmia or worsening heart failure in patients with HCM [26].

Fibrosis manifests in three forms, that are, reactive interstitial fibrosis, replacement fibrosis and infiltrative interstitial fibrosis. Replacement fibrosis occurs in response to an injury causing cardiomyocyte death, as in the case of myocardial infarction; a reparative response is activated in the heart, causing replacement of dead cells and formation of a collagen-based scar. In reactive interstitial fibrosis, the cardiac interstitial space expands without significant cardiomyocyte loss [28]. Infiltrative interstitial fibrosis is a subtype of fibrosis induced by the progressive deposit of insoluble proteins (amyloidosis) or glycosphingolipids (Anderson Fabry's disease) in the cardiac interstitium.

Being a very heterogeneous disease with variable outcome, a better characterization of HCM is needed, and especially related to the potential presence of fibrosis. In this context, cardiac imaging can provide significant information allowing to elucidate the factors explaining HCM evolution.

8.5 Cardiac magnetic resonance imaging with LGE-MRI

The clinical diagnosis of HCM is based on the demonstration of LV hypertrophy in the absence of another disease process that can reasonably account for the magnitude of hypertrophy present. Various imaging modalities can be used to assess fibrosis and to guide treatment, screening, and preclinical diagnosis. These imaging modalities include: echocardiography, nuclear imaging, cardiovascular magnetic resonance, and cardiac computed tomography. Traditionally, the diagnosis of HCM relies upon clinical assessment and transthoracic echocardiography (TTE). In recent years, MRI has become established as a useful adjunct to TTE owing to its unrestricted field of view, more accurate measurement of LV wall thickness, mass, volumes, and function and its ability to provide non-invasive assessment of myocardial fibrosis [19]. This work aims at characterizing myocardial fibrosis in cardiac magnetic resonance imaging (CMRI), focusing on fibrosis detection.

Cardiac magnetic resonance is the new gold standard to measure myocardial wall thickness and to diagnose HCM [21]. Also, it is used to characterize myocardial tissue enabling to quantify the fibrosis/scar extension [29]. It is now documented that approximately half of patients with HCM have LGE suggestive of areas of fibrosis [24].

Assessment of myocardial viability is performed using 5- to 20-minute delayed, gadolinium-enhanced MRI which involves intravenous administration of gadolinium-based contrast agent followed by the acquisition of T1-weighted images of the myocardium using an inversion recovery (IR) technique. IR technique is commonly employed to suppress the signal from viable myocardium by modifying the contrast of the acquired image between viable and nonviable myocardium. IR uses inversion pulses typically followed by a prescribed delay to allow recovery of the prepared magnetization before a spin echo or gradient echo pulse sequence used to "read out" the MR signal. The associated delay is known as the time after inversion (TI) [26].

Gadolinium is an extracellular contrast agent; thus, it distributes from vascular sector to the interstitial sector and never enters the cellular sector. The intensity of the gadolinium enhancement depends upon (i) tissue perfusion and (ii) the volume in which the gadolinium is distributed.

On delayed enhanced CMRI (DE-CMRI), there is a relatively decreased washout of the gadolinium contrast agent in areas of myocardium that have been replaced by fibrosis or scar. In normal viable myocardium, the gadolinium contrast agent washes out more rapidly than it does from the fibrosis or scar. Since the difference between normal and abnormal myocardium is based on washout kinetics, images after contrast injection will optimally depict the fibrosis or scar.

There are two types of DE-CMRI: early gadolinium enhanced (EGE) and LGE. EGE- and LGE-CMRI are essentially the same, but the timing of the acquisition following intravenous administration of the contrast agent is a distinguishing factor, being greater than 10 min for LGE-CMRI. In a typical LGE exam, 10–12 breath hold slices are acquired in short axis orientation, followed by long axis and 4-chamber views when clinically indicated.

In [29], a direct correlation between the percentage of LGE and percentage of histologic collagen in an explanted HCM heart was obtained. LGE has been reported in up to 75% of patients with HCM in whom the vast majority have patchy mid-wall-type enhancement which is typically most pronounced within the segments most severely affected by hypertrophy. LGE most often involves the interventricular septum, particularly the anteroseptal mid to basal segments and right ventricular insertion points [19].

In this work, we focus on LGE-MRI. There are two types of LGE sequences: inversion recovery CMR (IR-CMR) and phase sensitive inversion recovery CMR (PSIR-CMR).

The IR sequence acquires the magnitude of the magnetization at the time of inversion (TI). This time must be carefully selected to null viable myocardium which increases the range of gray levels between viable and non-viable tissue [30].

In contrast, PSIR-CMR is less sensitive to the selection of TI because it takes account of the sign of the magnetization at the time of data acquisition; therefore, the dynamic range of IR signal intensity levels increases.

8.6 The assessment of cardiac fibrosis detection in LGE-MRI: a brief state-of-the-art

Several studies have shown the relevance of LGE in cardiovascular MRI in the location and the assessment of myocardial fibrosis [31]. The accurate estimation of the transmural extent (from the endocardic to the epicardic zone) of the hyper-enhanced regions is crucial to estimate functional myocardial recovery after reperfusion therapy that is a medical treatment after the patient has suffered from myocardial infarction, or heart attack.

Also, the degree of improvement in global wall-motion and ejection fraction is significantly related to the transmural extent of LGE. In ischemic cardiomyopathy, the transmural extent of LGE is predictive of adverse LV remodeling. At the clinical level, infarct size is an independent prognostic factor for heart failure, arrhythmic events and cardiac mortality [32,33]. Patterns of fibrosis may be also used to differentiate

HCM from secondary causes of LV hypertrophy such as aortic stenosis or severe hypertension [34].

Earlier studies performed visually assessment of the transmural extent of gadolinium enhanced and the amount of functional recovery using semiquantitative Likert scales. For example, in [35], the segmental transmurality of the scar was graded on the conventional five-class scale:

- 0: no hyper-enhancement.
- 1: hyper-enhancement extending from 1% to 25% of LV wall thickness.
- 2: hyper-enhancement extending from 26% to 50% of LV wall thickness.
- 3: hyper-enhancement extending from 51% to 75% of LV wall thickness.
- 4: hyper-enhancement extending from 76% to 100% of LV wall thickness.

An automated segmental scoring of infarct extent begins with the detection of the infarct on the images. Several methods based on the tuning of signal intensity thresholds with manual interaction of the user [35–38] or automated definition of the infarcted zones using morphological operators [39,40] have been developed to this end.

An overview of previously published scar detection, quantification, and segmentation methods in LGE-CMR is presented in [41] where a standardized evaluation benchmarking framework for algorithms segmenting fibrosis and scar in left atrium (LA) myocardium from LGE-CMR images is also presented. Table 8.1 shows the overview reported in [41] in terms of: (a) the model or type of data evaluated (canine or human), (b) the number of datasets, (c) the structure of interest: LV or LA, (d) the method employed, and finally (e) the evaluation measures used.

Most methods employed comprise: (i) simple standard deviation (SD) thresholding taking as reference a base healthy tissue intensity value [37,40,42–45], (ii) the full-width-at-half-maximum (FWHM) used to identify the infarct boundaries into an initial region that include all pixels with signal intensity (SI) greater than 50% of maximum [37], (iii) expectation–maximization (EM) fitting of a mixture model consisting in the Rayleigh distribution to represent the darker healthy myocardium, and the Gaussian distribution to represent the bright late enhanced regions [46], (iv) the graph-cuts method which combines the intensity and boundary information to separate infarct from healthy myocardium, and (v) methods based on clustering that avoid the choice of gray level thresholds [47–49].

The visualization of infarcted regions can be performed by using the maximum intensity projection (MIP) of the intensities from the MR images into an anatomically derived cardiac surface [50].

In clustering methods, the fuzzy c-means is an unsupervised approach providing each voxel with a level of membership to both, LGE and non-LGE classes, describing the belongingness of the voxel to the class. The level of membership is a number between 0 and 1. This is advantageous in the case of fibrosis quantification because, excepted for highly enhanced pixels with a bright gray intensity, the enhancement between fibrotic and not fibrotic tissues remains unprecise.

The 6 standard deviations above the mean signal of the remote myocardium (the region with no contrast enhancement and normal wall thickening) method was

Table 8.1 Overview of previously published scar detection, quantification, and segmentation methods presented in [41]

Reference	Model	*n*	LV/LA	Method	Evaluation
[42]	Canine	26	LV	SD	Infarct size, ex vivo
[37]	Animal	13	LV	SD, FWHM	Bland–Altman, Infarct volume
[40]	Human	23	LV	SD	Percentage scar, Bland–Altman
[47]	Human	15	LV	Clustering	Percentage scar
[43]	Human	144	LV	SD	Percentage scar
[44]	Human	47	LV	SD	Infarct size
[46]	Human	21	LV	EM fitting	Percentage scar, Bland–Altman
[45]	Human	81	LA	SD	Percentage scar
[49]	Human	15	LV	Clustering	Infarct size
[51]	Human	20	LV	Otsu thresholding	Dice
[50]	Human	7	LA	MIP	Percentage scar
[52]	Human	10	LV	Graph-cuts	Infarct size and Bland–Altman

SD, simple standard deviation thresholding; FWHM, full-width-at-half-maximum;
MIP, maximum intensity projection; EM, expectation–maximization.

previously shown to best correlate with visual LGE assessment and was used in several large studies analyzing the relation between fibrosis and clinical events in patients with HCM [53,54].

In [55], a comparison of various methods for quantitative evaluation of myocardial infarct volume from LGE-CMR data is presented. The necrosis volumes were quantified using: (1) manual delineation, (2) automated fuzzy c-means method, and (3) +2 to 6SD thresholding approaches. The fuzzy c-means method proved appropriate correlations with biochemical myocardial infarct (scar) quantification as well as LV function parameters.

Segmentation of fibrosis or scar in LGE-CMR is challenging due to multiple causes including contrast variation due to inversion time, signal-to-noise ratio, motion blurring and artefacts [56]. The inversion time choice can generate the appearance of more or less scar, and change the appropriate scar threshold. Motion blurring also reduces the appearance of scar [41]. In the next section, we present our proposed approach that is based on dictionary learning for the detection of fibrosis without the needs of tuning threshold parameters.

8.7 The proposed method

We follow the idea of the framework for clustering datasets that are well represented in the sparse modeling framework with a set of learned dictionaries [57]. In this framework, a set of K clusters is used to learn K dictionaries for representing the data, and then associate each signal to the dictionary for which the best sparse decomposition is obtained.

The proposed approach in LGE-MRI is applied in each LGE-SAX image (from basal to apical planes) to detect enhanced and non-enhanced regions by splitting each image in several patches. Based on the DL framework, first, an initial dictionary is constructed with learning samples from 2 clusters (LGE and non-LGE regions). Second, the sparse coefficients of the learning data are computed and then used to train a K-nearest neighbor (*K-NN*) classifier. Finally, the label (LGE/Non-LGE) of a test patch is obtained with its respective sparse coefficients obtained over the learned dictionary and using the trained *K-NN* classifier. The zones of fibrosis can be finally detected in the myocardium delimited by the endo- and epicardial contours. Figure 8.2 resumes the proposed fibrosis detection procedure. The process is thus divided in 4 stages as described next [58].

8.7.1 Feature extraction

The ability to compare image regions (patches) has been the basis of many approaches to core computer vision problems, including object, texture and scene categorization [59].

First, in order to normalize the intensity differences across slices and subjects, each LGE-SAX image is normalized. For that purpose, each pixel in each LGE-SAX image is set to $(I_{i,j} - \mu)/\sigma$, where, $I_{i,j}$ is the interpolated pixel intensity value, μ and σ are, respectively, the mean and the standard deviation of the LGE-SAX image [60].

Second, from different patients, random non-overlapping patches covering enhancing and non-enhancing regions from LGE-SAX images in the medial plane are extracted. The random extraction procedure was guided by the selection of approximately one half of the extracted patches with high gray level intensity values and the other half with low gray level intensity values. We have performed several experiments varying the number of patches and found out that selecting 1,184 patches from four random patients achieves a good trade-off between complexity and visual detection performance. Thus the learning samples take 296 patches from each one of the selected patients.

Figure 8.3 shows an example of the feature extraction from four random LGE-SAX images. The non-labeled extracted patches can belong to different regions: LV and RV cavities, fibrosis and other regularly enhancing and non-enhancing structures inside and outside the heart.

The similarity among the extracted patches is then calculated by using a Gaussian (radial basis function RBF) kernel with bandwidth σ. The RBF kernel on two samples $\boldsymbol{x}$ and $\boldsymbol{x}'$, represented as feature vectors in some input space, is defined as:

$$K(\boldsymbol{x},\boldsymbol{x}') = \exp\left(\frac{-\|\boldsymbol{x}-\boldsymbol{x}'\|_2^2}{\sigma^2}\right) \tag{8.6}$$

Since the value of the RBF kernel decreases with distance and ranges between zero (in the limit) and one (when $\boldsymbol{x} = \boldsymbol{x}'$), it has a ready interpretation as a similarity measure.

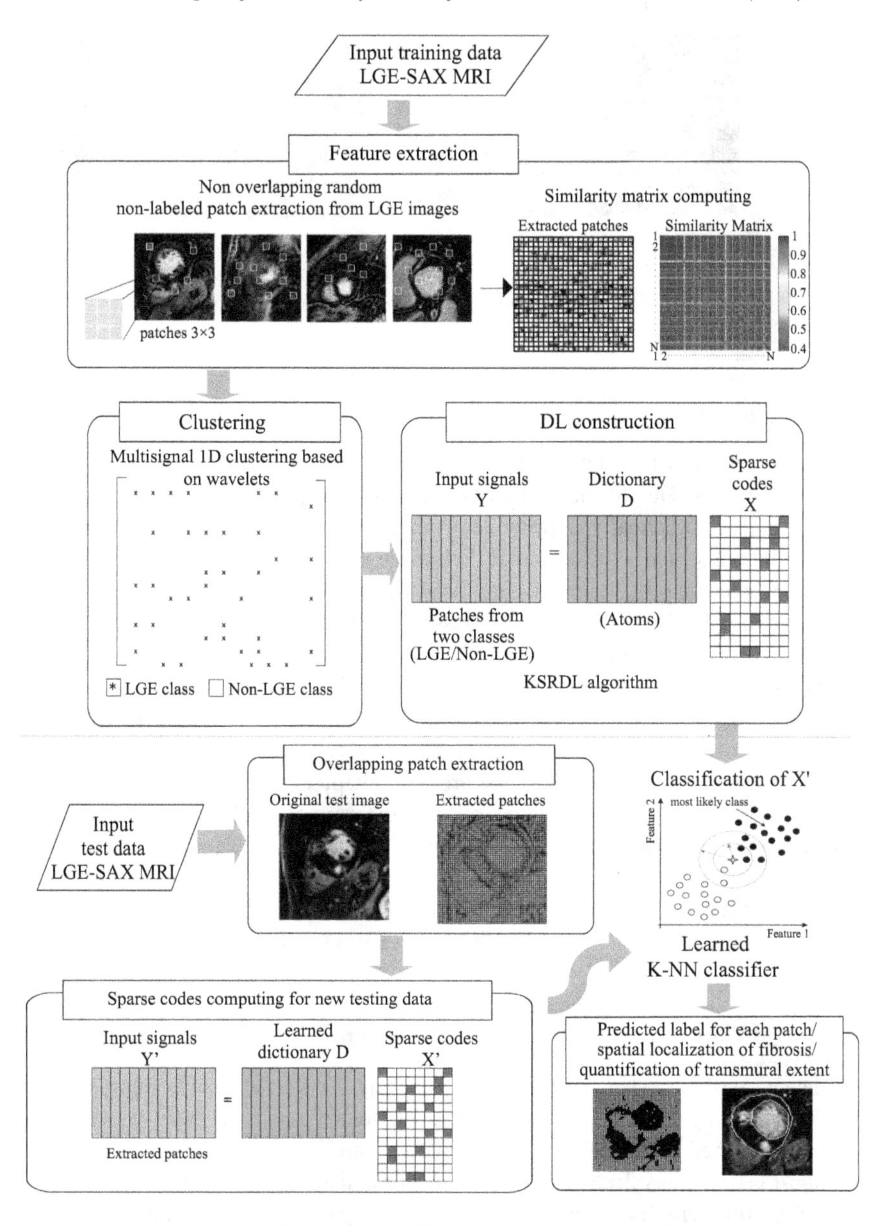

Figure 8.2 Overview of the proposed fibrosis detection method applied in each LGE-SAX image (from basal to apical planes) to detect enhanced and non-enhanced regions. Four main stages are identified: feature extraction, clustering, DL-based classification: training stage, DL-based classification: training stage: testing.

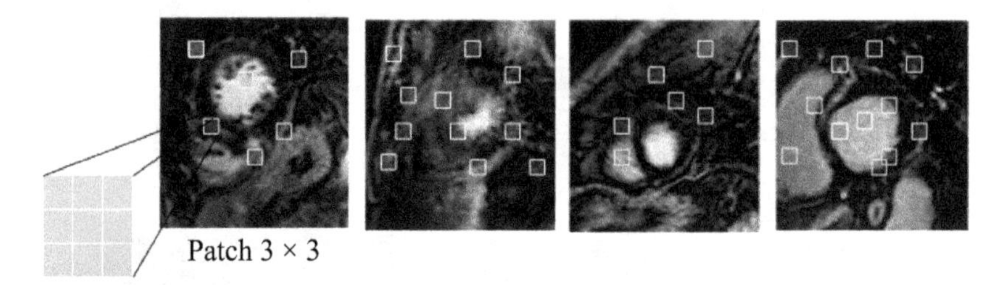

Figure 8.3 Feature extraction example from LGE-SAX images

8.7.2 Clustering

The initialization of the dictionary is very important for the success of the fibrosis detection process. Due to the cost associated with the procedure, repeating random initializations is practically impossible. Thus a "smart" initialization is needed. We propose the construction of an initial dictionary with two classes based on an unsupervised clustering process over the similarity measures among patches. Specifically, the aim is to split the patches in two classes, each one associated respectively with LGE and non-LGE regions.

For simplicity, we apply a clustering approach based on wavelets [61] which constructs clusters from a hierarchical cluster tree. This clustering approach was used in the construction of the clustering curve parameter (Cl) in the LV wall motion classification problem presented in the previous chapter achieving good performance results. In this case, the input matrix of the clustering algorithm corresponds to the similarity matrix among patches. Each row of this matrix is decomposed using the DWT function with the Haar Wavelet. A binary hierarchical cluster tree is constructed with the first level coefficients of the decomposition. Pairs of objects that are in close proximity are linked using the Euclidean distance. Data is then partitioned into two clusters.

8.7.3 DL-based classification: training stage

The detection of fibrosis is performed by adapting the Kernel Sparse Representation DL (KSRDL) algorithm [5] with an initial dictionary resulting from the clustering process described before where training patches are identified in two classes: LGE and non-LGE patches. In the KSRDL algorithm, also applied for the assessment of LV wall motion with local parametric evaluation, sparse representation is introduced from a Bayesian viewpoint assuming Gaussian prior over the atoms of the dictionary. The KSRDL model is defined as follows:

$$\min_{\boldsymbol{D},\boldsymbol{X}} \frac{1}{2}\|\boldsymbol{Y} - \boldsymbol{D}\boldsymbol{X}\|_F^2 + \frac{\alpha}{2}\textbf{trace}\left(\boldsymbol{D}^T\boldsymbol{D}\right) + \lambda \sum_{i=1}^{N} \|\boldsymbol{x}_i\|_1, \tag{8.7}$$

where the input signals $\boldsymbol{Y} \in \mathbb{R}^{n\times N}$ represent a data matrix of patches where each column is a vectorized patch (n is the signal size, N is the number of input signals or patches). $\boldsymbol{D} = [\boldsymbol{d}_1, \boldsymbol{d}_2, \boldsymbol{d}_3, \ldots, \boldsymbol{d}_K] \in \mathbb{R}^{n \times K}$ with K atoms being the dictionary to be learned and $\boldsymbol{X} = [\boldsymbol{x}_1, \boldsymbol{x}_2, \ldots, \boldsymbol{x}_N] \in \mathbb{R}^{K \times N}$ are the estimated sparse codes of input signals $\boldsymbol{Y}$. A *K-NN* classifier is the constructed over the sparse training coefficients matrix $\boldsymbol{X}$ before to perform the classification based on DL.

8.7.4 DL-based classification: testing stage

The class label of new p test instances can be predicted using the classifier obtained in the training step and the learned dictionary $\boldsymbol{D}$. As the selected classifier is trained based on the sparse coefficients of the input data, the test data needs to be represented in the same space of representation (sparse coefficients) over the learned dictionary. To this end, the sparse coefficients matrix $\boldsymbol{X}$ for the new test instances can be obtained by solving the non-negative quadratic problem (*NNQP*):

$$\min_{X} \sum_{i=1}^{p} \frac{1}{2} \boldsymbol{x}_i^T \boldsymbol{H} \boldsymbol{x}_i + \boldsymbol{g}_i^T \boldsymbol{x}_i \quad s.t. \quad \boldsymbol{X} \geq 0 \tag{8.8}$$

where $\boldsymbol{H}_{k \times k} = \boldsymbol{D}^T \boldsymbol{D}$ and $\boldsymbol{g} = -\boldsymbol{D}^T \boldsymbol{Y}$. As the optimization of the above problems only require inner products between the data, the sparse coding problem is solved by replacing inner products to a radial basis function (Gaussian) kernel.

Each LGE-SAX test image is represented as input by a grid of overlapping feature patches of dimension [3× 3]. The sparse coefficients of each patch (LGE and non-LGE) are obtained as described previously with the learned dictionary and then, the label of each patch is obtained using the trained *K-NN* classifier. By the overlapping of patches, each pixel is categorized as LGE or non-LGE pixel. Finally, the LGE pixels corresponding to myocardium fibrosis zones are selected by adding a spatial constraint given with the endo- and epicardial borders of the myocardium.

8.8 First experiments and results

8.8.1 Study population

MRI from 30 HCM data patients were acquired at CHU-Pontchaillou in Rennes, France, in collaboration with CIC-IT 804. This study was part of a systematic database review conducted according to the Declaration of Helsinki and approved under the CNIL (National Commission on Informatics and Liberty of France) number 909378. HCM was defined as recommended by recent guidelines [21].

8.8.1.1 Patient selection

The inclusion and exclusion criteria were:

- Inclusion criterion: Subjects carrying a primitive non-obstructive HCM with left ventricular ejection fraction (LVEF) greater than 60%, in sinus rhythm.
- Exclusion criteria: Prospective subjects with one of the following characteristics were disqualified from inclusion in this study: under legal age, contraindication to MRI (in particular, patients with an implanted cardiac stimulator/defibrillator

at the moment of MRI acquisition), history of coronary artery disease, permanent atria fibrillation, and left ventricular systolic dysfunction (LVEF $\leq$ 60%).

All patients had a clinical examination, a resting arterial blood pressure measurement (Dinamap Procare Auscultatory 100), a resting 12-lead electrocardiogram, a transthoracic echocardiography (Vivid 7, General Electric Healthcare, Horten, Norway) and a cardiac MRI (Philips Achieva 3T).

8.8.1.2 Cardiac magnetic resonance imaging

CMR images were performed with a 3T Achieva® clinical imager (Philips Medical Systems, Best, The Netherlands), using cardiac SENSE Coil (multicoil). Cardiac synchronization was performed using a four-electrode vectocardiogram. Scout images were acquired initially to identify the cardiac axes. CMR acquisitions included cine-SAX, cine-LAX (4CH, 2CH), LGE-SAX, and LGE-LAX (4CH, 2CH). Among the LGE-CMR acquisitions, IR and PSIR sequences were acquired, retrospective ECG-gated acquisition at mid-diastole and breath-hold volumetric SAX, 4CH, and 2CH. Typical parameters for these acquisitions were:

- IR acquisition: Turbo field echo (IR-TFE). TR/TE/FA=4.01 ms/1.23 ms/15°, IR prepulse delay = 280 ms.
- PSIR acquisition: TR/TE/FA = 4.50 ms/2.19 ms/15°, IR prepulse delay = 280 ms.
- IR, PSIR resolution: 256×256 pixels (in-plane) with 16 and 12 slices for SAX and LAX, respectively. In-plane pixel size=1.25 × 1.25 mm^2, spacing between slices=5 mm, acquisition slice thickness=10 mm. Given that the spacing between slices is half of the slice thickness, image slices overlap in a half of their thickness. This resulted in an output image volume with slice thickness=5 mm. All IR and PSIR acquisitions were acquired following this procedure (slice overlapping). This aimed at improving the observability of small portions with fibrosis.

For this study, LGE-SAX images in inversion recovery (IR) from 11 patients in the set of 30 HCM patients were retained (patients with exploitable LGE images).

8.8.2 Results

In the training stage, a set of 1,184 non-overlapping patches from four random inter-patients LGE-SAX IR images at mid-diastole and at mid-cavity plane are extracted in order to construct the initial dictionary. The first stage of clustering process splits the patches in two clusters (LGE and non-LGE) of size 952 and 232, respectively. Then, the KSRDL algorithm [5] is applied in order to obtain the sparse codes of the training data that is used in the *K-NN* classifier.

Several experiments were performed by reducing the number of atoms employed to represent each testing patch in the learned dictionary. The results presented here were obtained by using K=15 atoms from the initial dictionary to train the classifier and to represent each patch. The DL regularization parameter λ and the parameter α for the Gaussian kernel were tuned by heuristic search in a mesh from 0.01 to 10 with a step of 0.01. The final values used in these experiments were: $\lambda = 0.001$ and α=0.1.

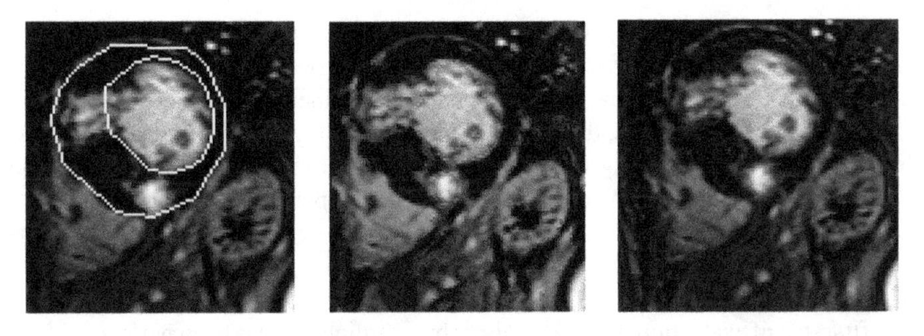

Figure 8.4 Examples of resulting fibrosis detection using the proposed approach with patches of size [3×3] (left), [5×5] (middle), and [7×7] (right)

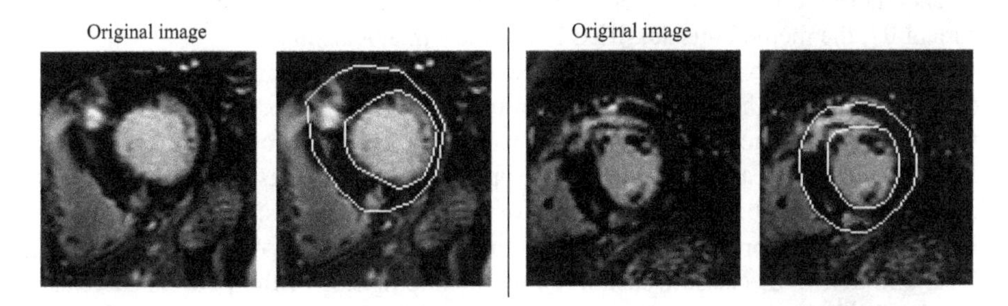

Figure 8.5 Examples of resulting fibrosis detection using the proposed approach: patients 01 and 02

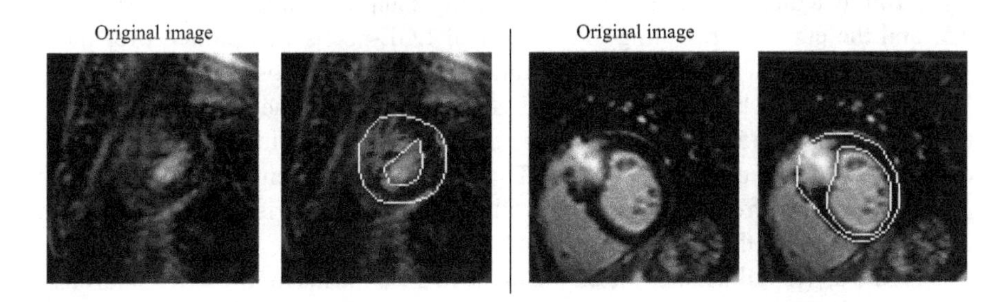

Figure 8.6 Examples of resulting fibrosis detection using the proposed approach: patients 07 and 08

Several experiments have been realized varying the size of the feature patch. Figure 8.4 shows an example of an LGE-SAX image with the detection of structures varying the patch size for [3×3], [5×5], and [7×7].

As it has been observed and as we can see visually in these images, a good detection of enhanced pixels is achieved by selecting patches of a dimension [3×3]. Figures 8.5 and 8.6 show LGE-SAX images at mid-cavity plane for 2 HCM patients of the database and the fibrosis detection using the proposed approach. In these figures, the original image for each patient is shown in the left part and the detected fibrosis is represented on the right part in colors inside the myocardium delimited by endo- and epicardial boundaries that have been manually delineated by a cardiologist.

About computing times, the proposed method employs about 59.38 s in the clustering stage and for training the classifier about 83.46 s. For testing, the method employs about 26.35 s for the detection of pixels LGE per slice and per patient.

8.8.3 Evaluation

In a first step of evaluation, a visual analysis by a cardiologist has been performed. The proposed method is able to detect fibrosis in 9 of 11 patients. The method misclassified LGE pixels in all the slices of the patient 04, classifying 100% of the pixels as LGE pixels due to the low contrast observed between myocardium and the LV cavity. In patient 07, the method misclassified LGE pixels in apical slices.

In a second step of evaluation, our method has been compared with one method of the literature that has been proposed for the detection of fibrosis in LGE-MRI. The fuzzy c-means method proposed by [48] has been retained to compare the results.

The fuzzy c-means method is an unsupervised method classifying the pixels in the myocardium as belonging to one out of two possible classes: LGE pixels of non-LGE pixels. This approach provides each pixel with a level of membership to both, LGE and non-LGE classes, describing the belongingness of the pixel to the class. The level of membership is a number between 0 and 1. The approach includes a defuzzification procedure to obtain a binary description of those pixels being part or not of LGE class (fibrosis for pixels inside the myocardium). Then, for the entire myocardium, a threshold of the LGE-class membership was varied between 0.25 and 0.5, and the curve representing the number of LGE-pixels was plotted over these varying threshold. Then the threshold value providing the most stable output (the longest portion in which the number of LGE-pixels remains the same) was selected as the optimal one [55]. This threshold is then used to get a binary image from the output of the fuzzy c-means approach. This approach has been applied in our data and were obtained from [26]. Figure 8.7 illustrates the comparison of this approach and our method for three patients.

Figure 8.7(top) illustrates the output of the fuzzy c-means for the LGE-pixel class for the pixels into the myocardium. Figure 8.7(middle) shows the resulting detection of fibrosis for the patients on the top after the defuzzification procedure. Figure 8.7(bottom) shows the resulting detection of fibrosis using our proposed approach. It can be noted that the fibrotic zones are identified in both methods, for those regions presenting a high concentration of pixels with LGE, however, in general the number of pixels LGE detected by the two methods is different.

To illustrate, Table 8.2 shows the quantification of pixels LGE for the patients shown in Figure 8.7 by the two methods and for each slice. The values shown in this table represent, for each slice, the percentage of pixels LGE with respect to the total number of pixels in the myocardium for the three patients. It can be seen that for the patient 02, both methods detect the same percentage of pixels LGE. For the other two patients, our method detect in general more LGE pixels than with the fuzzy c-method.

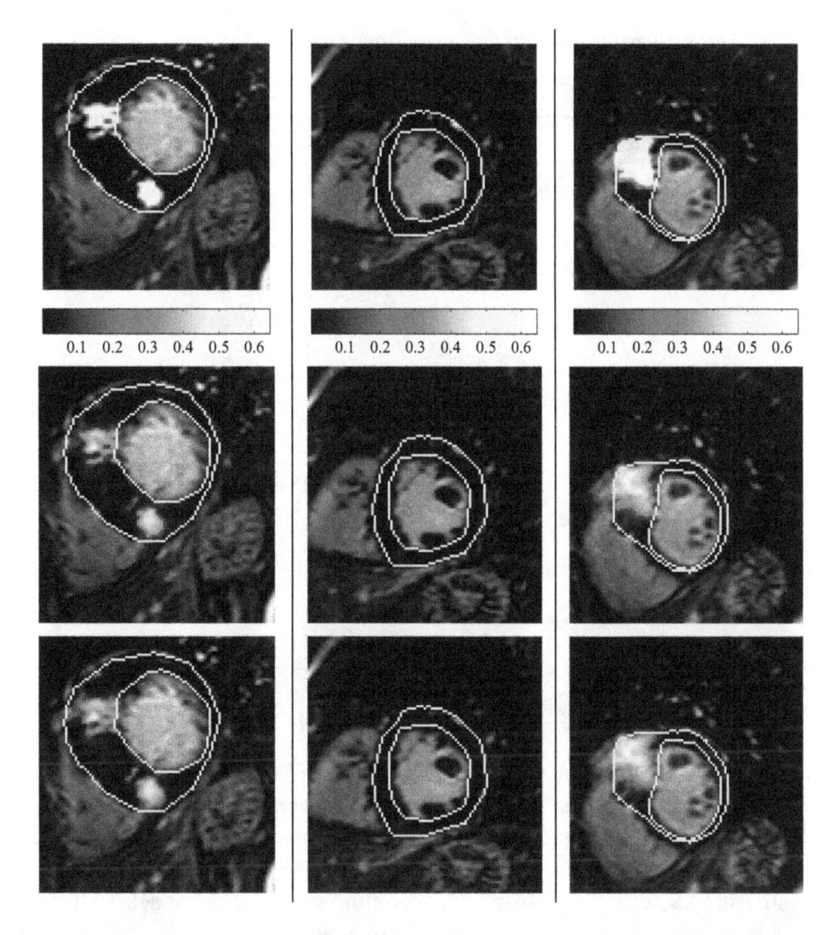

Figure 8.7 Comparison of results illustrated for three patients (P01–P02–P08): top: the output of the fuzzy c-means with the membership map of the LGE-class. Middle: the final result with fuzzy c-mean approach. Bottom: fibrosis detection using the proposed approach.

In apical slices (1–5), both methods detect a small number of pixels LGE [58]. These first quantitative results need to be completed by more experiments.

8.9 Qualification and quantification of myocardial fibrosis: a first proposal

The importance of detecting myocardial fibrosis in patients with HCM lies in the potential prognostic implication of this finding [62]. Previous studies have shown

Table 8.2 Quantification of cardiac fibrosis per slice level in patients 01, 02, and 08

Plane	Slice	P01	P02	P08
		Fuzzy–DL	Fuzzy–DL	Fuzzy–DL
Apical	1	0.00–3.82	0.00–0.00	0.00–0.55
	2	0.06–5.01	0.00–0.00	0.00–0.75
	3	0.00–2.72	0.00–0.00	0.12–2.86
	4	1.60–3.26	0.79–0.79	2.00–5.36
	5	0.16–5.82	1.89–7.86	5.14–7.27
Mid-cavity	6	0.15–5.38	3.49–16.98	5.00–7.11
	7	7.59–10.08	12.30–22.10	7.78–13.68
	8	4.52–5.72	19.93–23.02	4.00–7.14
	9	1.09–6.56	24.32–19.12	4.55–7.97
	10	1.62–6.91	25.02–15.81	10.00–12.95
	11	1.44–8.06	11.59–11.91	10.29–13.75
Basal	12	3.07–13.51	**9.18–9.18**	15.00–14.25
	13	4.13–15.13	10.23–6.92	10.21–13.35
	14	5.97–19.81	1.84–6.51	24.16–37.77
	15	**16.18–21.04**	1.87–2.17	**36.29–47.30**
	16	12.57–21.78	0.00–0.50	29.72–37.54

that the amount of myocardial fibrosis and the degree of cardiomyocyte degeneration are inversely related to both systolic and diastolic left ventricular (LV) functions [63].

For the detection and quantification of the fibrotic area, we perform a spatial localization of the myocardial fibrosis according the AHA 17 model representation. To identify the LV anatomical segments according to the AHA model, first, the LV centroid is automatically calculated using the contour of the endocardium delineated by a cardiologist. Second, the anterior intersection between the right and left ventricles is manually positioned by the user. By using a radial reference line traced from the LV centroid to the anterior intersection between LV and RV, the myocardium is divided in different anatomical segments at the different level slices (basal, mid-cavity, and apical): 6 segments for basal and mid-cavity slices using an angular variation of 60° and 4 segments for the apical slices using an angular variation of 90°. This procedure allows the localization of fibrosis by anatomical segments. Figure 8.8 shows an illustrative example of this procedure.

To illustrate, Figure 8.9 shows the result of the spatial localization of fibrosis LGE-SAX images at the mid-cavity plane in three patients. In Figure 8.9(left), fibrosis is localized in the anteroseptal and inferoseptal regions. For the patient in the Figure 8.9(middle), fibrosis is present in the anterior and anteroseptal segments and also in the inferior segment. Finally, for the patient in Figure 8.9(right), fibrosis can be observed in the anterior and anteroseptal regions.

The quantification of fibrosis is performed by calculating the percentage of LGE pixels in one particular anatomical segment with respect to the total number of pixels

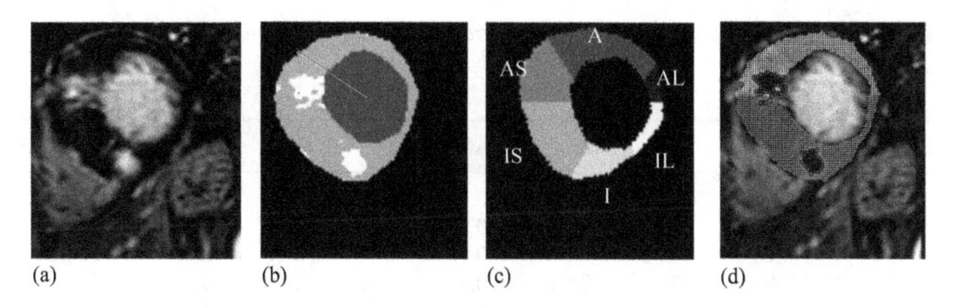

Figure 8.8 Framework for spatial localization of fibrosis using the proposed approach over an LGE-SAX image at mid-cavity plane: (a) the original LGE-SAX image; (b) a reference line from the LV centroid to the anterior intersection between RV and LV superimposed over the LV myocardium with the respective zones of fibrosis; (c) the 6 anatomical segments identified in the figure at the left: AL = antero septal, A = anterior, AS = antero septal, IS = infero septal, I = inferior, IL = infero lateral; (d) the spatial localization of fibrosis according the LV anatomical segments

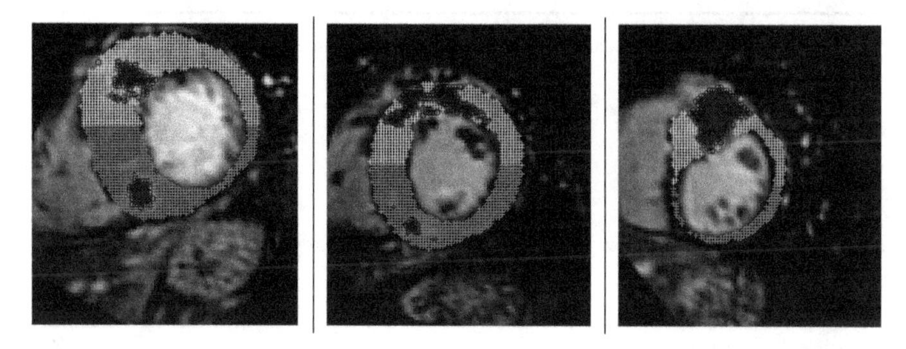

Figure 8.9 Spatial localization of fibrosis using the proposed approach for patients 01, 02, and 08

in the myocardium. Table 8.3 shows the quantification of fibrosis of the patient 02 shown in the middle of Figure 8.9.

As we can see, the highest concentration of LGE pixels is quantified in the medial plane, with 23.02% of fibrosis presented in slice 8. In this slice, broader areas of fibrosis are observed in the anterior and antero-septal segments with 12.38% and 4.36%, respectively.

Table 8.4 shows the quantification of fibrosis in each anatomical segment, calculating the percentage of LGE pixels in one particular anatomical segment with respect to the total number of pixels in that anatomical segment.

As we can see, the highest concentration of LGE pixels is quantified in the medial plane (slice 8), in which the Antero Septal segment presents 54.25% of fibrosis, the Anterior segment presents 51.89% and the Inferior segment presents 30.79% of

Table 8.3 Global quantification of cardiac fibrosis per segment and slice level in patient 02

Plane	Slice	AL	A	AS	IS	I	IL	(%) Total
Apical	1	0.00	0.00	0.00	0.00	0.00	0.00	0.00
	2	0.00	0.00	0.00	0.00	0.00	0.00	0.00
	3	0.00	0.00	0.00	0.00	0.00	0.00	0.00
	4	0.11	0.00	0.11	0.00	0.00	0.56	0.79
	5	1.88	0.00	1.44	0.00	0.00	4.54	7.86
	6	0.39	9.46	3.22	0.00	3.90	0.00	**16.98**
	7	3.57	13.10	0.00	2.76	2.67	0.00	**22.10**
Mid-cavity	8	0.96	12.38	4.36	0.00	5.32	0.00	**23.02**
	9	0.73	13.86	0.49	2.27	1.78	0.00	**19.12**
	10	0.41	11.63	2.38	1.06	0.33	0.00	**15.81**
	11	0.00	6.30	5.61	0.00	0.00	0.00	**11.91**
	12	0.00	2.69	6.49	0.00	0.00	0.00	9.18
	13	0.00	1.68	5.24	0.00	0.00	0.00	6.92
Basal	14	0.00	3.86	2.65	0.00	0.00	0.00	6.51
	15	0.00	1.79	0.38	0.00	0.00	0.00	2.17
	16	0.00	0.25	0.25	0.00	0.00	0.00	0.50

AL, antero septal; A, anterior; AS, antero septal; IS, infero septal; I, inferior; and IL, infero lateral.

Table 8.4 Local quantification of cardiac fibrosis per segment and slice level in patient 02

Plane	Slice	AL	A	AS	IS	I	IL
Apical	1	0.43	1.47	0.00	8.20	2.30	2.67
	2	0.40	0.46	0.00	0.00	0.00	1.81
	3	0.00	0.00	0.00	2.78	0.00	0.00
	4	1.20	9.05	3.83	0.00	3.28	0.00
	5	0.51	26.14	21.15	0.00	13.92	0.00
	6	0.00	33.95	41.42	0.92	23.43	0.00
	7	0.52	42.09	50.39	0.00	29.52	0.67
Mid-cavity	8	0.94	51.89	54.25	0.00	30.79	0.00
	9	0.00	51.55	44.44	2.00	23.76	0.00
	10	0.00	45.08	43.23	0.66	9.15	0.00
	11	0.51	34.39	37.85	0.38	0.37	0.00
	12	0.00	18.69	37.55	0.00	0.83	0.00
	13	0.00	10.42	31.85	0.00	0.00	0.00
Basal	14	0.00	12.23	34.83	0.00	0.60	0.00
	15	0.00	8.15	15.03	0.00	0.00	0.70
	16	1.11	1.78	12.90	2.56	2.30	2.14

AL, antero septal; A, anterior; AS, antero septal; IS, infero septal; I, inferior; IL, infero lateral.

fibrosis. Results are consistent with the observations in [19,64] in which LGE most often involves the interventricular septum, particularly the anteroseptal mid to basal segments and right ventricular insertion points.

8.10 Conclusion

We have presented a method for the detection of fibrosis in LGE-SAX images using a DL-based clustering approach. The detection approach has been applied on a set of 11 patients with HCM from which LGE-SAX images at 16 different slices were processed. The proposed method allows the detection of fibrosis inside the myocardium using the endo- and epicardial boundaries manually delineated by a cardiologist. The method has been evaluated by a visual evaluation and by comparing with the results of one method of the literature. The method has been able to successfully detect fibrosis in 9 of the 11 patients. By using the boundaries and a manual localization of the anterior intersection between the right and left ventricles, the region of fibrosis can be localized in different anatomical segments according to the AHA representation. The method could also be applied without the endo- and epicardial contours resulting in a segmentation approach of different structures in the MRI image that has to be analyzed. The proposed method based on DL has resulted in a promising technique for the detection of fibrosis in LGE-MRI that deserves more extensive validation. The method should be evaluated in a quantitative way on a set of more HCM patients with the fibrotic zones delineated by an expert. It must be deepened with broader experiments in the clustering and DL stages, for example, varying the number of patches for the initialization of the dictionary and performing a manual selection of LGE/Non LGE pixels. An analysis of the retained atoms used to represent each testing patch is also required.

References

[1] Aharon M, Elad M, and Bruckstein A. K-SVD: an algorithm for designing overcomplete dictionaries for sparse representation. *IEEE Transactions on Signal Processing*. 2006;54(11):4311–4322.

[2] Elad M and Aharon M. Image denoising via sparse and redundant representations over learned dictionaries. *IEEE Transactions on Image Processing*. 2006;15(12):3736–3745.

[3] Wright J, Ma Y, Mairal J, *et al*. Sparse representation for computer vision and pattern recognition. *Proceedings of the IEEE*. 2010;98(6):1031–1044.

[4] Rubinstein R, Zibulevsky M, and Elad M. Double sparsity: learning sparse dictionaries for sparse signal approximation. *IEEE Transactions on Signal Processing*. 2010;58(3):1553–1564.

[5] Li Y, and Ngom A. Sparse representation approaches for the classification of high-dimensional biological data. *BMC Systems Biology*. 2013;7(Suppl 4):S6.

[6] Tosic I and Frossard P. Dictionary learning. *IEEE Signal Processing Magazine*. 2011;28(2):27–38.
[7] Olshausen BA and Field DJ. Sparse coding with an overcomplete basis set: a strategy employed by V1? *Vision Research*. 1997;37(23):3311–3325.
[8] Lewicki MS and Sejnowski TJ. Learning overcomplete representations. *Neural Computation*. 2000;12(2):337–365.
[9] Engan K, Aase SO, and Husfy JH. Method of optimal directions for frame design. In: *1999 IEEE International Conference on Acoustics, Speech, and Signal Processing*, 1999, vol. 5, 1999. p. 2443–2446.
[10] Pati YC, Rezaiifar R, and Krishnaprasad PS. Orthogonal matching pursuit: recursive function approximation with applications to wavelet decomposition. In: *Proceedings of 27th Asilomar Conference on Signals, Systems and Computers*, vol. 1, 1993. p. 40–44.
[11] Tropp J and Gilbert A. Signal recovery from random measurements via orthogonal matching pursuit. *IEEE Transactions on Information Theory*. 2007;53:4655–4666.
[12] Gorodnitsky IF and Rao BD. Sparse signal reconstruction from limited data using FOCUSS: a re-weighted minimum norm algorithm. *IEEE Transactions on Signal Processing*. 1997;45(3):600–616.
[13] Mairal J, Elad M, and Sapiro G. Sparse representation for color image restoration. *IEEE Transactions on Image Processing*. 2008;17(1): 53–69.
[14] Abolghasemi V, Ferdowsi S, and Sanei S. Sparse multichannel source separation using incoherent K-SVD method. In: *2011 IEEE Statistical Signal Processing Workshop (SSP)*, 2011. p. 477–480.
[15] Ataee M, Zayyani H, Babaie-Zadeh M, *et al.* Parametric dictionary learning using steepest descent. In: *2010 IEEE International Conference on Acoustics Speech and Signal Processing (ICASSP)*, 2010. p. 1978–1981.
[16] Rubinstein R, Bruckstein AM, and Elad M. Dictionaries for sparse representation modeling. *Proceedings of the IEEE*. 2010;98(6):1045–1057.
[17] Ribhu R and Ghosh D. Dictionary design for sparse signal representations using K-SVD with sparse Bayesian learning. In: *2012 IEEE 11th International Conference on Signal Processing (ICSP)*, vol. 1, 2012. p. 21–25.
[18] Maisch B, Noutsias M, Ruppert V, *et al.* Cardiomyopathies: classification, diagnosis, and treatment. *Heart Failure Clinics*. 2012;8(1):53–78.
[19] Hoey ETD, Elassaly M, Ganeshan A, *et al.* The role of magnetic resonance imaging in hypertrophic cardiomyopathy. *Quantitative Imaging in Medicine and Surgery*. 2014;4(5):397–406. Available from: http://www.ncbi.nlm.nih.gov/pmc/articles/PMC4213427/.
[20] Maron B, Gardin J, Flack J, *et al.* Prevalence of hypertrophic cardiomyopathy in a general population of young adults. Echocardiographic analysis of 4111 subjects in the CARDIA study. Coronary artery risk development in (young) adults. *Circulation*. 1995;92(4):785–789.
[21] Members WC, Gersh BJ, Maron B, *et al.* 2011 ACCF/AHA guideline for the diagnosis and treatment of hypertrophic cardiomyopathy: a report of the American College of Cardiology Foundation/American Heart Association Task

Force on Practice Guidelines. *Circulation*. 2011;124(24):e783–e831. Available from: http://circ.ahajournals.org/content/124/24/e783.
[22] Hughes SE. The pathology of hypertrophic cardiomyopathy. *Histopathology*. 2004;44(5):412–427.
[23] Al-Tubaikh JA. *Internal Medicine: An Illustrated Radiological Guide*. New York, NY: Springer Science & Business Media, 2010.
[24] 24 Maron M, Appelbaum E, Harrigan C, *et al*. Clinical profile and significance of delayed enhancement in hypertrophic cardiomyopathy. *Circulation Heart Failure*. 2008;1(3):184–191.
[25] Olivotto I, Cecchi F, Casey S, Dolara A, Traverse J, and Maron B. Impact of atrial fibrillation on the clinical course of hypertrophic cardiomyopathy, *Journal of Circulation*. 2001;104(21):2517–2524. doi:10.1016/S1062-1458(02)00656-6.
[26] Betancur J, Simon A, Halbert E, *et al*. Registration of dynamic multiview 2D ultrasound and late gadolinium enhanced images of the heart: application to hypertrophic cardiomyopathy characterization. *Medical Image Analysis*. 2015;11:28.
[27] Baum J and Duffy HS. Fibroblasts and myofibroblasts: what are we talking about? *Journal of Cardiovascular Pharmacology*. 2011;57(4):376–379. Available from: http://www.ncbi.nlm.nih.gov/pmc/articles/PMC3077448/.
[28] Krenning G, Zeisberg EM, and Kalluri R. The origin of fibroblasts and mechanism of cardiac fibrosis. *Journal of Cellular Physiology*. 2010;225(3): 631–637.
[29] Moon JCC, Reed E, Sheppard MN, *et al*. The histologic basis of late gadolinium enhancement cardiovascular magnetic resonance in hypertrophic cardiomyopathy. *Journal of the American College of Cardiology*. 2004;43(12): 2260–2264.
[30] Simonetti OP, Kim RJ, Fieno DS, *et al*. An improved MR imaging technique for the visualization of myocardial infarction. *Radiology*. 2001;218(1): 215–223.
[31] Ordovas KG and Higgins CB. Delayed contrast enhancement on MR images of myocardium: past, present, future. *Radiology*. 2011;261(2):358–374.
[32] Mewton N, Liu CY, Croisille P, *et al*. Assessment of myocardial fibrosis with cardiovascular magnetic resonance. *Journal of the American College of Cardiology*. 2011;57(8):891–903.
[33] Adabag AS, Maron BJ, Appelbaum E, *et al*. Occurrence and frequency of arrhythmias in hypertrophic cardiomyopathy in relation to delayed enhancement on cardiovascular magnetic resonance. *Journal of the American College of Cardiology*. 2008;51(14):1369–1374.
[34] Rudolph A, Abdel-Aty H, Bohl S, *et al*. Noninvasive detection of fibrosis applying contrast-enhanced cardiac magnetic resonance in different forms of left ventricular hypertrophy relation to remodeling. *Journal of the American College of Cardiology*. 2009;53(3):284–291.
[35] Kim R, Wu E, Rafael A, *et al*. The use of contrast-enhanced magnetic resonance imaging to identify reversible myocardial dysfunction. *New England Journal of Medicine*. 2000;343(20):1445–1453.

[36] Gerber BL, Garot J, Bluemke DA, *et al.* Accuracy of contrast-enhanced magnetic resonance imaging in predicting improvement of regional myocardial function in patients after acute myocardial infarction. *Circulation*. 2002;106(9):1083–1089.

[37] Amado LC, Gerber BL, Gupta SN, *et al.* Accurate and objective infarct sizing by contrast-enhanced magnetic resonance imaging in a canine myocardial infarction model. *Journal of the American College of Cardiology*. 2004;44(12):2383–2389.

[38] Schuijf JD, Kaandorp TAM, Lamb HJ, *et al.* Quantification of myocardial infarct size and transmurality by contrast-enhanced magnetic resonance imaging in men. *The American Journal of Cardiology*. 2004;94(3):284–288.

[39] Hsu LY, Natanzon A, Kellman P, *et al.* Quantitative myocardial infarction on delayed enhancement MRI. Part I: Animal validation of an automated feature analysis and combined thresholding infarct sizing algorithm. *Journal of Magnetic Resonance Imaging*. 2006;23(3):298–308.

[40] Kolipaka A, Chatzimavroudis GP, White RD, *et al.* Segmentation of non-viable myocardium in delayed enhancement magnetic resonance images. *The International Journal of Cardiovascular Imaging*. 2005;21(2-3):303–311.

[41] Karim R, Housden RJ, Balasubramaniam M, *et al.* Evaluation of current algorithms for segmentation of scar tissue from late gadolinium enhancement cardiovascular magnetic resonance of the left atrium: an open-access grand challenge. *Journal of Cardiovascular Magnetic Resonance*. 2013;15(1):105.

[42] Kim R, Fieno D, Parrish T, *et al.* Relationship of MRI delayed contrast enhancement to irreversible injury, infarct age, and contractile function. *Circulation*. 1999;100(19):1992–2002.

[43] Yan AT, Shayne AJ, Brown KA, *et al.* Characterization of the peri-infarct zone by contrast-enhanced cardiac magnetic resonance imaging is a powerful predictor of post-myocardial infarction mortality. *Circulation*. 2006;114(1):32–39.

[44] Schmidt A, Azevedo CF, Cheng A, *et al.* Infarct tissue heterogeneity by magnetic resonance imaging identifies enhanced cardiac arrhythmia susceptibility in patients with left ventricular dysfunction. *Circulation*. 2007;115(15):2006–2014.

[45] Oakes RS, Badger TJ, Kholmovski EG, *et al.* Detection and quantification of left atrial structural remodeling with delayed-enhancement magnetic resonance imaging in patients with atrial fibrillation. *Circulation*. 2009;119(13):1758–1767.

[46] Hennemuth A, Seeger A, Friman O, *et al.* A comprehensive approach to the analysis of contrast enhanced cardiac MR images. *IEEE Transactions on Medical Imaging*. 2008;27(11):1592–1610.

[47] Positano V, Pingitore A, Giorgetti A, *et al.* A fast and effective method to assess myocardial necrosis by means of contrast magnetic resonance imaging. *Journal of Cardiovascular Magnetic Resonance: Official Journal of the Society for Cardiovascular Magnetic Resonance*. 2005;7(2):487–494.

[48] Kachenoura N, Redheuil A, Herment A, *et al.* Robust assessment of the transmural extent of myocardial infarction in late gadolinium-enhanced MRI studies

using appropriate angular and circumferential subdivision of the myocardium. *European Radiology*. 2008;18(10):2140–2147.

[49] Detsky JS, Paul G, Dick AJ, *et al.* Reproducible classification of infarct heterogeneity using fuzzy clustering on multicontrast delayed enhancement magnetic resonance images. *IEEE Transactions on Medical Imaging*. 2009;28(10):1606–1614.

[50] Knowles BR, Caulfield D, Cooklin M, *et al.* 3-D visualization of acute RF ablation lesions using MRI for the simultaneous determination of the patterns of necrosis and edema. *IEEE Transactions on Bio-Medical Engineering*. 2010;57(6):1467–1475.

[51] Tao Q, Milles J, Zeppenfeld K, *et al.* Automated segmentation of myocardial scar in late enhancement MRI using combined intensity and spatial information. *Magnetic Resonance in Medicine: Official Journal of the Society of Magnetic Resonance in Medicine/Society of Magnetic Resonance in Medicine*. 2010;64(2):586–594.

[52] Lu Y, Yang Y, Connelly KA, *et al.* Automated quantification of myocardial infarction using graph cuts on contrast delayed enhanced magnetic resonance images. *Quantitative Imaging in Medicine and Surgery*. 2012;2(2):81–86.

[53] Małek ŁA, Werys K, Kłopotowski M, *et al.* Native T1-mapping for non-contrast assessment of myocardial fibrosis in patients with hypertrophic cardiomyopathy – comparison with late enhancement quantification. *Magnetic Resonance Imaging*. 2015;33(6):718–724. Available from: http://www.sciencedirect.com/science/article/pii/S0730725X15000995.

[54] Spiewak M, Malek LA, Misko J, *et al.* Comparison of different quantification methods of late gadolinium enhancement in patients with hypertrophic cardiomyopathy. *European Journal of Radiology*. 2010;74(3):e149–153.

[55] Baron N, Kachenoura N, Cluzel P, *et al.* Comparison of various methods for quantitative evaluation of myocardial infarct volume from magnetic resonance delayed enhancement data. *International Journal of Cardiology*. 2013;167(3):739–744.

[56] Peters DC, Wylie JV, Hauser TH, *et al.* Detection of pulmonary vein and left atrial scar after catheter ablation with three-dimensional navigator-gated delayed enhancement MR imaging: initial experience. *Radiology*. 2007;243(3):690–695.

[57] Sprechmann P and Sapiro G. Dictionary learning and sparse coding for unsupervised clustering. In: *2010 IEEE International Conference on Acoustics Speech and Signal Processing (ICASSP)*, 2010. p. 2042–2045.

[58] Mantilla J, Paredes JL, Bellanger JJ, *et al.* Detection of fibrosis in LGE-cardiac MRI using Kernel DL-based clustering. In: *2015 Computing in Cardiology Conference (CinC)*, 2015, in press. p. 357–360. doi:10.1109/CIC.2015.7408660.

[59] Shakhnarovich G. Learning task-specific similarity [Thesis]. Massachusetts Institute of Technology, 2005.

[60] Lu Y, Radau P, Connelly K, *et al.* Pattern recognition of abnormal left ventricle wall motion in cardiac MR. In: *Proceedings of the 12th International*

Conference on Medical Image Computing and Computer-Assisted Intervention: Part II. MICCAI '09. Berlin, Heidelberg: Springer-Verlag, 2009. p. 750–758.

[61] Misiti M, Misiti Y, Oppenheim G, *et al*. Clustering signals using wavelets. In: Sandoval F, Prieto A, Cabestany J, *et al*., editors, *Computational and Ambient Intelligence. No. 4507 in Lecture Notes in Computer Science*. Berlin: Springer Berlin Heidelberg, 2007. p. 514–521.

[62] O'Hanlon R, Grasso A, Roughton M, *et al*. Prognostic significance of myocardial fibrosis in hypertrophic cardiomyopathy. *Journal of the American College of Cardiology*. 2010;56(11):867–874.

[63] Hein S, Arnon E, Kostin S, *et al*. Progression from compensated hypertrophy to failure in the pressure-overloaded human heart: structural deterioration and compensatory mechanisms. *Circulation*. 2003;107(7):984–991.

[64] Noureldin RA, Liu S, Nacif MS, *et al*. The diagnosis of hypertrophic cardiomyopathy by cardiovascular magnetic resonance. *Journal of Cardiovascular Magnetic Resonance*. 2012;14(1):17. Available from: http://www.jcmr-online.com/content/14/1/17/abstract.

Chapter 9

Enhancing physical performance with machine learning

Trinh C.K. Tran[1,2], Yongtai Raymond Wang[1,2], Alexandria Remus[1], Ivan Cherh Chiet Low[2,3], Dean Ho[1] and Jason Kai Wei Lee[1,2,3]

9.1 Introduction

Artificial Intelligence (AI) and machine learning (ML) have unarguably become fashionable terms within the discourse of healthcare research in the past decade. Like many novel technologies that have come to past, the hype is not substantiated by immediate success. However, AI/ML has made steady progress in clinical research. The area of human performance is not immune to the appeal of this technology. Like other industries, the question remains unanswered as to what extent AI/ML can be practically employed to enhance physical performance outcomes.

In recent years, there is a growing interest in using ML to model physical adaptations to training and predicting athletic performances, all of which will be further discussed. In these cases, ML was applied to transform raw physiological data to advanced, interpreted information based on which athletic coaches make informed decisions. However, the challenge remains to harness the true potential of this technology to impact physical performance of individuals as well as athletes in a measurable, verifiable, and scalable manner.

There is a need to bridge the gap between data scientists and human performance experts in order to translate the enthusiasm in AI into material transformation of physical performance. This chapter endeavours to do so by presenting the tenets of physical performance in ways that are intuitive for data scientists, such that they can find footholds to employ ML applications on problems that can enable improvement in physical performance. Vice versa, by suggesting steps to apply ML in physical performance, the chapter hopes to lower the barrier for sports and exercise experts to integrate data-centric approaches to training. Ultimately, the aim is not only to aid in

[1]The N.1 Institute for Health (N.1), National University of Singapore, Singapore
[2]Department of Physiology, Yong Loo Lin School of Medicine, National University of Singapore, Singapore
[3]Department of Physiology, Yong Loo Lin School of Medicine, National University of Singapore, Singapore

realising the goal of "higher, faster, and stronger" in competitive sports but also to improve quality of life through physical activity.

9.2 Physical performance and data science

9.2.1 Physical performance overview

An individual's physical performance is primarily determined by their physiological adaptations resulting from physical activity, exercise, and training. Examples of such body adaptations are the changes in endurance and maximal aerobic capacity in marathon training, and the changes in strength and muscle size in strength training. Physical performance can typically be assessed by physiological indicators and performance outcomes such as muscle size, gait characteristics, and race time. For one to achieve the highest performance outcomes, many factors need to be incorporated into the design of the exercise training such as intensity and duration of training, nutrition, and recovery [1–3].

Physical activity is defined on broad terms as "all body movements that increase energy use beyond resting levels" [4], which can occur spontaneously in leisure, work and transport or in an organised setting with targeted designs and outcomes. In general, there are two main forms of physical activity: exercise and training. Exercise is aimed primarily at improving health and physical capacity; while training is aimed primarily at increasing the individual's maximum physical capacity and performance [5].

Physical activity is highly recommended as it provides positive health benefits in physiological and psychosocial development, and one such benefit is the protection against aging-related declines in physical functions [3,6]. In addition, sporting participation as a youth is associated with greater knowledge of nutrition, exercise, and health aspects which in turn improves long-term life quality [4,7].

9.2.2 The role of data in physical performance

The yardstick for pushing the bounds of physical performance is constantly moving due to the advent of competitive sports. The resources put into competitive sports has given rise to an increasingly comprehensive physiological understanding of the human body and its response to training, rest, nutrition, and external factors, resulting in an insurmountable amount of data, as measured from devices such as wearable devices, video analysis systems, tracking systems, and questionnaires. While monitoring these numerous variables and their fluctuations can be arduous for humans, computers have proven to excel at these tasks. As such, with the growing importance of data science, there is also a growing demand for ML to manage, process and interpret these accumulating physiological data to further enhance and push the limits of human physical performance. Among the range of capabilities of data science, three applications are particularly relevant to enhancing physical performance: (i) precision, (ii) personalisation, and (iii) prediction.

9.2.2.1 The need for precision

As sports coaches and athletes pursue higher physical performance, they require a high level of precision monitoring of advanced physiological parameters. Essential parameters, such as level of exertion, hydration status, thermal discomfort, among others, are traditionally monitored by subjective questionnaires and ratings; but in recent years have been replaced or accompanied by data collected through wearable technologies with higher precision. Unlike subjective ratings, data from wearables are objective and not influenced by perception. To optimise the precision of these data, Internet-of-Things (IoT) systems, and ML algorithms can be used to account for the complex factors influencing human physiological responses [8]. As such, this higher precision data obtained enables coaches and athletes to better monitor their performance status and adaptive responses, resulting in safer, and more effective training regimes.

9.2.2.2 The need for personalisation

Personalisation in physical training is also needed for not only athletes but also the general population. One strong motivation for personalised programmes is to account for the high interpersonal variation in physiological responses due to genetic makeup [3,6,9]. A second motivation is to account for the dissimilarity in risk factors that expose an individual to injury in sports and physical activity participation. These risk factors, such as prior injuries, experience, skill level, and psychological status, should be taken into account to mitigate the risk of injury [4,10–12]. Another compelling motivation for personalisation is to account for the variation in lifestyles and time constraints. In general, the vast majority of the global population do not meet the recommended amount of exercise to maintain fitness due to lifestyle-related reasons [3]. For example, only half of the adult U.S. population engages in the recommended 150 min of Moderate and Vigorous Physical Activity per week, and less than 30% meet the guideline for strength training of at least 2 days per week [13]. Therefore, customisation of training dosage based on personal needs, goals, and availability is critically important to ensure high compliance and effectives training [3,6].

Although crucial, personalising physical training necessitates an extensive collection of historical data and ongoing monitoring and modification of physical routines, which presents a significant opportunity for leveraging AI applications. For example, in the context of strength training, AI can learn each individual's adaptations to training, and hence find the most efficient way to induce a desired response. For example, there are established algorithms/formulae that can be used to inform strength training programmes that are personalised relative to an individual's historical performance (e.g., Prilepin's table from a one repetition maximum); however, these methods typically do not factor in any additional physiological data or perceived performance metrics. With the multitude of factors that influence the prescription of exercise, ML stands out as a prime tool to facilitate training personalisation given its ability to incorporate and interpret these data effectively. One commercial example of how ML is being used to prescribe exercise is FitnessAI (https://www.fitnessai.com/). FitnessAI is an AI-driven personal trainer that integrates workout history with data from Apple's Health Kit and Apple watch to prescribe personalised training sessions to reach their fitness goals. While sports athletes generally have the ability to receive personalised coaching, this personalised training opportunity is largely inaccessible to the general population, so such applications of ML are invaluable to personalise physical activity for a population that typically would receive one-size-fits-all prescription.

9.2.2.3 The need for prediction

Prediction plays a vital role in enhancing physical performance. For instance, precise forecasting of athletic performance and injury risks not only benefits athletes but also informs the development of sports programs [12,14]. Being aware of an anticipated unfavorable outcome, such as injury or subpar performance, provides an opportunity for coaches and athletes to take proactive measures to increase the likelihood of success. Other examples of useful predictions in physical performance include, but are not limited to:

- Prediction of sporting performance
- Prediction of injury breakpoint
- Prediction of recovery duration

AI and ML offer the potential to generate accurate predictions by analyzing multifaceted and intricate data. In athletic performance, univariate historical data, such as training dose, serves as a foundation for projecting future outcomes. However, it is insufficient alone to make precise predictions due to the complex relationships among performance factors (e.g., training load and physiological adaptations) and external factors (e.g., environmental factors, nutritional status, psychological status). Thus, ML can be an invaluable tool for integrating these variables to provide more accurate predictions [8, 15-17]. As a result, predictive ML applications for physical performance outcomes are rapidly emerging and have the potential to significantly enhance the field by guiding training decisions, predicting adverse effects of training, and even project recovery trajectory.

9.2.3 Why ML?

9.2.3.1 The technological advancement of AI/ML

AI can be defined as defined as a system's ability to correctly interpret external data, to learn from such data, and to use those learnings to achieve specific goals and tasks through flexible adaptation [18]. Since the 1960s, AI has gone through a tremendous period of development. Generally, the technological development of AI can be broadly categorised into three stages: Artificial Narrow Intelligence (ANI), Artificial General Intelligence (AGI), and Artificial Super Intelligence (ASI) [19]. Figure 9.1 demonstrates these AI stages and their respective physical performance application.

1. ANI: AI is applied to one specific problem and can outperform/equal humans in solving that problem: prediction of physiological parameters, personalised recommendation of exercise programme.

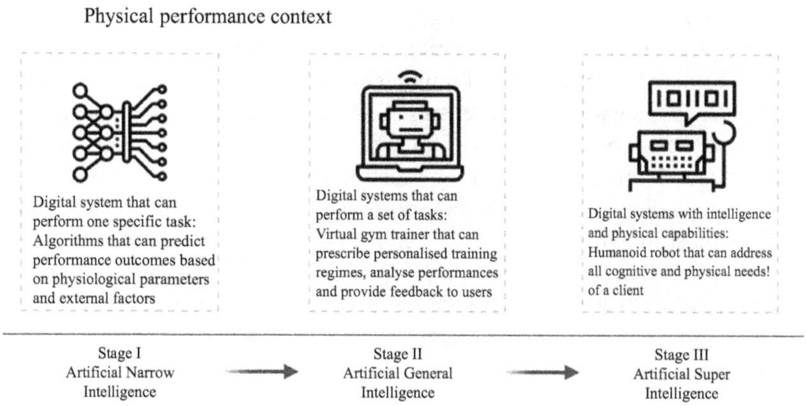

Physical performance context

Digital system that can perform one specific task: Algorithms that can predict performance outcomes based on physiological parameters and external factors

Digital systems that can perform a set of tasks: Virtual gym trainer that can prescribe personalised training regimes, analyse performances and provide feedback to users

Digital systems with intelligence and physical capabilities: Humanoid robot that can address all cognitive and physical needs! of a client

Stage I Artificial Narrow Intelligence → Stage II Artificial General Intelligence → Stage III Artificial Super Intelligence

Figure 9.1 Examples of three stages of AI development with contextualised physical performance examples. Adapted from [19, Table 1]

2. AGI: AI is applied to several areas and can outperform/equal humans in solving those problems: virtual gym trainer that can prescribe personalised training regimes, analyse performances and provide feedback.
3. ASI: AI is applied in any area and can solve any problems instantaneously with a better performance than humans: a humanoid robot that can replace a human physical trainer and address all questions and needs of a client, including emotional concerns.

Today, we are only in the first stage of AI. While all applications are classified as ANI, they are near ubiquitous to the modern world [19]. On the contrary, the ASI stage, which might be expressed as all-knowing gym robot trainer that replaces real human trainers, seems like a distant future. ANI often exists under the form of a ML model designed to perform a specific task with high accuracy. It is not too farfetched to imagine that these ML models that each solves a specific task would eventually be integrated into an ultra-sophisticated system, completed with robotics and virtual reality, to fulfil the expectations of AGI and ASI systems. But for now, the following sections of this chapter will focus on discussing ANI-level ML systems which have been applied widely across healthcare and other industries [8,18].

9.2.3.2 ML in physical performance

ML is best conceptualised as a subset of AI that uses computational algorithms with the aim of providing new insights on relationships between variables [18–20]. The 'learning' in its name refers to the process by which a computer system uses data to train itself to make better decisions through a feedback loop. The basis of ML is data science, statistics, and mathematics. Because of its ability to process and identify interdependency in large clusters of data, ML can be leaned on to analytically conceptualise non-linear dynamics of physiological responses and thus make predictive and classification calculations [12,18,20].

In recent times, the tool has emerged as a strategical area to exploit knowledge in exercise physiology to provide solutions in physical performance domain including, but not limited to, prediction of training outcomes, risks of injury, and recovery rates. As such, there have been precedents of ML algorithms being applied in various exercise and sports contexts [12,14,17,21,22]. Such applications include injury prediction in CrossFit training using survey-based epidemiological data [12] future sports performance prediction from historical sports achievement data [14], the identification of the most important injury risk factors and high injury risk athlete detection [17], early prediction of performance to support staff in play substitution strategies [15], and the classification and evaluation of various exercises [23–25]. Therefore, it is indisputable that ML has significant potential for further application in sports medicine and physical performance research.

9.2.3.3 Data from wearables and IoT

'What gets measured gets managed' a cliché quote with a lost origin. The axiom speaks to the rising interests in managing health and fitness that follows the increase in physiological data collected by wearable fitness bands such as Fitbit, Whoop, Apple watch, just to name a few. The advent and constant improvement of microchips gave rise to a myriad of wearable biosensors that are attached to human's skin or fabrics to monitor live physiological parameters and give users feedback on their physical status.

As wearables become increasingly common all around the world, the amount of data that are constantly generated by wearables is enormous. These data are often available for individuals and their coaches to quantify performance and physical fitness. In addition, IoT is an ecosystem of data communications that draw in external information such as environmental conditions and connects the wearables to the cloud with computing power to process data at extremely high speed. The combination of wearables and IoT presents sports and data scientists with the opportunity to better understand the physical performance outcomes and implications.

Digitalisation allows for record, transfer, study, cross-validation of physiological data for the first time in the history. This also opens the possibility of using mathematical methods to form indices derived from one or more parameters to represent complex physiological responses to training and exercise. One of such examples is the Physiological Strain Index (PSI) which presents individual thermal strain. In exercise physiology, the body's ability to regulate heat, which comes both from external environment and as a by-product of internal systems when exercising, is crucial to endurance, performance factors, and safety. PSI is an index to present this individual thermal strain, and its equations incorporate simultaneous measurements of the cardiovascular and thermoregulatory systems, presented by Heart Rate and Core Body Temperature, respectively, to monitor an individual's strain experienced from heat exposure and physical activity [26,27]. This simple PSI overcomes limitations of other indexes and allows for the evaluation of heat stress either in real-time online or later when data analysis is applied.

Wearable and sensor technology combined with big data analytics is revolutionising the way sports are played, analysed and enhanced in today's connected world [28]. Although the nature of competitive sports resides primarily in the sporting skills of athletes, combining the efforts of AI technology will result in a qualitative leap in thinking mode, training method, and evaluation means.

9.2.3.4 Growth in data appreciation in physical training and exercise

As more people have access to their physiological data through off-the-shelf wearable devices, there is an increasing awareness among the general population in using physiological parameters to quantitatively determine the effectiveness of their exercise and the long-term trend of their physical health. This is evidenced by the rapidly growing market for wearable technologies in which approximately 600 million wearable users in 2020 are predicted to expand to 1100 million users by the end of 2022 [29]. As a result, there is also a demand in understanding the explicit impact training and exercise have on their bodies. This can translate to positive motivation, especially for training where changes come slow and unnoticeable such as muscle growth in strength resistance training [3]. Sports and exercise experts are also becoming more cognizant of the utility of data. This fuels data-driven research and application is sports. The advantage conferred over traditional statistical methodologies is that this process allows interdependency of data points and facilitates the detection of previously hidden clusters, revealing more information about multivariable dynamics of physiological adaptations to training [8].

9.2.3.5 Developing ML solutions for a physical performance problem

Although there is no hard and fast rule, the development of ML generally follows the following pipeline, as depicted in Figure 9.2 [8,12,18,20]:

1. Define target: e.g. a physiological parameter
2. Data capture/pre-processing: *Capture of an adequate dataset of sufficient sample size and remove any erroneous, noise, or incomplete data.*
3. Feature engineering: *The generation/extraction and selection of suitable features for the model.*
4. ML modelling: *Apply a suitable model for the dataset and intended goals.*
5. Evaluation: *Common performance measurements of ML models are Mean Absolute Error (MAE), Root Mean Square Error (RMSE), Confidence Intervals (CI), accuracy, precision, confusion matrix, Area Under the Curve- Receiver Operation Characteristic (AUC-ROC), among others.*
6. Parameter tuning: *Optimise model by tuning hyperparameters of the given model and improve feature engineering.*

We will discuss the targeted outcomes of physical performance and suitable ML methods for these types of problems.

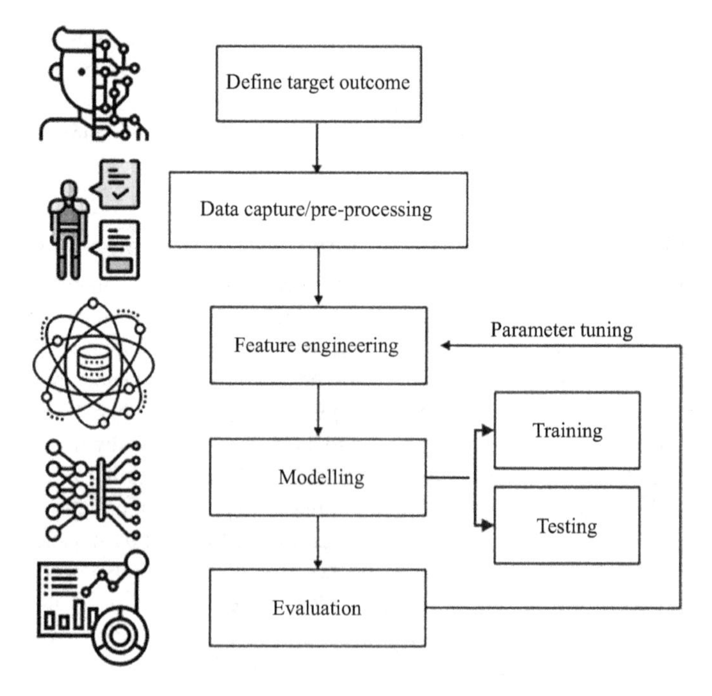

Figure 9.2 ML model development pipeline

9.3 Contextualise physical performance factors: ML perspectives

To dive deeper into how ML can be applied to physical performance, it is helpful to break down physical performance into its three foundational components: (i) Training, (ii) Nutrition, and (iii) Sleep and Recovery.

9.3.1 Training

9.3.1.1 Physical training: definition and determinants

Physical training can have two primary purposes: improving aerobic capacity and muscle-strengthening. Aerobic physical activity is the type of activity typically associated with stamina, fitness, and the biggest health benefits [29,30,31]. Muscle-strengthening physical activity, often referred to as "strength training" or "resistance training," is primarily intended to maintain or improve various forms of muscle strength and increase or maintain muscle mass [32].

Adaption to physical activity and training is a complex physiological process, but in the context of this paper, it can be simplified by a fundamental basic principle which assumes that physical activity disturbs the body's physiological balance, which the body then seeks to restore, all in a dose-related response relationship [4].

The overload principle states that if exercise intensity is too low, overload is not reached to induce desired physiological adaptations, whereas an intensity too high will result in premature fatigue and possibly overtraining. Thus, for adaptation to occur, greater than normal stress must be induced, interspersed with sufficient recovery periods for restoration of physiological balance [30].

9.3.1.2 Training indicators for ML

The degree of adaptation depends on many factors, including age, heredity, the environment, and diet [9,31]. The hereditary factor, genetics, may be the most critical for adaptation [9]. The degree of adaptation also depends on how the person in question trained previously; a well-trained athlete usually does not have the same relative improvement as an untrained individual. General recommendations for health may be stated, but individual predispositions make general training schedules for specific performance effects unpredictable. Effective exercise training should be individualised based on the purpose, goals, and circumstances [4]. Below we provide examples of key training indicators for both aerobic endurance and strength performance for consideration in ML applications.

- Aerobic endurance
  - Endurance performance indicators
    - VO_{2max}: *measure of the maximum amount of oxygen your body can utilise during exercise*
    - Running economy: *steady-state oxygen consumption (VO_2) during running at a submaximal intensity*
  - Anthropometric indications [32,33]
    - Height
    - Weight
    - Skinfold thickness
  - Training indications
    - Frequency of training
    - Intensity of training: e.g., *%VO_2, heart rate reserve*
    - Time/duration
    - Perceived efforts: e.g., *Borg's ratings of perceived exertion*
- Strength
    - Skeletal muscle indicators [31]
      - Size
      - Structure
      - Composition
      - Metabolic capacity
      - Contractile indices
      - Blood-based biomarkers of muscle status: e.g. *endocrine regulation of muscle adaptations, metabolic homeostasis such as anabolic–catabolic balance, protein and amino acid deficiencies, substrate availability* [34]
    - Anthropometric indications [35]
      - Height
      - Weight

    - Body composition: e.g., *regional lean masses, fat mass, free fat mass, arm, leg, and trunk length, and shoulder width*
- Training indications
    - Frequency of training
    - Intensity of training: e.g., *%VO_2, heart rate reserve*
    - Time/duration
    - Perceived efforts: e.g., *Borg's ratings of perceived exertion*

9.3.1.3 Examples of ML applications

The first seminal work in modelling training is the Fitness-Fatigue model (FFM) [36,37]. The FFM was developed based on a systems theory approach to model the relationship between fitness and fatigue – the two key outcomes of a training stimulus [37,38]. This relationship was originally described by a set of first-order differential equations and further simplified into the FFM depicted in Figure 9.3.

In this mathematical framework, the modelled performance outcome is presumed to be resultant of two components, fitness and fatigue. These two, thus, can be manipulated to induce the desired performance outcome. As helpful as the conceptual model has been in guiding basic understanding of physical responses to training, it has several limitations. In the current mathematical form, the model is commonly described as a linear time-invariant system. That is, the constants of basic level of performance, or magnitudes of fitness gains and fatigue do not change over time. However, physical adaptations do change over time, as they are indeed dependent on the accumulation of training [36,39]. In addition, physical performance training is multi-factorial that goes beyond linear gains and losses. Modern exercise physiology has brought attention to alternative factors that contribute to training outcomes, such as environmental factors, nutritional, and psychological status [16,39]. To this goal, Imbach *et al.* have introduced a multivariate framework by combining the conceptual FFM framework with ML models for predicting physical adaptation [16]. Hence, the use of ML in enhancing physical performance can be seen as an extension of this FFM model, in order to incorporate a complex framework that is aligned with the development of exercise physiology.

In a broader view of ML applications in physical training, some notable efforts include the prediction of ratings of perceived exertion (RPE) and Global Positioning System (GPS) training load variables using generalised estimating equations (GEEs) and Artificial Neural Network (ANN) analyses in Football players [40]. This was one of the first studies to assess the viability of ML and team sports, popularising the

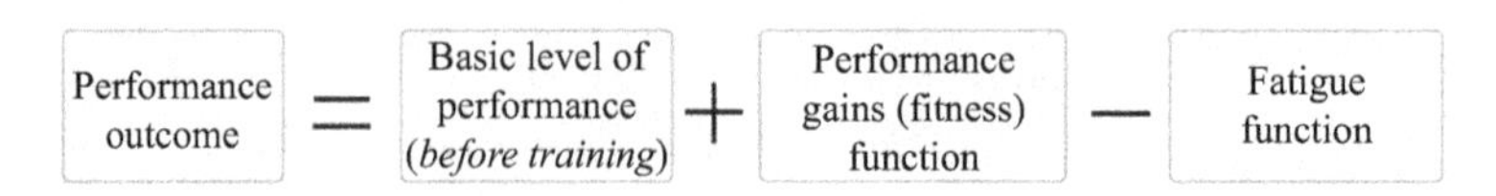

Figure 9.3 Simplified fitness-fatigue model to explain the performance outcome as a function of baseline level of performance and a positive change from performance games and negative change due to fatigue. Adapted from [36, equation 8]

application of ML for sports prediction. Based on 2,711 training sessions (using both pooled group and individual athlete data), the study showed that the distance covered during each session is most strongly associated with RPE. Further, ANN was able to take into consideration the between- and within-individual non-linearity inherent in the datasets, yielding a prediction error of 1.24 ± 0.41 (root mean square error [RMSE] ± standard deviation [SD]) for individualised ANN. In comparison to GEEs, with a prediction error of 1.58 ± 0.41.

Extending the previous study findings, Carey *et al.* enhanced the accuracy and predictive ability of the models. The study developed various models such as linear regression, multivariate adaptive regression splines, random forests, support vector machine (SVM), and neural networks [41]. These models were used to predict RPE from training load variables, such as GPS, HR monitors, accelerometers, and data obtained from wellness questionnaires. A more accurate, yet simpler model was developed using random forest (1.09 ± 0.05). Together with other similar studies [42,43], these findings suggest the possibility of non-linear ML to accurately predict athlete's RPE.

9.3.2 Nutrition

9.3.2.1 Nutrition: definition, significance, and determinants

Besides training, nutrition is a significant determinant of physical performance. Specific nutritional deficiencies are common in athletes such as vitamin D and iron, for which studies have reported deficiency rates in most athletes in both genders [44,45]. Moreover, less common nutritional deficiencies in nutrients such as folate, vitamin B12, or magnesium may result in reduced endurance work performance and muscle function.

Supplements can have direct effects on exercise performance and endurance activities. Research efforts have pointed to possible ergogenic effects of sport drinks and caffeine on physical performance [46–48]. The extent of the effect they can impose depends on a host of factors such as the substance, daily habits, physiological factors, and genetic factors [46–48]. A better understanding and control of these variables should be considered in future research into personalised nutritional strategies.

A crucial component of nutrition is water. Water is the most essential nutrient of the body undergoing continuous recycling, functioning as a solvent, and regulating cell volume, while playing a critical role in thermoregulatory and overall function. Water balance is mainly regulated by thirst and the antidiuretic hormone, known also as vasopressin, through its renal effect [49,50]. Acute decrease in body weight has been used as the gold standard to evaluate the degree of dehydration because it reflects mainly a decrease in total body water and not energy substrates (e.g., fat, protein) [49]. A hypo-hydrated state of greater than −2% body mass has been linked to decreases in exercise/sport performance, cognitive function, mood, and increases in risk of exertional heat illness or exertional heatstroke for individuals exercising in hot and humid environments [51].

During exercise, especially in the heat, most people tend to drink less than what they lose through sweating, resulting in involuntary dehydration. Because sweat is hypotonic, exercise-induced dehydration leads mainly to a decrement of the extracellular fluid volume. This state is described as hypertonic hypovolemia due to water loss from the plasma. Because of this, blood biomarkers of haemoconcentration have thus been used as an index of dehydration [34,52].

Individual nutritional needs depend largely on sport- and training-specific bioenergetic demands as well as on an athlete's metabolic tolerance, needs, and preferences. Frequent monitoring of macronutrient and micronutrient intake may help identify individual deficiencies and track changes, especially as training volume and nutritional demands increase [34,52].

9.3.2.2 Nutritional indicators for ML

Markers of nitrogen balance (e.g., urea nitrogen) are important for assessing the nutritional status of an athlete, while specific amino acids can reveal information about protein synthesis, nutrition, and fatigue [34]. As with all physiological parameters, nutritional indicators need to be contextualised with each individual's habitual diet in a dynamic fashion. In other words, absolute values for certain parameters may not direct action for a given athlete, but changes with training that coincide with reduced capacity to recover, and decreased performance should be monitored on an individual basis. This approach to monitoring will allow coaches and staff to better monitor groups of highly variable athletes who will inevitably have highly different diets and other behaviours that affect performance [34]. Moreover, red blood cell indices may provide early indications of nutritional deficiencies. For example, haematocrit, haemoglobin, and red blood cell indices may suggest iron, vitamin B12, or folate deficiency [53].

Researchers have studied indicators including urine, saliva, sweat, and even tears as possible biological samples which is used to measure hydration state. In addition, the hydration needs of an athlete is influenced by many factors, such as height, weight, body composition, genetic predisposition and metabolic rate, level of conditioning, exercise intensity and duration, environmental conditions, clothing worn, and heat acclimation [54–56].

Monitoring or detection tools have been developed over the years to assess the hydration levels in the body, ensuring measures to be taken in case of dehydration or over-hydration. Recent studies demonstrate interest in monitoring hydration level, with most relying on manual entry by the user using a mobile app [57]. Others have introduced bottle mount recorders to log the volume of water consumed [58,59]. However, these methods mentioned were simply a log of water intake, instead of assessing the actual hydration level in the body. Few have attempted to assess the actual hydration level in the human body due to the challenges in accurately measuring the proportion of water in the various tissue compartments of the body. The complexities associated with the regulation of body fluid and variations in tissue morphology, density, and distribution, pose a challenge in defining a superior method to determining the accurate hydration levels during daily activities [60,61].

9.3.2.3 Examples of ML applications

Few studies have examined the usage of AI and ML on macro- and micro-nutrient monitoring in health and fitness performance. A review by Sak and Suchodolska demonstrated the use of AI through means of ANN methodology on food composition and production of nutrients in food science research [62]. In a similar scope, ML algorithms were widely used to examine the influence of nutrients on the functioning of the human body in health and disease and in studies on gut microbiota. Deep learning algorithms prevailed in a group of research works on clinical nutrients intake. Sundaravadivel *et al.* proposed an IoT based fully automated diet monitoring solution using Bayesian algorithms and 5-layer perceptron neural network method for diet monitoring. The Bayesian network was used to determine nutrients features from food materials and for suggesting future meals or recipes, while the 5-layer deep learning model (perceptron neural network) was used to determine the nutritional balance after each meal is proposed. This prototype system registers the nutritional values of food intake autonomously for an individual, acting as a personal food log, potentially assisting individuals and nutritionists to track nutrition consumption effortlessly [63]. Although few studies have been conducted to examine the feasibility of ML algorithm on personalising macro-, and micro-nutrient intake, findings thus far have shown the potential in developing dietary systems using AI technology. This could lead to the creation of a global network that will be able to both actively support and monitor the personalised supply of nutrients [62].

ML methods have been suggested to apply to analyse and predict nutrition and hydration levels in conjunction with bioelectrical impedance techniques, galvanic coupled signal and microfluidic wearables. Asogwa *et al.* reported that bioelectrical impedance analysis (BIA) serves as a useful method to assess fluid level in the human body by applying the relationship between the volume of a conductor, and the length of the conductor, as well as the impedance of the conductor, and the material it is made of, soft tissues [57]. Similarly, studies have used a galvanic coupled signal propagation method to assess fluid changes on a particular area of the body [64]. Other commercial hydration sensors have been shown to allow direct conversion of measured skin impedance, or chemically sensed parameters to assess hydration levels [65].

In line with the hypothesis on fluid regulatory hormones and its influence on cardiovascular system physiology, another study examined the feasibility of predicting hydration status using electrocardiographic signals and various heart rate responses [66]. Specifically, the study aimed to evaluate the classification capability of time domain heart rate variability parameters using SVM algorithm and K-means, in the detection of dehydration. Participants were subjected to a protocol of dehydration, and alternations in electrocardiographic signal during rest, post-exercise and post-hydration were measured. The study indicated that the RR-interval obtained with the SVM method had the highest accuracy (a confusion matrix score of 0.6) in comparison to the other physiological parameters and K-means method. The viability of the use of RR interval to differentiate the states of the dehydration process can be argued from a physiological perspective. In the process of dehydration, the cardiac muscles respond via increase in heart rate to compensate the cardiac output decrease.

Others have attempted to validate this hypothesis further by studying the autonomic nervous system (ANS) response and electrocardiographic signals [67]. The study found that subspace ensemble of K-nearest neighbour classifiers and cubic SVM yields an accuracy of 91.2%. In contrast to the previous study [66], the differences in outcome could be attributed to the extent of dehydration. Posada-Quintero *et al.* performed the experiments with participants undergoing non-dehydrated and mild dehydration (via fluid restriction) conditions, while the former [66] had participants undergoing 37 min of cycling routine. Further differences are also accounted by the different indices used by Posada-Quinterol *et al.*, where ANS response was evaluated based on electrodermal activity and pulse rate variability. Nevertheless, the findings highlight the potential of examining the interaction of ANS and electrocardiographic signal response and dehydration through ML

Besides being utilised to monitor nutritional consumption and hydration levels, ML has also been used as an analytical tool to understand compensatory hydration habits in sports. Suppiah *et al.* employed K-means cluster analysis to identify unique subgroups of drinking characteristics amongst participants, namely through fluid consumption, sweat rate, and percentage body mass change [50]. The clustering analysis found notable hydration characteristics between sports, gender, and training intensity[50]. This effort has not only introduced a novel method for hydration management and interventional strategy but also the use of ML analytical capabilities in performance research.

9.3.3 Sleep and recovery

9.3.3.1 Sleep and recovery: definition, significance, and determinants

Recovery is essential to physical performance as it activates the mechanisms required for physical adaptations to training. Factors that are indicative of recovery are sleep, resting heart rate, appetite, estimated fatigue, and they interact with each other to influence optimal performance and recovery outcomes [68,69]. Given the effect of sleep deprivation on performance, mood, cognitive function, memory, learning, metabolism, illness and injury, sleep assessment in athletes has become increasingly popular over the years [70]. Numerous international consensus statements, position stands, and guidelines for managing and treating various sleep disorders are available, however, little is known on the measurement, analysis, feedback provision, and intervention for athletes with no clinical sleep disorder, and where monitoring is utilised as a tool in optimising sporting performance [71–73].

9.3.3.2 Sleep and recovery indicators for ML

There are well-validated markers related to sleep, fatigue, recovery, protein synthesis, or fuelling strategies, which are all major athlete concerns. Wearable commercial sleep technology (CSTs), such as Fitbit and Whoop, assesses the duration of sleep and wake over a sustained period. To track sleep quality, these devices tend to take in total sleep time and sleep efficiency as inputs [74]. These devices can be used at the comfort of the individual's usual environment and are relatively inexpensive, with additional sensors (e.g., oximetry, temperature) attached [75]. Though CSTs are widely available, algorithms utilised by these devices are proprietary and few published findings are available regarding sleep and wake detection [74]. It is worth noting that when compared these sleep indicators to Polysomnography (PSG), which is the gold standard for measuring sleep, there are deviations. The correlation between the two methods deteriorates further in the light and deep sleep calculation, detection of sleep stages and in individuals with short sleep times and disturbed sleep [74,76]. PSG typically consists of the assessment of eye movement, brain activity, heart rate, muscle activity, oxygen saturation, breathing rate, and body movement [77]. Therefore, data derived from wearables and algorithmic methods holds a promise in making sleep monitoring more scalable when correlation with PSG data is further enhanced.

Molecular biomarkers are also potential features for ML models aiming to evaluate recovery. After muscle-damaging exercise, the enzyme creatine kinase (CK) leaks from the muscle into the circulation [78,79]. Monitoring CK levels during training in comparison with baseline levels may help athletes to monitor muscle status. Creatine kinase levels peak approximately 24 h after damaging exercise such as heavy strength training but may remain elevated up to 7 days after exercise. Chronically elevated CK may indicate insufficient recovery. Because other components of muscle such as myoglobin may leak into circulation during muscle damage (peak 1–3 h after exercise), and urea nitrogen can indicate overall protein synthesis as compared to breakdown, using all three markers to determine an athlete's muscle status during training and recovery will be useful to athletes, coaches, and clinicians [34,80].

9.3.3.3 Examples of ML applications

Over the years, ML algorithm has demonstrated a huge potential to understand responses in sleep, and in particular sleep stage classification has been a focal point [81]. Typically, sleep staging requires manual inspection of electroencephalogram, electrooculogram, and electromyogram in 30-s epochs. As the procedure is labour intensive, AI has been harnessed to classify sleep stages in PSG data and to automate the PSG scoring in efforts to improve reliability when coupled with manual review by sleep technologists. Early efforts in sleep staging and AI examined the human–computer algorithm agreement compared to human interrater reliability [82–86].

The key challenge in applying ML in staging sleep is the subjectivity of sleep stage recognition. In the traditional sleep staging, different sleep technologist experts might not always arrive at the same result [81]. Recognising this, earlier ML efforts have explored statistical classifiers and supervised learning ML models are with human-defined labels. More recently, advanced neural network models have been applied to classify sleep stage based on eye and facial tracking [87]. The results of the study suggested ML could be used to distinguish levels of wakefulness in individuals, which could be beneficial in various safety outcomes in extreme sports and occupational health [87]. Other efforts in predicting sleep stage employed heart rate parameters, such as heart rate variability (HRV), and motion from triaxial accelerometery in multi-class Bayesian Linear Discriminant model with promising results in sleep–wake differentiation and sleep stage classification [88]. This model has the potential to scale widely as it employed data that are easy to capture accurately using wearable watches.

There have also been numerous examples of robust performance of ML algorithms across large, heterogenous patient populations [89–91]. The accuracy of these models to determine arousals and sleep stage transitions may have been enhanced by additional signal outputs not typically used to stage sleep, such as heart rate variability and respiratory dynamics [92]. The direction that ML studies is taking in this field may not only redefine sleep classification methods but also unearth new associations between complex signal patterns and sleep outcomes.

9.4 ML modelling for physical performance problems

ML is a fast-growing field in research and industry with new methodologies being developed. A ML algorithm, also called model, is a mathematical expression that represents data in the context of a business problem with the aim of going from data to insights. In applying ML in physical performance, it is important to choose the right ML model, which can be picked from the comprehensive of models available in Scikit, depending on the problems.

9.4.1 Choosing ML models for the right physical performance tasks

There are three recurring tasks which exist throughout the tenets of physical performance, representing the three primary challenges in bringing insights to physical performance data:

- Monitoring
- Individualisation
- Performance analysis
- Achievement prediction

9.4.1.1 Monitoring tasks

There are traditional physiological parameters of interests for exercise physiologists, such as VO_{2max}, hydration level and HRV which are expensive, invasive and difficult to monitor accurately. ML techniques present a potential to derive to those advanced parameters using data attained from minimally invasive and non-invasive parameters [52]. In comparison, prediction models using ML require less time and effort, and reducing health risks associated with exercise testing. Algorithmic approaches that can report individuals' physical, psychological, and physiological readiness and ability to carry out assigned tasks successfully can play a significant role in sports and other sectors.

In VO_{2max}, as an example, various ML and statistical approaches have been used in combination with different predictor variables. Gender, age, BMI, HRmax, and test duration were the most commonly used predictor variables, indicating their strong predictive power for VO_{2max} [93]. Algorithms that have been used to predict VO_{2max} based on different predictor variables include but are not limited to linear regression (LR) Random Forrest (RF), SVM, multilayer perceptron (MLP) and ANN; and all of the models demonstrated the ability to predict VO_{2max} with relatively high accuracy [93].

9.4.1.2 Individualisation tasks

The ability to individualise training and exercise has a profound impact on physical performance. For one, how a person feels emotionally towards a training program can affect training outcomes [94,95]. For more practical purposes, individuals have different needs in terms of physiological stress. Different groups of population have different physical performance needs which require different modality, intensity, and frequency of exercise [3]. These differences in physical activity needs derive from physical function continuum, age, health as well as culture [96]. Evidence suggests a dose–response relationship such that being active, even to a modest level, is superior to being inactive or sedentary. The need for individualisation calls for phenotypic and historic performance data, such that complex predictions and trend analysis can be achievable through data science and ML [6,97].

ML can perform such tasks by learning users' physical adaptations to exercise and recovery, from which a prediction of future adaptations can be made. This is the principle on which CURATE.AI, an AI platform, found to generate personalised training programs based on historical dose–response data [98]. Similarly, a mobile-based application called HRV4Training aims at individualising training using HRV as a risk factor. Using a ML algorithmic approach, HRV and phenotypic characteristics are utilised as input features to predict risks of injury. From this prediction, a customised training intensity is recommended for users [99].

9.4.1.3 Performance analysis tasks

Data-centric approaches have demonstrated its superiority to human's ability in evaluating physical performance. Cust *et al.* reported the feasibility of using ML to recognise sport-specific motions, which has profound implications for training as well as performance evaluation in sport competitions [97].

In training, mastering the correct form, posture, and motions not only improve performance outcomes but also alleviates risks of musculoskeletal injuries due to overloading of joints and muscles. The need for feedback on forms is universal in sports, such as gait analysis in endurance runners, lifting motions in strength training and even crucial scoring motions in golf and shooting in basketball. The primary way professional and non-professional athletes master these sport-specific forms is through feedback from coaches and trainers. However, the general public have limited access to professional guidance. ML has proved to be a competent tool in image recognition tasks, coupled with its ability to characterise motions based on accelerometers and gyroscopic data, thus ML models can provide individuals and their trainers correct information on movements. There have also been proposals to apply in sports that need a fair, objective, and precise movement evaluation such as figure skating [21].

Clustering analysis methods such as K-means are particularly inept at identifying patterns and complex dynamic relationships in unstructured, unlabelled data [50, 97]. For performance evaluation that requires visual analysis, such as in the case of form feedback or performance scoring in figure skating, the ML field of Computer Vision is most suitable. Computer Vision employed one of many types of Neural Network algorithms to identify, classify, and interpret digital images [20].

9.4.1.4 Performance prediction tasks

The task of predicting a performance outcome is incredibly challenging. Besides having to understand the athlete's current condition, the predictive model needs to also account for external factors. Changes in various natural conditions have an important impact on sports performance. The change of environmental conditions, such as temperature, humidity, atmospheric pressure, and wind, can affect an athlete's technical level, and hence sporting performance. Therefore, sports performance is related to many factors, and various factors interact with each other, which leads to very complicated changes in sports performance.

Deep neural network (DNN) is reported to be fitting in model representation and data processing ability for performance prediction [14]. DNN can produce results in data representation through complex network structure and deep network layers. Using the hierarchical structure of the DNN can realise the nonlinear intersection of features. This model shows the mapping relationship between athletes' performance and various factors. Furthermore, it improves the efficiency of modelling and the accuracy of performance prediction [14].

In a more applied manner, Dijkhuis *et al.* attempted to make prediction of the physical performance of individual soccer players after the first 15 min of a soccer match with the goal to support tactical substitution decision making process [15]. The players' physical performance is expressed in the variables of distance covered, distance in speed category, and energy expenditure in power category. Utilising a combination of Random Forest and Decision Tree models, the ML approach reported high precision in the classification of 'performance' and 'underperformance' [15,20]. The study proved that it is possible to predict physical performance of individual players in an early phase of the match. These findings have an implication beyond the sport of soccer, for now ML can offer predictive performance of players to support coaches in tactical decision making.

9.4.2 Contributing ML features and methods

9.4.2.1 Big data

ML systems are dynamic, which means that the accuracy can improve with more data. This puts ML into an advantageous position in addressing physical performance problems as there are vast datasets constantly being generated by fitness wearables. These data allow ML systems to identify patterns in similar populations of users and thus improve the system's predictive power. The ability to record an increasing richness of information, such as epigenetic modifications, along with better quantification of present metrics, such as training load, should assist in the production of valuable predictive models in the future, which, with the application of ML, will constantly evolve to increase predictive power with the increasing amounts of data being entered into the model [6]. Similar works using ML on big data have been done, such as on cancer pathology detection.

Hundreds of millions of sports data are generated each day from millions of schools, events, and communities, representing the volume feature [100–102]. The variety in sports big data is attributed to the various entities and relationships that it draws upon. It contains various features such as (1) physical fitness, such as high, weight, physical function; (2) exercise behaviours and their behaviour trajectory; (3) personal information, such as gender, age, ethnicity; and (4) competition outcomes [103].

9.4.2.2 Data quality

Data quality is imperative in ML. If a model is trained with erroneous data, the model will not generalise and will generate false conclusions – hence the phrase 'errors in, errors out' [8,20]. The quality of any predictive model depends on the ability to have effective and informational inputs, with an emphasis on collecting reliable and valid data [6]. As much of the data comes from biosensors, there is a need to pay attention to possible noise motion artefacts. It is evident that due to these factors, wearables have performed better in bed-bound patients in clinical settings as compared to free-living users. In all ML systems, it is important for data scientists to work on pre-processing stage more before applying ML model to mitigate data artefacts [8,20].

These data pre-processing may remove outliers and reduce noise using algorithm-driven methods, and below is a non-exhaustive list of filtering methods:

- Kalman filter: *A recursive two-phase process that uses a series of observed measurements over time (inclusive of noise) to provide estimates of some unknown variables by estimating a joint probability distribution of the variables for each time point [104].*
- Adaptive filtering: *Typically a digital filter with time-varying, self-adjusting characteristics which are frequently used when the filter response is not known a priori or when the nature of the operating environment is expected to change over time. They are capable of adjusting their filter coefficients automatically to adapt the input signal via an adaptive optimisation algorithm by minimising the mean squared error between filter output and a desired signal. The most common adaptive filters are: Recursive Least Square (RLS) and Least Mean Square (LMS) [105].*
- Empirical mode decomposition: *A data-adaptive multiresolution algorithm that decomposes time-series data signals into meaningful components while remaining in the same time domain with no assumptions about the data [106].*
- Ensemble of multiples denoising methods: A combination of methods may be selected.

9.4.2.3 Feature engineering

ML has the ability to pick out patterns between physiological indicators and performance outcomes that so far have not been studied or are undefined. It is possible that indicators which, in raw forms, have little correlation with desired targets but can markedly improve in correlation after being processed mathematically or combined with other features. This process is referred to as feature engineering in data science [20]. Feature engineering includes the generation/extraction (the creation of new features from the original raw data to perform better with in the ML algorithm) and selection (where only a subset of the original data is used in the ML model) of these features. In sports science studies, a subjective selection of features (e.g., a signal's maximal value) is traditionally used [8]. However, for an optimal and robust collection of features, it is recommended that sports scientists should combine objective data-driven approach and domain-specific knowledge to select and process these features to yield the best results [6,8,20]. As such, data-driven approaches are beginning to be adopted and different processes are being used. For example, Rossi *et al.* aimed to forecast injury in soccer from GPS data by first generating a list of 50+ features and then using a Recursive Feature Elimination with Cross-Validation (RFECV) process to determine the most relevant features to be used in their model during their data engineering phase [107]. Similarly, Rodas *et al.* used the LASSO (Least Absolute Shrinkage and Selection Operator) method to select significant features for their ML model to identify genomic predictors of tendinopathy in elite athletes [108]. Carey *et al.* and Ritcher *et al.* both used principal component analysis (PCA) to reduce the dimensionality of the data for their models [109]. The reduction of features and dimensionality of the feature space in the aforementioned examples also reduces the risk of overfitting and allows for an easier interpretation of the ML model [110].

9.4.3 Challenges

There is a sense of contradiction often seen in AI research. Despite the benefits of AI that has been touted for two decades, commercialisation has not picked up as expected and that speaks to the limited impact ML and AI currently have in clinical use as well as in the health and fitness industry. Looking into these areas gives sports scientists an idea of pitfalls to avoid in incorporating AI/ML to physical performance.

9.4.3.1 Doubts and lack of verifiability

Although many ML applications mentioned in the chapter thus far reported good accuracy, the lack of cross-evaluation and scalability are the prime reasons they have not found a path to commercial success. Even in commercial devices such as Apple Watch Blood Oxygen measurement, there are caveats to exclude the device from medical uses, including self-diagnosis or consultation with a doctor, emphasising that it is only designed for general fitness and wellness purposes. This is emblematic of ML-led approach in physical performance which casts doubt on the true level of accuracy and, hence, the commercial value.

The results of ML research efforts are also difficult to verify and replicate. Due to propriety and data sharing concerns, the dataset and the algorithms used are not made public, impeding any efforts to validate these findings [8]. If datasets are shared among sport scientists, it should be noted that it is highly probable that data captured from different laboratories are likely to have different noise characteristics or have been processed using different frameworks. Therefore, a data sharing framework is needed to facilitate this process [8].

There is significant impetus for improving how ML prediction algorithms are developed and validated, including using larger, richer, and more diverse data sources, improvements in model design, and fully reporting on the development process as well as the final model [22]. Guidelines and reporting standards of implementing ML algorithms could improve the utility of studies in this regard and future studies would benefit from attempting to evaluate potential impact and clinical utility [22].

9.4.3.2 Performance and interpretability

In ML, there is a fundamental trade-off between performance and algorithm interpretability. Practitioners often settle with easier interpretable models such as simple linear, logistic regression, decision trees, because they are easier to validate and explain. Although a deep learning model may result in a better performance, models that are explainable would have a higher chance of adaption in practice [20].

This interpretability is especially demanded in health care and human potential management, where human systems are concerned [18]. But, as one is trying to apply these predictive models to solve real-world problems with access to high-dimensional heterogeneous complex data sets interpretable models often fall short in terms of performance [8]. On the other hand, the DNN methods are difficult to interpret due to the complex, multi-layered decision-making process that is integrated in the model. This group of models is also referred to as 'Black box technology' to denote the difficulty in explaining the reasons behind the predicted outcomes. Therefore, practitioners need to strike a balance between performance and interpretability [8,20].

9.4.3.3 Data bias

Although the ML model algorithms themselves are not subject to social bias, the data that are fed into the models can hold certain biases [8,20]. An example of this is the variation in performance of optical heart rate sensors across skin tones, wearable devices, and modality of physical activity [111]. In addition, gender data bias due to a lack of female representation in database has also been reported in sports that are male dominated, resulting in lower ML performance for female subjects [8].

To mitigate bias in physical performance research, transparent and complete reporting is required to allow the reader to critically assess the presence of bias, facilitate study replication, and correctly interpret study results [8,22,112].

9.5 Limitation

Development of physical performance models is not the end goal, as they are eventually intended to be applied practice. Besides accuracy of prediction performance, a ML system should also be evaluated on economic and practical values. However, most of the ML models presented in this chapter are only exploratory academic research to elucidate the feasibility of ML application in physical performance.

Commercialisation is limited due to a myriad of factors. Before deployment to practical usage, extensive external validation is required to ensure robustness of the model outside the database used for development. However, external validation and individual predictions are difficult with ML practitioners not willing to share dataset and algorithm [112]. More work can be done in deployment of ML models in physical performance.

In addition, the fast pace and large volume of ML research makes it inevitable to cover comprehensively. Readers are encouraged to read on the newest trends and development of ML-led physical performance to keep up to date.

9.6 Conclusion

AI and ML help users and health professionals better understand the person's physiological responses to exercise and physical performance. ML can improve effectiveness and efficacy of physical activity and exercise through targeting training, nutrition, and recovery. In these areas of targets, various ML approaches have emerged as potential solutions to address monitoring, individualisation, performance analysis, and prediction.

A large amount of research and commercial efforts have been published in recent years across sports and fitness industries. With the ability to process and interpret large and complex data, ML helps to form nonlinear dynamic relationships between physiological parameters, external factors, and physical outcomes. This helps inform the decision-making process in sports and fitness. Hence, it is imperative for human performance researchers to appreciate data science in order to harness the power of ML to enhance physical performance.

Challenges remain for scalability and commercialisation of ML models in physical performance. The need to create standards for ML algorithmic devices remains, by establishing gold standard in literature or comparing to similar devices in the market. Guidelines and reporting standards of implementing ML algorithms could improve the utility of studies in this regard and future studies would benefit from attempting to evaluate potential impact and clinical utility [22].

References

[1] R. C. Hickson and M. A. Rosenkoetter, "Reduced training frequencies and maintenance of increased aerobic power," *Med Sci Sports Exerc,* vol. 13, no. 1, pp. 13–16, 1981.

[2] C. S. Bickel, J. M. Cross, and M. M. Bamman, "Exercise dosing to retain resistance training adaptations in young and older adults," *Med Sci Sports Exerc,* vol. 43, no. 7, pp. 1177–1187, 2011, doi: 10.1249/MSS.0b013e318207c15d.

[3] B. A. Spiering, I. Mujika, M. A. Sharp, and S. A. Foulis, "Maintaining physical performance: the minimal dose of exercise needed to preserve endurance and strength over time," *J Strength Condit Res,* vol. 35, no. 5, pp. 1449–1458, 2021, doi: 10.1519/jsc.0000000000003964.

[4] C. Malm, J. Jakobsson, and A. Isaksson, "Physical activity and sports-real health benefits: a review with insight into the Public Health of Sweden," *Sports (Basel),* vol. 7, no. 5, p. 127, 2019, doi: 10.3390/sports7050127.

[5] L. National Heart and B. Institute, "What is physical activity," Retrieved from http://www.nhlbi.nih.gov/health/health-topics/topics/phys, 2011.

[6] C. Pickering and J. Kiely, "The development of a personalised training framework: implementation of emerging technologies for performance," *J Funct Morphol Kinesiol,* vol. 4, no. 2, p. 25, 2019. Available: https://www.mdpi.com/2411-5142/4/2/25.
[7] J. S. Brenner, "Sports specialization and intensive training in young athletes," *Pediatrics,* vol. 138, no. 3, 2016, doi: 10.1542/peds.2016-2148.
[8] C. Richter, M. O'Reilly, and E. Delahunt, "Machine learning in sports science: challenges and opportunities," *Sports Biomech,* pp. 1–7, 2021, doi: 10.1080/14763141.2021.1910334.
[9] A. C. Venezia and S. M. Roth, "Recent research in the genetics of exercise training adaptation," *Med Sport Sci,* vol. 61, pp. 29–40, 2016, doi: 10.1159/000445239.
[10] D. G. Behm and J. C. Colado Sanchez, "Instability resistance training across the exercise continuum," *Sports Health,* vol. 5, no. 6, pp. 500–503, 2013/11/01 2013, doi: 10.1177/1941738113477815.
[11] T. M. H. Eijsvogels, K. P. George, and P. D. Thompson, "Cardiovascular benefits and risks across the physical activity continuum," *Curr Opin Cardiol,* vol. 31, no. 5, pp. 566–571, 2016, doi: 10.1097/hco.0000000000000321.
[12] S. Moustakidis, A. Siouras, K. Vassis, I. Misiris, E. Papageorgiou, and D. Tsaopoulos, "Prediction of injuries in CrossFit training: a machine learning perspective," *Algorithms,* vol. 15, p. 77, 2022, doi: 10.3390/a15030077.
[13] C. Harris, K. Watson, S. Carlson, J. Fulton, J. Dorn, and L. Elam-Evans, "Adult participation in aerobic and muscle-strengthening physical activities – United States, 2011," *Morbidity Mortality Weekly Rep,* vol. 62, pp. 326–330, 2013.
[14] Q. Zhou, "Sports achievement prediction and influencing factors analysis combined with deep learning model," *Sci Program,* vol. 2022, p. 3547703, 2022, doi: 10.1155/2022/3547703.
[15] T. B. Dijkhuis, M. Kempe, and K. Lemmink, "Early prediction of physical performance in elite soccer matches – a machine learning approach to support substitutions," *Entropy (Basel),* vol. 23, no. 8, 2021, doi: 10.3390/e23080952.
[16] F. Imbach, N. Sutton-Charani, J. Montmain, R. Candau, and S. Perrey, "The use of fitness-fatigue models for sport performance modelling: conceptual issues and contributions from machine-learning," *Sports Med – Open,* vol. 8, no. 1, p. 29, 2022, doi: 10.1186/s40798-022-00426-x.
[17] H. Van Eetvelde, L. D. Mendonça, C. Ley, R. Seil, and T. Tischer, "Machine learning methods in sport injury prediction and prevention: a systematic review," *J Exp Orthopaed,* vol. 8, no. 1, p. 27, 2021, doi: 10.1186/s40634-021-00346-x.
[18] S. A. Bini, "Artificial intelligence, machine learning, deep learning, and cognitive computing: what do these terms mean and how will they impact health care?," *J Arthroplasty,* vol. 33, no. 8, pp. 2358–2361, 2018, doi: https://doi.org/10.1016/j.arth.2018.02.067.
[19] A. Kaplan and M. Haenlein, "Siri, Siri, in my hand: Who's the fairest in the land? On the interpretations, illustrations, and implications of artificial

intelligence," *Business Horizons,* vol. 62, no. 1, pp. 15–25, 2019, doi: https://doi.org/10.1016/j.bushor.2018.08.004.

[20] J. Grus, *Data Science from Scratch: First Principles with Python*. Sebastopol, CA: O'Reilly Media, 2019.

[21] C. Xu, Y. Fu, B. Zhang, Z. Chen, Y.-G. Jiang, and X. Xue, *Learning to Score the Figure Skating Sports Videos*. arXiv.

[22] W. Wang, M. Kiik, N. Peek, *et al.*, "A systematic review of machine learning models for predicting outcomes of stroke with structured data," *PLoS One,* vol. 15, no. 6, p. e0234722, 2020, doi: 10.1371/journal.pone.0234722.

[23] M. A. O'Reilly, D. F. Whelan, T. E. Ward, E. Delahunt, and B. M. Caulfield, "Classification of deadlift biomechanics with wearable inertial measurement units," *J Biomech,* vol. 58, pp. 155–161, 2017/06/14/ 2017, doi: https://doi.org/10.1016/j.jbiomech.2017.04.028.

[24] D. F. Whelan, M. A. O'Reilly, T. E. Ward, E. Delahunt, and B. Caulfield, "Technology in rehabilitation: evaluating the single leg squat exercise with wearable inertial measurement units," *Methods Inf Med,* vol. 56, no. 2, pp. 88–94, 2017, doi: 10.3414/me16-02-0002.

[25] M. A. O'Reilly, D. F. Whelan, T. E. Ward, E. Delahunt, and B. Caulfield, "Classification of lunge biomechanics with multiple and individual inertial measurement units," *Sports Biomech,* vol. 16, no. 3, pp. 342–360, 2017, doi: 10.1080/14763141.2017.1314544.

[26] D. S. Moran, A. Shitzer, and K. B. Pandolf, "A physiological strain index to evaluate heat stress," *Am J Physiol – Regulat Integr Comparat Physiol,* vol. 275, no. 1, pp. R129–R134, 1998, doi: 10.1152/ajpregu.1998.275.1.R129.

[27] C. Byrne and J. K. W. Lee, "The Physiological Strain Index modified for trained heat-acclimatized individuals in outdoor heat," *Int J Sports Physiol Perform,* vol. 14, no. 6, pp. 805–813, 2019, doi: 10.1123/ijspp.2018-0506.

[28] N. P. Mali, "Modern technology and sports performance: an overview," *Int J Physiol Nutr Phys Educ,* vol. 5, no. 1, pp. 212–216, 2020.

[29] H. Tankovska, "Global connected wearable devices 2016–2022 Statista," February 2019. Available: https://www.statista.com/statistics/487291/global-connected-wearable-devices/

[30] T. R. Baechle and R. W. Earle, *Essentials of Strength Training and Conditioning*. Champaign, IL: Human kinetics, 2008.

[31] J. P. Ahtiainen, S. Walker, H. Peltonen, *et al.*, "Heterogeneity in resistance training-induced muscle strength and mass responses in men and women of different ages," *Age (Dordr),* vol. 38, no. 1, p. 10, 2016, doi: 10.1007/s11357-015-9870-1.

[32] A. L. Arrese and E. S. Ostáriz, "Skinfold thicknesses associated with distance running performance in highly trained runners," *J Sports Sci,* vol. 24, no. 1, pp. 69–76, 2006, doi: 10.1080/02640410500127751.

[33] P. Bale, D. Bradbury, and E. Colley, "Anthropometric and training variables related to 10km running performance," *Br J Sports Med,* vol. 20, no. 4, pp. 170–173, 1986, doi: 10.1136/bjsm.20.4.170.

[34] E. C. Lee, M. S. Fragala, S. A. Kavouras, R. M. Queen, J. L. Pryor, and D. J. Casa, "Biomarkers in sports and exercise: tracking health, performance, and recovery in athletes," *J Strength Cond Res,* vol. 31, no. 10, pp. 2920–2937, 2017, doi: 10.1519/JSC.0000000000002122.

[35] S. T. Stanelle, S. F. Crouse, T. R. Heimdal, S. E. Riechman, A. L. Remy, and B. S. Lambert, "Predicting muscular strength using demographics, skeletal dimensions, and body composition measures," *Sports Med Health Sci,* vol. 3, no. 1, pp. 34–39, 2021, doi: https://doi.org/10.1016/j.smhs.2021.02.001.

[36] E. W. Banister, T. W. Calvert, M. V. Savage, and T. Bach, "A systems model of training for athletic performance," *Aust J Sports Med,* vol. 7, no. 3, pp. 57–61, 1975.

[37] T. Busso, R. Candau, and J.-R. Lacour, "Fatigue and fitness modelled from the effects of training on performance," *Eur J Appl Physiol Occupat Physiol,* vol. 69, no. 1, pp. 50–54, 1994/01/01 1994, doi: 10.1007/BF00867927.

[38] T. W. Calvert, E. W. Banister, M. V. Savage, and T. M. Bach, "A systems model of the effects of training on physical performance," *IEEE Trans Syst Man Cybernet,* vol. SMC-6, pp. 94–102, 1976.

[39] T. Busso, "Variable dose–response relationship between exercise training and performance," *Med Sci Sports Exerc,* vol. 35, no. 7, pp. 1188–1195, 2003, doi: 10.1249/01.Mss.0000074465.13621.37.

[40] J. D. Bartlett, F. O'Connor, N. Pitchford, L. Torres-Ronda, and S. J. Robertson, "Relationships between internal and external training load in team-sport athletes: evidence for an individualized approach," *Int J Sports Physiol Perform,* vol. 12, no. 2, pp. 230–234, 2017.

[41] D. L. Carey, K. Ong, M. E. Morris, J. Crow, and K. M. Crossley, "Predicting ratings of perceived exertion in Australian football players: methods for live estimation," *Int J Comput Sci Sport,* vol. 15, no. 2, pp. 64–77, 2016.

[42] A. Rossi, E. Perri, A. Trecroci, M. Savino, G. Alberti, and F. M. Iaia, "GPS data reflect players' internal load in soccer," in *2017 IEEE International Conference on Data Mining Workshops*, 2017, pp. 890–893.

[43] A. Jaspers, T. Op De Beéck, M. S. Brink, *et al.*, "Relationships between the external and internal training load in professional soccer: what can we learn from machine learning?," *Int J Sports Physiol Perform,* vol. 13, no. 5, pp. 625–630, May 1 2018.

[44] N. W. Constantini, R. Arieli, G. Chodick, and G. Dubnov-Raz, "High prevalence of vitamin D insufficiency in athletes and dancers," *Clin J Sport Med,* vol. 20, no. 5, pp. 368–371, 2010, doi: 10.1097/JSM.0b013e3181f207f2.

[45] G. Dubnov and N. W. Constantini, "Prevalence of iron depletion and anemia in top-level basketball players," *Int J Sport Nutr Exerc Metab,* vol. 14, no. 1, pp. 30–37, 2004, doi: 10.1123/ijsnem.14.1.30.

[46] D. McCartney, M. J. Benson, B. Desbrow, C. Irwin, A. Suraev, and I. S. McGregor, "Cannabidiol and sports performance: a narrative review of relevant evidence and recommendations for future research," *Sports Med – Open,* vol. 6, no. 1, p. 27, 2020, doi: 10.1186/s40798-020-00251-0.

[47] G. L. Martins, J. P. L. F. Guilherme, L. H. B. Ferreira, T. P. de Souza-Junior, and A. H. Lancha, "Caffeine and exercise performance: possible directions for definitive findings," *Front Sports Active Living,* vol. 2, 2020, doi: 10.3389/fspor.2020.574854.
[48] C. Heneghan, R. Perera, D. Nunan, K. Mahtani, and P. Gill, "Forty years of sports performance research and little insight gained," *Br Med J,* vol. 345, p. e4797, 2012, doi: 10.1136/bmj.e4797.
[49] B. M. Popkin, K. E. D'Anci, and I. H. Rosenberg, "Water, hydration, and health," *Nutr Rev,* vol. 68, no. 8, pp. 439–458, 2010, doi: 10.1111/j.1753-4887.2010.00304.x.
[50] H. Suppiah, E. Ling Ng, J. Wee, *et al.*, "Hydration status and fluid replacement strategies of high-performance adolescent athletes: an application of machine learning to distinguish hydration characteristics," *Nutrients,* vol. 13, p. 4073, doi: 10.3390/nu13114073.
[51] M. N. Sawka, S. N. Cheuvront, and R. W. Kenefick, "Hypohydration and human performance: impact of environment and physiological mechanisms," *Sports Med,* vol. 45, no. Suppl 1, pp. S51–60, Nov 2015, doi: 10.1007/s40279-015-0395-7.
[52] D. Sim, M. C. Brothers, J. M. Slocik, *et al.*, "Biomarkers and detection platforms for human health and performance monitoring: a review," *Adv Sci,* vol. 9, no. 7, p. 2104426, 2022, doi: https://doi.org/10.1002/advs.202104426.
[53] H. C. Lukaski, "Vitamin and mineral status: effects on physical performance," *Nutrition,* vol. 20, no. 7–8, pp. 632–644, 2004, doi: 10.1016/j.nut.2004.04.001.
[54] N. R. Singh and E. M. Peters, "Artificial neural networks in the determination of the fluid intake needs of endurance athletes," *AASRI Proc,* vol. 8, pp. 9–14, 2014.
[55] M. N. Sawka, L. M. Burke, E. R. Eichner, R. J. Maughan, S. J. Montain, and N. S. Stachenfeld, "American College of Sports Medicine position stand. Exercise and fluid replacement," *Med Sci Sports Exercise,* vol. 39, no. 2, pp. 377–390, 2007.
[56] M. N. Sawka, S. N. Cheuvront, and R. Carter, 3rd, "Human water needs," *Nutr Rev,* vol. 63, no. 6, pp. S30–S39, 2005.
[57] C. O. Asogwa and D. T. H. Lai, "A review on opportunities to assess hydration in wireless body area networks," *Electronics,* vol. 6, no. 4, p. 82, 2017.
[58] T. A. Howell, A. Hadiwidjaja, P. P. Tong, C. D. Thomas, and C. Schrall, "Method and apparatus for hydration level of a person," Patent Appl. US Patent 16/270,773, 2019.
[59] N. E. Lee, T. H. Lee, D. H. Seo, and S. Y. Kim, "A smart water bottle for new seniors: Internet of Things (IoT) and health care services," *Int J Bio-Sci Bio-Technol,* vol. 7, pp. 305–314, 2015.
[60] L. E. Armstrong, "Assessing hydration status: the elusive gold standard," *J Am Coll Nutr,* vol. 26, no. 5 Suppl, pp. 575S–584S, 2007.

[61] A. Bak, A. Tsiami, and C. Greene, "Methods of assessment of hydration status and their usefulness in detecting dehydration in the elderly," *Curr Res Nutr Food Sci,* vol. 5, no. 2, pp. 43–54, 2017.
[62] J. Sak and M. Suchodolska, "Artificial Intelligence in nutrients science research: a review," *Nutrients,* vol. 13, no. 2, p. 322, 2021. Available: https://www.mdpi.com/2072-6643/13/2/322.
[63] P. Sundaravadivel, K. Kesavan, L. Kesavan, S. P. Mohanty, and E. Kougianos, "Smart-Log: a deep-learning based automated nutrition monitoring system in the IoT," *IEEE Trans Consumer Electron,* vol. 64, no. 3, pp. 390–398, 2018, doi: 10.1109/TCE.2018.2867802.
[64] C. O. Asogwa, S. F. Collins, P. Mclaughlin, and D. T. H. Lai, "A galvanic coupling method for assessing hydration rates," *Electronics,* vol. 5, no. 39, 2016.
[65] X. Huang, H. Cheng, K. Chen, *et al.*, "Epidermal impedance sensing sheets for precision hydration assessment and spatial mapping," *IEEE Trans Biomed Eng,* vol. 60, no. 10, pp. 2848–2857, 2013.
[66] A. Alvarez, E. Severyn, J. Velasquez, S. Wong, G. Perpinan, and M. Heuerta, "Machine learning methods in the classification of the athletes dehydration," in *2019 IEEE Fourth Ecuador Technical Chapters Meeting*, 2019.
[67] H. F. Posada-Quintero, N. Reljin, A. Moutran, *et al.*, "Mild dehydration identification using machine learning to assess autonomic responses to cognitive stress," *Nutrients,* vol. 12, no. 1, p. 42, 2019.
[68] S. L. Halson, "Recovery techniques for athletes," *Sports Sci Exchange,* vol. 26, no. 10, pp. 1–6, 2013.
[69] H. H. Fullagar, S. Skorski, R. Duffield, D. Hammes, A. J. Coutts, and T. Meyer, "Sleep and athletic performance: the effects of sleep loss on exercise performance, and physiological and cognitive responses to exercise," *Sports Med,* vol. 45, no. 2, pp. 161–186, 2015.
[70] S. L. Halson, "Sleep in elite athletes and nutritional interventions to enhance sleep," *Sports Med,* vol. 44 Suppl 1, pp. S13–S23, 2014.
[71] D. Riemann, C. Baglioni, C. Bassetti, *et al.*, "European guideline for the diagnosis and treatment of insomnia," *J Sleep Res,* vol. 26, no. 6, pp. 675–700, 2017.
[72] A. Qaseem, D. Kansagara, M. A. Forciea, M. Cooke, T. D. Denberg, and P. Clinical Guidelines Committee of the American College of, "Management of Chronic Insomnia Disorder in Adults: A Clinical Practice Guideline From the American College of Physicians," *Ann Internal Med,* vol. 165, no. 2, pp. 125–133, 2016.
[73] J. A. Douglas, C. L. Chai-Coetzer, D. McEvoy, *et al.*, "Guidelines for sleep studies in adults – a position statement of the Australasian Sleep Association," *Sleep Med,* vol. 36 Suppl 1, pp. S2–S22, 2017.
[74] B. P. Kolla, S. Mansukhani, and M. P. Mansukhani, "Consumer sleep tracking devices: a review of mechanisms, validity and utility," *Expert Rev Med Dev,* vol. 13, no. 5, pp. 497–506, 2016.
[75] M. T. Bianchi, "Sleep devices: wearables and nearables, information and interventional, consumer and clinical," *Metabolism,* vol. 84, pp. 99–108, 2018.

[76] C. Sargent, M. Lastella, S. L. Halson, and G. D. Roach, "The validity of activity monitors for measuring sleep in elite athletes," *J Sci Med Sport,* vol. 19, no. 10, pp. 848–853, 2016.

[77] S. Roomkham, D. Lovell, J. Cheung, and D. Perrin, "Promises and challenges in the use of consumer-grade devices for sleep monitoring," *IEEE Rev Biomed Eng,* vol. 11, pp. 53–67, 2018.

[78] A. J. Koch, R. Pereira, and M. Machado, "The creatine kinase response to resistance exercise," *J Musculoskelet Neuronal Interact,* vol. 14, no. 1, pp. 68–77, 2014.

[79] V. Mougios, "Reference intervals for serum creatine kinase in athletes," *Br J Sports Med,* vol. 41, no. 10, pp. 674–678, 2007, doi: 10.1136/bjsm.2006.034041.

[80] C. Z. Hong and I. N. Lien, "Metabolic effects of exhaustive training of athletes," *Arch Phys Med Rehabil,* vol. 65, no. 7, pp. 362–365, 1984.

[81] C. A. Goldstein, R. B. Berry, D. T. Kent, *et al.*, "Artificial intelligence in sleep medicine: background and implications for clinicians," *J Clin Sleep Med,* vol. 16, no. 4, pp. 609–618, 2020.

[82] P. Anderer, H. Danker-Hopfe, S. -L. Himanen, *et al.*, "An E-health solution for automatic sleep classification according to Rechtschaffen and Kales: validation study of the Somnolyzer 24 x 7 utilizing the Siesta database," *Neuropsychobiology,* vol. 51, no. 3, pp. 115–133, 2005.

[83] P. Anderer, G. Gruber, S. Parapatics, *et al.*, "Computer-assisted sleep classification according to the standard of the American Academy of Sleep Medicine: validation study of the AASM version of the Somnolyzer 24 x 7," *Neuropsychobiology,* vol. 62, no. 4, pp. 250–264, 2010.

[84] L. Fraiwan, K. Lweesy, N. Khasawneh, M. Fraiwan, H. Wenz, and H. Dickhaus, "Classification of sleep stages using multi-wavelet time frequency entropy and LDA," *Methods Inf Med,* vol. 49, no. 3, pp. 230–237, 2010.

[85] S. F. Liang, C. E. Kuo, Y. H. Hu, and Y. S. Cheng, "A rule-based automatic sleep staging method," *J Neurosci Methods,* vol. 205, no. 1, pp. 169–176, 2012.

[86] N. Schaltenbrand, R. Lengelle, M. Toussaint, *et al.*, "Sleep stage scoring using the neural network model: comparison between visual and automatic analysis in normal subjects and patients," *Sleep,* vol. 19, no. 1, pp. 26–35, 1996.

[87] M. S. Daley, D. Gever, H. F. Posada-Quintero, Y. Kong, K. Chon, and J. B. Bolkhovsky, "Machine learning models for the classification of sleep deprivation induced performance impairment during a psychomotor vigilance task using indices of eye and face tracking," *Front Artif Intell,* Original Research vol. 3, 2020, doi: 10.3389/frai.2020.00017.

[88] P. Fonseca, T. Weysen, M. S. Goelema, *et al.*, "Validation of photoplethysmography-based sleep staging compared with polysomnography in healthy middle-aged adults," *Sleep,* vol. 40, no. 7, p. zsx097, 2017.

[89] A. Patanaik, J. L. Ong, J. J. Gooley, S. Ancoli-Israel, and M. W. L. Chee, "An end-to-end framework for real-time automatic sleep stage classification," *Sleep,* vol. 41, no. 5, pp. 1–11, 2018.

[90] H. Sun, J. Jia, B. Goparaju, *et al.*, "Large-scale automated sleep staging," *Sleep,* vol. 40, no. 10, p. zsx139, 2017.
[91] S. Biswal, H. Sun, B. Goparaju, M. B. Westover, J. Sun, and M. T. Bianchi, "Expert-level sleep scoring with deep neural networks," *J Am Med Inf,* vol. 25, no. 12, pp. 1643–1650, 2018.
[92] L. Citi, M. T. Bianchi, E. B. Klerman, and R. Barbieri, "Instantaneous monitoring of sleep fragmentation by point process heart rate variability and respiratory dynamics," In: *Annual International Conference of the IEEE Engineering in Medicine and Biology Society,* vol. 2011, pp. 7735–7738, 2011.
[93] A. Ashfaq, N. Cronin, and P. Müller, "Recent advances in machine learning for maximal oxygen uptake (VO2 max) prediction: a review," *Inf Med Unlocked,* vol. 28, p. 100863, 2022, doi: https://doi.org/10.1016/j.imu.2022.100863.
[94] M. E. Jung, J. E. Bourne, and J. P. Little, "Where Does HIT Fit? An examination of the affective response to high-intensity intervals in comparison to continuous moderate- and continuous vigorous-intensity exercise in the exercise intensity-affect continuum," *PLoS One,* vol. 9, no. 12, p. e114541, 2014, doi: 10.1371/journal.pone.0114541.
[95] P. Ekkekakis, "Let them roam free?," *Sports Med,* vol. 39, no. 10, pp. 857–888, 2009/10/01 2009, doi: 10.2165/11315210-000000000-00000.
[96] C.-N. Shin, Y.-S. Lee, and M. Belyea, "Physical activity, benefits, and barriers across the aging continuum," *Appl Nurs Res,* vol. 44, pp. 107–112, 2018, doi: https://doi.org/10.1016/j.apnr.2018.10.003.
[97] E. E. Cust, A. J. Sweeting, K. Ball, and S. Robertson, "Machine and deep learning for sport-specific movement recognition: a systematic review of model development and performance," *J Sports Sci,* vol. 37, no. 5, pp. 568–600, 2019, doi: 10.1080/02640414.2018.1521769.
[98] A. Blasiak, J. Khong, and T. Kee, "CURATE.AI: optimizing personalized medicine with artificial intelligence," *SLAS TECHNOL: Transl Life Sci Innov,* vol. 25, no. 2, pp. 95–105, 2020, doi: 10.1177/2472630319890316.
[99] M. Altini and O. Amft, "HRV4Training: large-scale longitudinal training load analysis in unconstrained free-living settings using a smartphone application," *EMBC*, pp. 2610–2613, 2016.
[100] M. Du and X. Yuan, "A survey of competitive sports data visualization and visual analysis," *J Visual,* vol. 24, no. 1, pp. 47–67, 2021.
[101] Y. Zhang, Y. Zhang, X. Zhao, Z. Zhang, and H. Chen, "Design and data analysis of sports information acquisition system based on internet of medical things," *IEEE Access,* vol. 8, pp. 84792–84805, 2020.
[102] Z. Yin and W. Cui, "Outlier data mining model for sports data analysis," *J Intell Fuzzy Syst,* vol. 22, 2020.
[103] Z. Bai and X. Bai, "Sports big data: management, analysis, applications, and challenges," *Complexity,* vol. 2021, pp. 1–11, 2021.
[104] Y. Kim and H. Bang, *Introduction to Kalman Filter and Its Applications Croatia: IntechOpen*, 2018.
[105] P. S. R. Diniz, "Algorithms and practical implementation," in *Adaptive Filtering*, vol. XXIV, 3 ed., New York, NY: Springer, 2008, p. 627. Originally

published as volume 694 in the series: The International Series in Engineering and Computer Science.

[106] M. Usman, M. Zubair, H. S. Hussein, *et al.*, "Empirical mode decomposition for analysis and filtering of speech signals," *IEEE Can J Electr Comput Eng,* vol. 44, no. 3, pp. 343–349, 2021, doi: 10.1109/ICJECE.2021.3075373.

[107] A. Rossi, L. Pappalardo, P. Cintia, F. M. Iaia, J. Fernàndez, and D. Medina, "Effective injury forecasting in soccer with GPS training data and machine learning," *PLoS One*, vol. 13, no. 7, p. e0201264, 2018, doi: 10.1371/journal.pone.0201264.

[108] G. Rodas, L. Osaba, D. Arteta, R. Pruna, D. Fernández, and A. Lucia, "Genomic prediction of tendinopathy risk in Elite team sports," *Int J Sports Physiol Perform,* pp. 1–7, 2019, doi: 10.1123/ijspp.2019-0431.

[109] C. Richter, E. King, S. Strike, and A. Franklyn-Miller, "Objective classification and scoring of movement deficiencies in patients with anterior cruciate ligament reconstruction," *PLoS One*, vol. 14, no. 7, p. e0206024, 2019, doi: 10.1371/journal.pone.0206024.

[110] D. W. Gareth James, T. Hastie, and R. Tibshirani, *An Introduction to Statistical Learning: with Applications in R*, 1 ed. (Springer Texts in Statistics). New York, NY: Springer, 2013.

[111] B. Bent, B. A. Goldstein, W. A. Kibbe, and J. P. Dunn, "Investigating sources of inaccuracy in wearable optical heart rate sensors," *NPJ Digit Med,* vol. 3, pp. 18–18, 2020, doi: 10.1038/s41746-020-0226-6.

[112] O. Q. Groot, P. T. Ogink, A. Lans, *et al.*, "Machine learning prediction models in orthopedic surgery: a systematic review in transparent reporting," *J Orthopaed Res,* vol. 40, no. 2, pp. 475–483, 2022, doi: https://doi.org/10.1002/jor.25036.

Chapter 10

Wearable electrochemical sensors and machine learning for real-time sweat analysis

Matthew Douthwaite[1] and Pantelis Georgiou[1]

10.1 Electrochemical sensors: the next generation of wearables

Wearable technology can greatly improve our ability to monitor of health and physical activity. Using wearable sensors in medicine gives patients more freedom and comfort, while allowing physicians to obtain measurements continuously over longer periods of time and in alternate environments and test conditions. For example, early non-invasive monitoring in a clinical setting used ECG monitoring patches to replace restricting cables [1,2].

Over the past decade, continuous and non-invasive monitoring of body chemistry has emerged as a popular next objective for wearable sensor research [3]. Our body chemistry is affected by everything from our lifestyle choices to our physical and mental health, and can often reveal problems before we even notice them ourselves. For this reason, blood tests form a vital part of modern medical diagnoses and treatments, as they allow a doctor to obtain a picture of a patient's internal physiological condition. However, these tests require blood to be extracted through the skin, which can not only cause discomfort for the patient and poses an infection risk in certain scenarios but also limits the analysis to discrete samples and often requires lab-processing which can result in long delays [4]. Consequently, there is a demand to monitor body chemistry in a non-invasive and continuous fashion so that physicians can obtain a more comprehensive insight into a patient's state-of-the-health. This would enable more efficient treatments, faster reactions to change and longer observation periods, both in the hospital and during ambulatory care. Numerous applications can be found outside of medicine, such as monitoring the vital signs and stress levels of workers in hazardous environments, such as military personnel or firefighters, or the metabolism and electrolyte balance of athletes in sports science and recreational exercise.

For a non-invasive assessment of body chemistry, research has focused on biofluids other than blood. Possible alternatives frequently considered are tears, saliva, urine, interstitial fluid (the fluid which surrounds subdermal cells) and sweat. Each

[1]Centre for Bio-Inspired Technology, Department of Electrical and Electronic Engineering, Imperial College London, UK

of these has their own unique advantages and challenges, with whole research fields exploring their use. However, this chapter focuses solely on sensors and devices which analyse sweat, primarily because it is arguably most accessible and least invasive option [5,6]. In comparison, continuous monitoring of tears or saliva requires placing sensors in a subject's eyes or mouth, which introduces significant technological and safety challenges. Analysing urine is not a suitable option for continuous monitoring, and could still be classed as invasive depending on the method of sampling. Interstitial fluid, which surrounds cells beneath the epidermis, may be probed using micro-needles, which are small enough to be painless. However, this poses bio-compatibility and manufacturing challenges from a technical standpoint [6]. In contrast, sweat is produced by glands which cover almost all of the human body [7]. The coverage of sweat glands provides flexibility in the choice of measurement site, and the glands themselves aid collection by actively pumping sweat out onto the skin.

Sweat monitoring has been historically used as the gold standard test to diagnose cystic fibrosis in new-born babies [8,9] and more recently it has been used in some lab-based sports science tests [10]. However, it is only in the past decade that technology has reached a stage where sweat analysis can be carried out on the body. This has been made possible by advances in the miniaturisation and integration of sensors and electronics and by improvements in materials and manufacturing techniques. These developments have been driven by consumer electronics and research into point-of-care and portable lab-on-chip applications [3]. Now that the potential of analysing sweat on the body has been rediscovered along with the technology that makes it possible, the interest in the field of wearable electrochemical sensors has exploded.

One of the underlying technologies which has enabled the advancement of wearable sensors is integrated semiconductor-based electronics. Complementary metal-oxide-semiconductor (CMOS) technology facilitates low-power, scalable and reliable processing in small physical areas, which has enabled the world of smart-phones and portable computers that we know today. Chemical sensors have already benefited greatly from CMOS technology. The ion-sensitive field-effect transistor (ISFET) is a CMOS-compatible sensor that converts chemical signals to electrical signals [11], bringing significant benefits to lab-on-chip devices. Arrays of thousands of ISFETs can now be fabricated on a single microchip for ion sensing and imaging. As a result, processes such as DNA detection and sequencing have followed the same trend as computers; going from slow, expensive processes requiring bulky equipment to using only bench-top, and in some cases portable, devices that perform highly parallelised functions at a rapidly decreasing cost [12].

Since the latest wearables are built around microchips, it is logical to explore whether ISFETs in CMOS can bring the lab-on-a-chip onto the body. While the majority of wearable chemical sensors are based on optical methods or use screen-printed electrodes, a fully integrated CMOS solution could bring numerous advantages. Not only are there reductions in size and power consumption, but CMOS offers the potential for versatile detection of a wide range of bio-markers simultaneously on a single silicon substrate. On top of that, integrating custom instrumentation at the sensor front-end can bring improvements in signal-to-noise ratio (SNR) [13], which is important in a noisy wearable environment where the signal is often small. Furthermore, microcontrollers can be incorporated along with the sensor frontend, which can enable signal processing and data analysis 'at the edge'. With the latest developments of edge-AI in popular microcontrollers, this opens up the possibilities of wearable ISFETs with machine learning (ML).

While ML is increasingly being used in other types of wearables, it has yet to be significantly investigated in the field of wearable electrochemical sensing [14]. There is a great scope to use these techniques to improve the quality of information and expand the inferences that can be made from such devices. Wearable sweat monitoring devices are ideal candidates for enhancement through ML, and CMOS ISFETs are well placed to facilitate monolithic implementation with numerous benefits.

As such, this chapter first aims to provide the reader with a foundation of knowledge in sweat analysis and the existing progress towards wearable sweat sensor development in Section 10.2. Section 10.5 will be introduce the ISFET and discuss this sensor in the context of this application. Finally, recent uses of machine learning for electrochemical sensors, sweat analysis and ISFETs will be reviewed in Section 10.6 to identify opportunities for research and development in this space.

10.2 The mechanisms and content of sweat

The application of on-body sweat analysis poses a different set of challenges to typical lab-on-chip measurements. Sweat is a particularly complex bio-fluid and so understanding its components and their inter-dependencies is key to designing a wearable sweat sensor. Much of sweat gland biology is still not fully understood, particularly how the contents of sweat relates to physiology [15]. This section aims to provide an overview of the different analytes of interest in sweat and what needs to be considered when making sweat measurements on the body.

Sweat is generated naturally through a variety of means, primarily for cooling through evaporation when body temperature rises due to the environment or physical activity but also due to psychological factors. Sweating can also be induced chemically using iontophoresis, a process where a small electrical current is used to drive charged chemicals that stimulate sweat glands through the skin [5].

There are in fact three main types of sweat produced by humans: eccrine, apocrine and apoeccrine [8]. In this chapter, we will focus solely on eccrine sweat, as apocrine and apoeccrine glands are inconveniently located around the axilla and groin and produce sweat with a greater concentration of bacteria and oil which could cause measurement errors [6]. In contrast, eccrine sweat glands are found almost everywhere else on the human body and eccrine sweat contains a wealth of biological components which have great potential to provide insight into the physiology of the subject [6,7]. As a result, all references to sweat hereafter refer to eccrine sweat.

The eccrine sweat gland is mainly located in the dermis, a structure filled with ISF. Figure 10.1 shows a diagram of the sweat gland which is made up of the secretory coil, the dermal duct and the upper coiled duct. Analytes primarily partition into sweat by passing into the secretory coil and duct through active transport channels (transcellularly) or in between cells (paracellularly), as shown in Figure 10.1(a) [6].

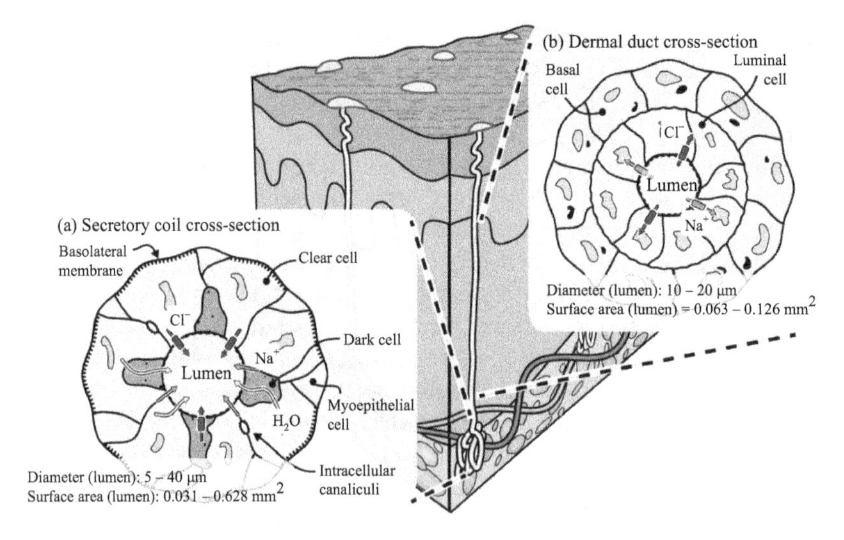

Figure 10.1 Diagram of a sweat gland showing cross section and the entry and exit processes of Na^+ and Cl^-. Reprinted from [7] with the permission of AIP Publishing.

Table 10.1 Key components in sweat and associated applications

Component	Concentration range (mmol) [15]	Proposed to monitor
Sodium (Na^+)	10–90	Hydration [16,17], sweat rate [18]
Chloride (Cl^-)	10–90	Cystic fibrosis [19,20], sweat rate [18]
pH (H^+)	3–8 [5]	Skin health [21]
Potassium (K^+)	2– 8	Exercise [22–24]
Ca^{2+}	0.2– 2	Bone health [25]
Zn^{2+}	$0.1–20 \times 10^{-3}$	Muscular stress & fatigue [26,27]
Glucose	$10–200 \times 10^{-3}$	Diabetes [22,28–32]
Lactate	5–40	Exercise [22,29,33]
Ethanol	0–7	Blood alcohol [28,34]
Cortisol	$0.6–7 \times 10^{-6}$	Stress [35,36]

Although sweat is 99% water, there are a great number of different components in the remaining 1% [7]. An in-depth list of the ions, metabolites and proteins can be found in literature [7,15]. Sonner *et al.* [7] present a comprehensive overview of sweat as a measurement target for biomedical devices and discusses each of the key analytes in detail. The most abundant and relevant components are presented in Table 10.1, along with their reported concentrations and proposed uses for each.

Overall, many processes and dependencies of the components of sweat are still not fully understood, but it is clear there is a lot of potential to obtain physiological information. Currently, the most realistic target applications are measuring $Na^+$and $Cl^-$for cystic fibrosis diagnosis, ethanol for tracking blood alcohol levels and glucose for monitoring blood sugar [5]. However, since ions such as Na^+, $K^+$and H^+ are very popular targets, due to their abundance and ease of detection, and substances like lactate and cortisol are gaining popularity, due to their physiological significance elsewhere in other bio-fluids, there is a need to thoroughly investigate the potential relationships between all components of sweat and physiological states [15]. Gaining a better understanding of these relationships will require not only robust measurement devices but also carefully designed sample collection methods and analysis in order to obtain results which reflect the actual content of sweat. The next section will discuss some of the factors which should be considered when analysing sweat on the body.

10.3 Considerations for on-body sweat analysis

This section will discuss the key aspects that should be considered when designing a system for on-body sweat analysis, including the possible techniques for obtaining a sweat sample and the different locations on the body where it can be collected.

10.3.1 Sweat gland densities and sweat rates

Sweat collection location is an important factor to consider, both for practical reasons, such as comfort of the wearer or available space, and because the volume of sweat produced varies across the body. Taylor *et al.* [37] provide a comprehensive review of the regional variation in both sweat gland densities and sweat rate per unit area for different sites across the body.

A particularly challenging aspect of performing sweat analysis on the body is low sweat rates per unit area. While it may seem like a typical person produces a lot of sweat, this is due to the large number of sweat glands across the body. It is reported that whole body sweat loss typically ranges from about 0.2 to $3\,l\,h^{-1}$ during a variety of sports [38], with a mean of $1.2\,l\,h^{-1}$ for an adult, and the total number of sweat glands across the body is estimated to be around 2 million [37]. A simple calculation therefore shows that an individual gland will produce sweat at a rate of around $10\,nl\,min^{-1}$ on average. Collecting sweats from multiple glands is therefore essential, but since a wearable device should ideally be small, collection areas are limited.

To maximise sample volume it is important to choose a sampling area with both high sweat gland density and local sweat rate, however practicality for the user should also be ensured. The areas of the body with the highest sweat gland density are the palm of the hand, fingers and bottom of the foot [37]; however, these locations are likely to be obtrusive to the wearer or dislodged during daily activity. The head is also an area with a generally high sweat gland density, in particular the forehead and scalp have densities of over 150 glands cm^{-2}. While the scalp is difficult due to hair, sampling from the forehead in the form of a head band is feasible, though the device must be reasonably small and lightweight to be unobtrusive. The upper arm,

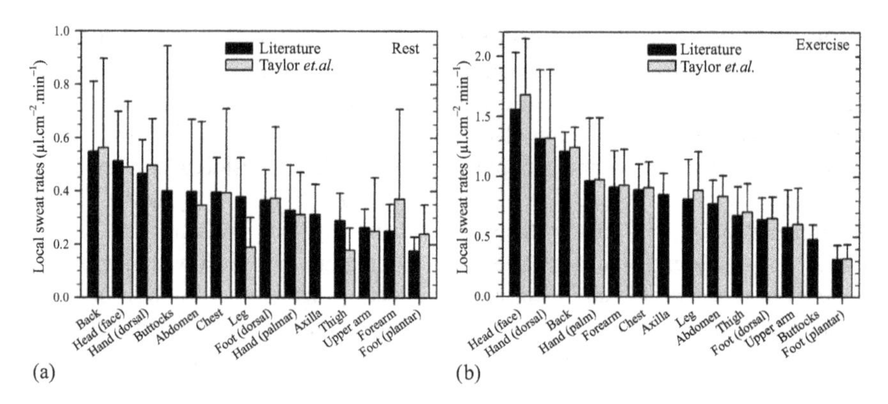

Figure 10.2 Local sweat rates per unit area for different regions of the body a) at rest (passive sweating) and b) during exercise (intense cycling). Data from literature and from study from the lab of Taylor et al., showing mean values and 95 % confidence intervals. Adapted with permission from [37].

forearm, chest and back all have densities of $\sim$ 100 glands cm^{-2} on average [37] and are consequently popular collection sites.

The rate of sweat production also varies across the body. Taylor *et al.* summarise the literature and their own measurements of local sweat rate for 14 aggregated sites across the body, for sweat generated by heat while the user is at rest and by exercise [37]. The results in Figure 10.2 show that the back, head (face) and hand (dorsal) make up the top three in both scenarios. This data shows that different sites could be considered depending on whether the intended application will involve the user at rest or exercising. Some sites, like the back, work well for monitoring both passive and active sweating, claiming the highest sweat rate across the body in the former case (0.55 μlcm^{-2}min^{-1}) and the third highest for the latter (1.2 μlcm^{-2}min^{-1}).

Researchers designing wearable monitoring devices tend to use the torso or arms as collection sites. As well as having a reasonable gland density, the back and chest are frequently used as they offer a large and relatively flat space to mount a wearable device [23,33], and the arm is used because of ease of access and user comfort [22,25,28,39].

10.3.2 Sweat collection techniques and challenges

When conducting analysis of sweat both on and off the body, it is very important for measurement integrity to consider how sweat should be obtained from the skin [15,40]. It has been shown that measured results can be affected by factors such as evaporation and skin contamination [5,40]. Furthermore, as discussed in Section 10.3.1, since the volume of sweat produced per unit is very low, researchers often rely on collecting samples from a larger area to obtain a measurement, which can increase the likelihood of evaporation and contamination errors occurring. As a result, it is important to consider the different available collection techniques that can be used and their advantages and disadvantages.

Absorbent patches were used in early studies as a method to collect sweat and maintain an adequate volume for reliable measurements [22,25,41]. However, it has been suggested that using an absorbent patch is not suitable for long-term measurements as, once saturated, old sweat remains and contaminates the new sample. This can artificially increase the analyte concentration [9].

An alternative method used in sweat studies is to seal off an area of the skin and allow sweat from a larger area to collect on the sensor through surface run-off, which reduces losses due to evaporation [42]. However, if the skin is not cleaned adequately, the sweat samples can be contaminated by dirt or dead skin cells [40]. This method can be improved by adding oil or Vaseline to the skin to protect run-off sweat from being contaminated and a centrifuge is also suggested to separate out cell-free sweat [42].

Modern efforts are moving towards creating microfluidic channels to extract sweat as soon as it is produced and then direct it over the sensing area, with an outlet to allow old samples to be replaced by new samples [43,44]. Sweat glands can actually help in this case by actively pumping sweat into the microfluidic cell, as they have been found to generate pressure in the order of 70 kNm^{-2} [7]. This can be thought of as positive pressure. Additionally, negative pressure techniques can be used, utilising materials and careful channel design to wick sweat from the skin passively. By removing sweat from the skin and allowing it to flow over the sensor and then out into the air, contamination from skin and evaporation losses are minimised. Using microfluidics is certainly the best option to enable sweat sensing with micro-litre or even nano-litre volumes, the ultimate goal of wearable sweat sensing [5], however design is more challenging. Many working prototypes have been designed, ranging from PDMS structures to paper based solutions improving wicking [45].

10.4 Current trends in wearable electrochemical sweat sensors

The popularity of sweat monitoring both in research and industry soared in recent years. In particular, since 2015 there has been a significant jump in publications and citations which have been growing each year. Figure 10.3 shows the exponential rise in publications with topics related to "wearable sweat sensors" or "on-body sweat analysis" up to 2021.* This is mainly thanks to advances in microelectronics, electrode fabrication techniques and microfluidics that have allowed devices for electrochemical analysis to become conformable and compact enough to be worn.

A number of trends have begun to emerge in the design of these devices. Some commonly described trends cover the device form-factor, for example "tattoo-based" [28], textile-based [29,46] and "electronic skin" [47]. Other common trends are in the features of proposed devices. For example, it is becoming increasingly common to

*Full search criteria were: "On-Body Sweat Analysis" or "Wearable Sweat Sensors".

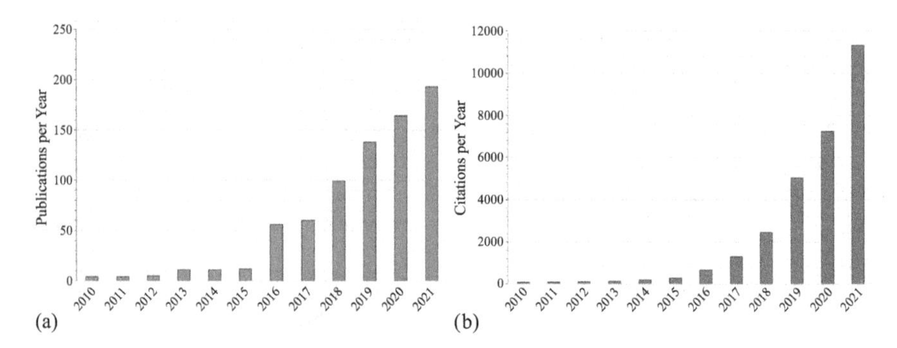

Figure 10.3 Results of a Web of Science search for publications on the topic of "wearable sweat sensors" or "on-body sweat analysis", showing increase in (a) publications and (b) citations from 2010 to 2021

measure multiple biochemical analytes or to monitor sweat in conjunction with other modes of physiological measurements. Another common goal is to create batteryless sensors to reduce size and cost of the wearables. Published works demonstrating these features are summarised in Table 10.2, and the following section describes these trends to provide context to the field and highlight the range of information sources which are being targeted and are available for deeper analysis with ML algorithms.

10.4.1 Common features of wearable sweat sensors

In 2016, a publication in *Nature* by Gao *et al.* began the trend of multi-analyte sensing [22]. They presented a wearable device for monitoring Na^+, K^+, glucose and lactate in sweat, using an array of six flexible screen-printed ion-selective electrodes (ISEs) on a PET (polyethylene terephthalate) substrate. Measurements were also taken on-body using a battery-powered, flexible PCB with bluetooth connectivity. This marked the start of both multi-analyte sensing and the integration of electronic systems which were presented over the following years. The same group followed up by also demonstrating measurements of calcium ($Ca2^+$) and pH [25], as shown in Figure 10.4(a), and further devices with flexible PCBs and sensors were subsequently presented [29,31,32,59]. By monitoring multiple analytes simultaneously, these researchers aim to build up a more detailed picture of the user's physiology in a range of applications. Almost all devices now presented target more than one analyte. The largest array at the time of writing was presented by He *et al.* and contains 15 electrodes on a PET substrate, measuring the six analytes [29]. The patch itself is approximately 5 $\times$ 5 cm, with a lot of area required by the connecting wires. This suggests fabricating arrays of ISEs in this way may not be scalable.

Since the primary reason to research sweat analysis is to gain a deeper insight into a subject's physiology, a logical next step is to combine these measurements with existing sources of information. As such, a more recent emerging trend is to combine electrochemical measurements with other modes of signal. For example,

Table 10.2 A comparison of existing wearable electrochemical sweat sensors

Trend: Battery-less	Trend: Multi-Analyte	Trend: Multi-Modal	Year	Detection target(s)	Type of sensor	Full wearable system?	Device form factor	Wireless communication	On-Body validation?	Collection location	Duration of test	Reference
x			2014	Na^+	ISE	Yes	Flexible Patch	RFID	-	Arm (proposed)	40 mins (in-vitro)	[16]
	x		2016	Na^+, K^+, Lac, Gl, T	ISE	Yes	Flexible sensors and circuit board	Bluetooth	Yes	Arm	100 min	[22]
		x	2016	Lac, ECG	ISE	Yes	Small rigid PCB with flexible electrodes	-	Yes	Chest	30 min	[48]
	x		2016	Ca2+, pH, T	ISE	Yes	Flexible sensors and circuit board	Bluetooth	Yes	Forehead	30 min	[25]
x			2017	pH	ISFET	-	Bench-top Circuit Board	-	-	N/A	N/A	[49]
		x	2017	pH, Temperature	ISFET	-	Plastic Strip	-	Yes	Neck	4 min	[50]
	x		2017	Na^+, pH, Lac, T	ISE	Yes	Flexible patch	Bluetooth	Yes	Lower Back	~115 min	[51]
	x		2018	Na^+, K^+	ISE	Yes	Flexible case adhered to back	Bluetooth	Yes	Lower Back	33.33 min	[17]
	x		2018	Gl, Eth	ISE	Yes	Small rigid PCB with tattoo electrode sensors	-	Yes	Arm	60 min	[28]
x	x	x	2018	Na^+, Cl^-, SR	Electrodes	Yes	Flexible patch with electronics and μfluidics	NFC	Yes	Forehead and Forearm	29 min	[52]
		x	2018	pH, HR, SP02	Optical	Yes	Patch with PCBs. Serpentine electronics	-	Yes	Lower Back	30 min	[53]
		x	2018	pH, T	Graphite electrode	-	Custom flexible electronics and antenna with PCB	NFC	-	N/A	1.5 min	[54]
x		x	2019	pH, HR	Au electrodes	Yes	Flexible circuits adhered to body	NFC	Yes	Neck	20 min	[55]
	x		2019	Na^+, K^+, pH, T	ISEs, RTD	Yes	Flexible PCB with Sensors attached	Bluetooth	-	Forehead (Proposed)	15 mins	[56]
	x		2019	Na^+, K^+, Gl, Lac, Ascorbic Acid, Uric Acid	CNT ISE	Yes	Flexible PCB	Bluetooth	Yes	Upper Arm	30 min	[29]
	x		2019	K^+, pH	AlGaN/ GaN HEMTs	-	Sensors stuck to skin	-	Yes	Neck	~25 min	[57]
	x		2019	Na^+, K^+, Gl	Au electrodes	Yes	Flexible sensors and PCB	Wireless	Yes	Wrist, ankle	120 min	[30]
x		x	2019	pH, T	ISE	Yes	Kapton flexible device with components on top	UHF	Yes	Arm	30 min	[58]
	x		2019	Na^+, K^+, pH, Ca2+	ISE	-	Flexible patch	-	Yes	Back	60 min	[23]
		x	2019	pH, Cor, EEG	ISE	Yes	Ear plug	BLE	Non Chemical Only	Inner Ear	N/A	[35]
x		x	2019	pH, ECG, T	ISE, Electrodes, Microphone	Yes	Flexible PCB and antenna	NFC	Yes	Chest	20 min	[59]
	x		2019	Na^+, K^+	ISE	Yes	PDMS μfluidics w/ PCB	BLE	Yes	Lower Back	33.3 min	[60]
x	x		2019	Na^+, K^+, pH, Ca^{2+}	ISFET	Yes	Plaster-like patch	NFC	-	N/A	N/A	[24]
x			2019	Gl	Au electrodes	Yes	Flexible Watch	-	Yes	Wrist (on-body) forehead (off-body)	120 min	[31]
	x		2020	Na^+, pH, Lac, NH_4^+	Carbon ink PE thread ISE	Yes	Plaster-like patch w/ PCB	Bluetooth	Yes	Upper Arm	30 min	[46]
	x	x	2020	pH, Gl, Lac, T	ISE	Yes	Watch sized PCB connected to flexible printed electrodes	-	-	Wrist (proposed)	30 min	[32]
	x	x	2020	pH, Gl, Z	RuS_2/PDMS	Yes	Flexible patch connected to PCB	Bluetooth	Yes	Wrist	10 min	[61]
x	x	x	2020	pH, Gl, NH_4^+, Urea, T, MC	ISE, Pt Electrodes	Yes	Electronic skin w/ electronics	BLE	Yes	Forehead	50 min	[47]

Key: Na^+, sodium; K^+, potassium; SR, sweat rate; Z, skin impedance; NH_4^+, ammonium; Gl, glucose; Lac, lactate; Cor, cortisol; Ca2+, calcium; T, temperature; HR, heart rate; CNT, carbon nanotubes; AlGaN, aluminium gallium nitride; HEMT, high-electron-mobility transistor; Au, gold; Pt, platinum; BLE, bluetooth low-energy; NFC, near-field communication; PE, polyester; MC, muscle contraction.

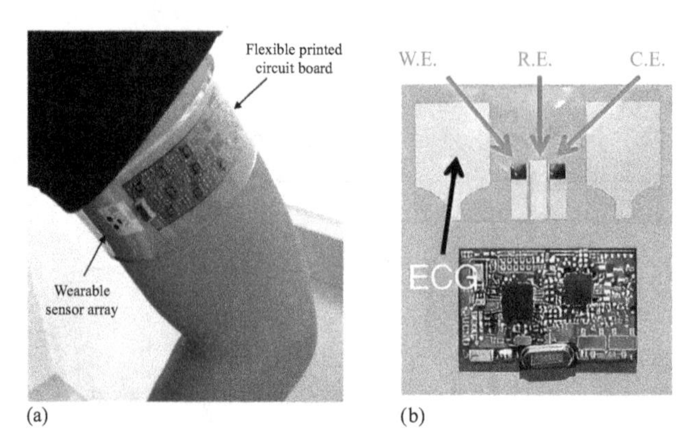

Figure 10.4 Examples of (a) a device for multi-analyte detection (reprinted with permission from [25]. Copyright 2016 American Chemical Society), and (b) a combined electrochemical electrophysical sensor monitoring sweat lactate and ECG, reprinted with permission from [48].

Figure 10.4(b) shows a combined heart rate (ECG) with lactate concentration [48]. There have also been devices combining heart rate monitoring with measurements of sweat pH [55,59], and brain activity with sweat pH and cortisol [35]. These devices currently only report the multi-modal measurements independently, but it is likely that in future ML algorithms will be applied to combine the data and allow more accurate predictions to be made about a user's state of health.

Another key area of research focuses on how wearable sweat sensors could be powered. While the simplest solution is to use a battery, there are several drawbacks to this option. Including a battery adds additional weight, size and cost to the device. Consequently, it is becoming increasingly common to see researchers opt for alternative solutions to batteries. Rose *et al.* presented the first batteryless sweat monitoring device in 2014, using the radio-frequency communication (RFID) protocol to wirelessly power and transmit data [16]. The device was presented as a disposable patch in the form factor of a medical plaster, and was aimed at measuring sweat sodium concentration with built in ISEs and an off-the-shelf potentiostat [16]. Other works have shown batteryless devices to minimise cost and size using near-field communication (NFC) [24,33,52,54,59], ultra-high frequency (UHF) powering [58] and photo-voltaic (PV) cells [31]. The feasibility of using body heat as a power source for ISFETs has also been demonstrated [49]. Most recently, proposals to use sweat itself as a bio-fuel source have been realised [47].

10.4.2 Opportunities for ISFETs and machine learning in wearable sweat sensing

As the field develops, there is a clear consensus among researchers that wearable sweat monitoring devices should detect multiple biomarkers, have a low-cost and be unobtrusive to the wearer. In this respect, ISFETs have great potential to advance the field. ISFETs require very little physical area for a complete system and potentially thousands of sensors can be fabricated in a few square millimetres, enabling the detection of multiple different analytes simultaneously. To achieve comfort, most works have focused on flexibility, but other important factors are minimising size and reducing unnecessary components, such as batteries. ISFET-based systems in CMOS can be designed to consume very little power, in the order of micro-watts or less, meaning that it is realistic to operate them without a battery. The additional processing that can be incorporated into a CMOS IC can provide improved signal quality and multiplexed inputs for multi-modal analysis. All of these features align well with the trends described in this chapter and the ultimate goal to increase our understanding of sweat with a robust, compact wearable device. The advantages and challenges of using ISFETs as a sensor for sweat monitoring are discussed in more detail in Section 10.5.

Interestingly, an area that has received little attention in the state-of-the-art so far is the use of machine learning when analysing the data obtained from sweat sensors. This is most likely because data sets are not yet large enough. The problem of making sense out of multiple data sources lends itself well to neural networks and other ML techniques. Sweat is a complex fluid, with a large number of different biomarkers with dependencies that are not fully understood. Edge ML could be used to help improve sensor selectivity or to classify data in order to reduce the amount of information that needs to be transmitted. For example, Veeralingam *et al.* [61] recently demonstrated a wearable sweat sensor targeting skin hydration, sweat glucose and sweat pH which incorporated an microcontroller running a machine learning algorithm. The algorithm helped to classify data from the different sensors simultaneously and provide a clear real-time readout to a smartphone application.

While edge-ML commonly makes use of a general-purpose microcontroller to perform machine learning, an alternative strategy is to develop custom hardware to tackle specific problems. By combining advanced CMOS-based sensors able to measure multiple biomarkers simultaneously with ML to analyse data for patterns, an unparalleled insight into real-time changes in physiology could be achieved.

As sweat sensors become more reliable and are distributed to a large number of subjects, large data datasets will become available, enabling the use of more complex analysis techniques such as deep neural networks. This is important to improve the reliability of classifications between different subjects and increase diagnostic potential of on-body sweat analysis.

Due to all these advantages, it is predicted that ML will be a key area of development in the future of wearable sweat sensor research [5]. In Section 10.6, we will review the existing research into ML for electrochemical sensors and presents an example of how machine learning at the sensor node could improve the performance of ISFETs for multi-analyte sensing.

10.5 The ion-sensitive field-effect transistor

ISFETs have been used in a range of applications for scalable, low-cost analysis of biological samples, and it is hoped that the advantages they bring can also be harnessed in the field of wearable electrochemical sensing. In particular, ISFETs can be fabricated in large arrays [12], creating huge quantities of data that is ideal for the application of ML. When considering that these ISFET arrays are directly integrated on-chip, it makes the prospect of ML at the edge both exciting and valuable. To understand the possibilities of the ISFET, it is first necessary to review the theory of operation and the main features and limitations of the sensor. This section provides a brief introduction to the fundamentals of the ISFET sensor and how it can be fabricated in CMOS in Sections 10.5.1 and 10.5.2, then Section 10.5.3 discusses the advantages of using this sensor for on-body sweat analysis and the practical implications of limitations of the device.

10.5.1 The fundamental theory of ISFETs

The ISFET sensor was originally proposed in the 1970s for the application of neurochemical recordings [62]. It was originally a discrete sensor, however almost 30 years later it was shown that they could be fully-integrated into CMOS [63]. This development opened up many new benefits and possible applications, the most popular of which has been DNA detection and sequencing [12,13].

An ISFET based on the same principle as a regular metal-oxide-semiconductor field-effect transistor (MOSFET). In the case of a MOSFET, an electrical signal is typically applied to the input (gate) terminal and this induces a current to flow between two other terminals (the source and drain). For detailed description of MOSFET operation, the reader is referred to the literature [64]. In the case of a discrete ISFET, the metal input terminal is replaced with a liquid and the electrical connection is made to a reference electrode which creates a bias point in the solution, as shown in Figure 10.5.

This liquid is in contact with an insulator surface, which in this case is typically and oxide, such as silicon dioxide, SiO_2. At the boundary between solid and liquid, a reaction occurs due to hydrogen ions (protons) binding with the surface. This is known as the cite binding model [65], which states that several layers of charge will build up, creating what is known as a double-layer capacitance [66]. The result is that an electrical surface potential builds up which couples to the FET channel as in a regular MOSFET, inducing the flow of current.

The surface potential is now dependent on the concentration of hydrogen ions, or the pH of the solution. As a result, it can be said that the ISFET threshold voltage becomes dependent on the pH of the solution [11]. The relationship between pH and ISFET threshold voltage is given by (10.1):

$$V_{th(ISFET)} = \underbrace{E_{ref} - \psi_0 + \chi^{sol} - \frac{\phi_M}{q}}_{V_{chem}} + V_{th(MOSFET)} \tag{10.1}$$

where E_{ref} is the constant potential of the reference electrode, ψ_0 is the pH dependent surface potential of the insulator, χ^{sol} is the surface dipole potential of the solution, ϕ_M is the work function of the MOSFET metal contact and $V_{th(MOSFET)}$ is the threshold voltage of the standard MOSFET. The first four terms are often grouped

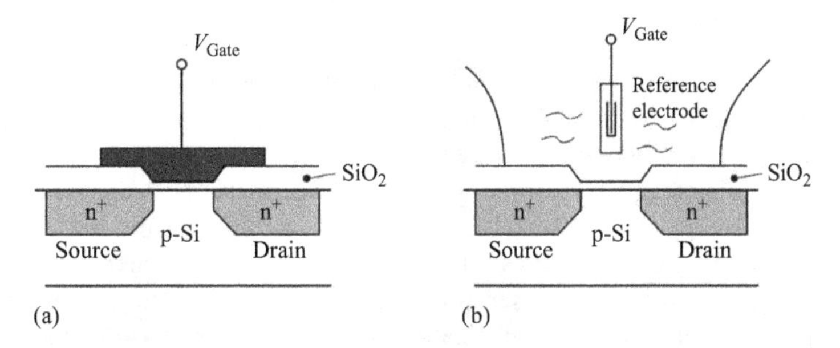

Figure 10.5 Diagrams of (a) a MOSFET and (b) an ISFET, showing the differences in gate connection. Reprinted from [11] Copyright 2003, with permission from Elsevier.

together into a single pH-dependent chemical potential, V_{chem} [67]. Since E_{ref} is constant and χ^{sol} is dependent on the composition of the electrolyte rather than pH, V_{chem} is only dependent on pH variations through ψ_0. ψ_0 is derived from the Nernst equation [67] and given by (10.2):

$$\psi_0 = 2.302\frac{kT}{q}.\alpha.\Delta pH \tag{10.2}$$

$$\alpha = \frac{1}{\frac{2.302kTC_{DL}}{q^2\beta_{int}} + 1} \tag{10.3}$$

where k is the Boltzmann constant, T is the temperature in kelvin, q is the absolute electron charge and $\frac{kT}{q} = U_T$, the thermal voltage. A potentiometric sensor has an ideal (or Nernstian) sensitivity of $S_N = 2.302\frac{kT}{q} = 59.5$ mV/decade at 300 K. α is a factor value between 0 and 1 representing the deviation from the ideal sensitivity, given by (10.3). It can be seen from this equation that ISFET sensitivity is closest to Nernstian when the surface material has a high buffer capacity, β_{int} and the double layer capacitance, C_{DL}, is small. The buffer capacity represents the change in net occupied sites for a change in proton surface concentration, measured in number sites per decade [67].

Now that the expression for $V_{th,ISFET}$ is known, it can be incorporated into the standard MOSFET equations. The general equation for the drain current, I_D, of a non-saturated ISFET is therefore given by (10.4):

$$I_D = \mu C_{ox}\frac{W}{L}(V_{GS} - V_{th(ISFET)} - \frac{1}{2}V_{DS})\, V_{DS} \tag{10.4}$$

where V_{GS} is the remote gate to source voltage, V_{DS} is the drain to source voltage, μ is the mobility of the charge carriers (i.e. electrons for n-channel), C_{ox} is the capacitance per unit area of the gate oxide and W and L are the device width and length respectively.

10.5.2 ISFETs in CMOS

CMOS technology forms the basis of almost all of our modern electronics. It allows circuits and systems to be fabricated on a sub-micrometre scale and consume a very little power. Additionally, due to existing infrastructure, manufacturing CMOS microchips at scale can be achieved at a low unit cost. ISFETs can be fabricated in CMOS to make use of these advantages. CMOS ISFETs have been adopted for several applications since their creation [11]. In addition to their success in enabling rapid DNA sequencing as discussed previously [12,13], there are now many publications focusing on portable amplification of DNA on-chip for infectious disease diagnosis [68], ion-imaging and multiple ion detection [24,69,70].

An ISFET can be constructed in unmodified CMOS technology by extending the gate connection of a MOSFET to the top metal layer of the chip through a via stack. No other electrical connections are made to this node, so the terminal is now a floating gate (FG). The top metal layer forms the sensing area of the ISFET and is capacitively coupled to the electrolyte through the die passivation layer, which now becomes the surface material discussed in Section 10.5.1. In unmodified CMOS, this is usually dual layer of silicon nitride (Si_3N_4) followed by silicon dioxide (SiO_2). This means that there is now an additional series capacitance between the reference electrode and the channel referred to as C_{pass}.

A cross-section of an ISFET in CMOS is shown in Figure 10.6 along with the final simplified ISFET macromodel, first derived in [71]. When using this macromodel, the chemical potential is usually subtracted from the remote gate voltage, V_G, as shown in 10.5, rather than threshold voltage as discussed in Section 10.5.1. However, as MOSFET operation is mainly dependent on overdrive voltage which includes V_{th}, i.e. $V_{GS} - Vth$, this does not change (10.4).

$$V'_G = V_G - V_{chem} = V_G - (\gamma + \alpha S_N pH) \tag{10.5}$$

The implications of this additional passivation capacitance in the stack are an attenuation of the signal due to a capacitive division between the ISFET surface and floating gate. The final equation for the drain current of an ISFET in CMOS is given by (10.6), where the attenuation factor due to the ratio of passivation capacitance C_{pass} and total capacitance of the floating gate node C_T is added to (10.6). Note that an additional potential representing an non-ideal effect, known as trapped charge, V_tc, is left out of the equation for simplicity. This effect will be discussed in the following section:

$$\begin{aligned} I_D &= I'_0 \exp\left(\frac{V_{FG}}{nU_T}\right) \\ &= I'_0 \exp\left(\frac{C_{pass}}{C_T}\left(\frac{V_{GS} - (\gamma + \alpha S_N \, \text{pH})}{nU_T}\right)\right) \end{aligned} \tag{10.6}$$

Since Si_3N_4 has a low buffer capacity, the achieved sensitivity for ISFETs fabricated in unmodified CMOS is at most 30 mV/pH [72]. Consequently, some users of ISFETs choose to add additional processing steps after fabrication to replace the surface material of the chip with a more sensitive material, though this adds time and cost

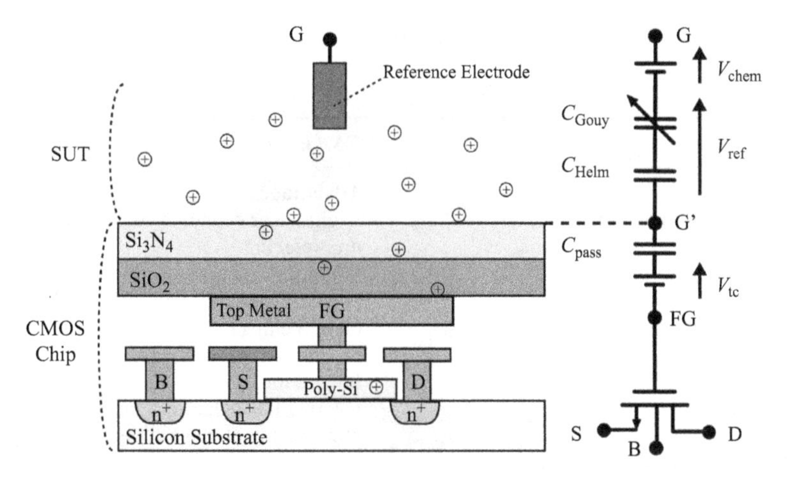

Figure 10.6 A cross-section of an ISFET in CMOS and the target protons in a solution-under-test (SUT) along with the associated macromodel [73]

to the fabrication process. Since a key-requirement of wearable chemical sensors is to be low-cost, as they have a relatively short life, unmodified CMOS is still used in some cases, both for ease of manufacture and to show a best-case performance in this low-cost set-up. The choice whether or not to re-passivate is based on the required cost versus performance.

10.5.3 ISFETs in CMOS for sweat sensing

Like many electrochemical sensors, ISFETs suffer from a number of non-idealities which need to be understood and taken into account when designing systems and evaluating measurements. A detailed review of ISFETs in CMOS and the non-ideal effects which are attributed to them has been published in literature [13,72]. In this section, the non-ideal effects will be briefly summarised in the context of the application of wearable sweat sensing.

Table 10.3 shows a summary of the specifications for this application and corresponding attributes of ISFETs in CMOS which can either bring advantages or pose challenges.

10.5.3.1 Advantages of using ISFETs for sweat sensing

There are several advantages of using CMOS integrated ISFETs in wearable devices. First, with a small sensing area and well-established fabrication processes, a high sensor density can be achieved. With each ISFET requiring an area in the order of square micrometres, potentially thousands of sensors can fit on a single microchip. By studying approaches taken in optical imaging ICs, researchers design ISFET 'pixels' which can be replicated into dense arrays [13]. Using arrays brings advantages of spatial averaging to reduce noise and robustness through redundancy. For monitoring

Table 10.3 Key specifications for a wearable sweat sensor and related ISFET attributes which provide advantages or challenges

Specification	CMOS ISFET feature
	Advantages
Analysis of multiple biomarkers	High sensor density for arrays and multiplexing
Sample volume of up to 10s μl	Small sensing area (<10s μm^2) allows practical integration with microfluidics
Comfortable and conformal to wear	CMOS chips typically a few mm^2 can be built into a patch
Good sensitivity and low noise	Integration of front-end amplification and filtering through custom circuits can improve signal magnitude and quality
Low-cost & disposable sensing element	Low unit cost with economies of scale
Low-cost power source with a small form factor	Ultra-low power operation possible with conventional techniques. Feasible to use without batteries
	Challenges
Robust to temperature variations	ISFETs are highly temperature dependent. Compensation and monitoring of local temperature is required
Automatic calibration	Variations in threshold voltage due to trapped charge make this challenging. Use of wide range bias voltages and calibration algorithms required
Measurement times up to hours	Challenging due to drift in $V_{th\text{-}ISFET}$ over time. Compensation required
Integrated reference electrode	Miniaturisation of reference electrode requires post-processing of materials and membranes to the chip surface, close to the sensing area
Absolute measurements	ISFET output will only show a difference in ion-concentration, so calibration against a known solution is required

multiple components of sweat, it has been shown that ion-selective membranes can be applied to the surface of ISFET arrays to detect multiple ions simultaneously. With individually addressable pixels, arrays can be customisable and calibrated in-pixel once membranes have been applied [69]. Additional benefits of a small individual sensor size are that operation is possible with very low sample volume, as required due to low sweat generation rates, and the overall device size can be smaller, improving wearability.

Non-integrated sensors require a signal to travel along a wire before reaching an amplifier. For example, if flexible electrodes are used in a wearable device, this could introduce noise due to bending of wires and additional parasitics. In contrast, the ISFET itself provides immediate gain at the signal source to minimise input-referred noise. The level of integration offered by CMOS technology allows custom signal processing and control systems to be included as close to the signal source as possible [64]. With low-noise, tunable amplification and filtering, an ISFET-based system could be suitable to detect the wide range of analyte concentrations present in sweat [74]. Advanced processing can also integrated such as algorithms to detect signal patterns and improve power management. Integrating the full system on a single IC also allows the smallest possible form factor for the wearable device, minimising size and weight to be completely unobtrusive to the user. While many existing devices rely on flexible electronics to be conformable, a typical microchip has an area of a few square millimetres and so can still be mounted onto a patch flexible enough to be worn on most areas of the body. Finally, by minimising the number of components, the manufacturing complexity and therefore cost is reduced.

In a wearable device, maintaining a low power consumption is a key factor to ensure wearability, a long operational life and low-cost, all through minimising battery size. We saw in Section 10.4 that many devices are now forgoing a battery altogether in favour of energy harvesting or wireless powering. Reducing cost is a particularly important since most electrochemical sensors have a short usable life before they need to be replaced, due to surface degradation [4]. The use of CMOS technology allows very low-power consumptions to be achieved, and thus battery-less operation is feasible. Creating a device without a battery would reduce both the monetary and environmental cost of the disposable wearable sensor. Finally, since CMOS foundries are optimised for mass manufacture, making use of this infrastructure can bring a very low cost per chip.

10.5.3.2 Challenges of sweat sensing with ISFETs

The main challenge for sweat sensors based on CMOS ISFETs is drift. Drift appears as a slow variation in ISFET threshold voltage over time and is thought to be in part due to the hydration of the sensor surface layer [13]. The concentrations of components in sweat change slowly over long periods of time, with reported rates of change for sweat pH ranging from 0.2 to 0.075 pH/min [25,39]. As a result, depending on the sensitivity of the ISFET, drift rate can be comparable to the signal making it difficult to identify and quantify a concentration change [72]. On top of that, if the target application is exercise monitoring, activities can often last over an hour. This means the ISFET sensor is required to stay in the correct operating region for this length of time, however drift can cause the threshold voltage to shift such that the device switches off or saturates. Any device constructed for long-term sensing needs to be able to compensate or recalibrate to overcome this issue, either in hardware or software. Drift can be thought of as low-frequency chemical noise which is difficult to eliminate with techniques such as chopping as it typically exists in similar frequency bands as the signal [72]. To help reduce drift, pre-hydrating of the sensing surface, operating in dark conditions and repassivation of the sensing surface have all been shown to be effective [11,24].

An additional challenge is trapped charge. Since the ISFET is a floating gate device, charge can become trapped in the passivation or gate metal during fabrication or handling, inducing offsets in threshold voltage that can vary between pixels in an array. In standard CMOS, the spread can be quite wide and require either positive or negative compensation voltages [72]. While some applications can use a reset switch at the floating gate to remove charge, this solution is not suitable for low-frequency signals such as the pH change of sweat due to current leakage in the switch. The wider the range of possible voltage biases which can be applied, the more likely it is that the ISFET array can be calibrated. However, in a wearable device only, the battery supply voltage is available. The bias voltage range can be extended using a charge pump IC or other circuit techniques, however this requires additional power. It has been shown that post-processing steps such as UV exposure [75], reactive ion etching (RIE) [76] or re-passivation [24] can significantly reduce trapped charge.

Temperature dependence of ISFETs becomes more complex when used outside of the lab. An ISFET has all the temperature-dependent parameters of a MOSFET plus that of the chemical system, for example in the surface potential. In a wearable device, sensors can be affected by both the temperature of the skin surface and the environment, both of which are variable, albeit within a limited range. Skin temperature variation depends on the location on the body, with the core having a smaller variation that the extremities. Webb recorded the skin temperature at different locations across the body for males at rest in different ambient temperatures [77]. The lower back ranged from between 30 and 38°C in cold (15°C) and hot (45°C) ambient temperatures respectively. The ambient temperature depends on a number of factors, for example whether the user is inside or outside and the climate. A final consideration is the temperature of the sweat sample itself, which could be the main dictator of sensor temperature in mild ambient conditions due to the thermal conductivity of water. Thus a device should have a method for measuring and compensating for such variations.

The integration of the reference electrode into the package also needs to be considered carefully. An external connection and wire would undermine the small form factor that is desired, and would also make it difficult for a sweat sample to make a reliable bridge between electrode and the sensing surface. A likely solution to this is to add a micro Ag/AgCl electrode to the chip surface in a post processing step, which minimises the area a sweat sample would need to cover. For this solution, Ag/AgCl paste could be used. It should be noted, that when using a pseudo-reference electrode such as this, an additional solid-state membrane is needed to provide a constant chloride reference in a complex solution like sweat [78,79].

10.5.4 Existing ISFET-based wearable sweat sensors

Having observed the general trends in sweat sensing devices in Section 10.4.1, it is clear that the most popular choice of sensor is an ISE. However, over the past 7 years, ISFETs have become increasingly popular choice for this application. This section will now focus on the devices which use ISFETs as a sweat sensor and how they align with the trends discussed previously.

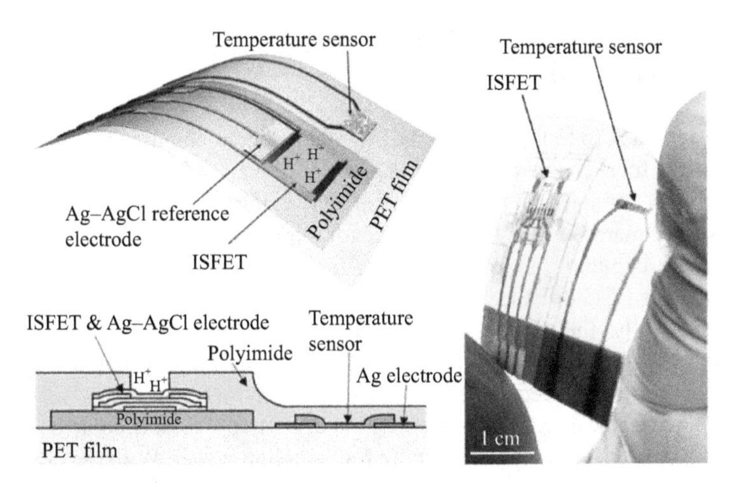

Figure 10.7 An example of a sweat monitoring device using a discrete InGaZnO-based ISFET printed on a flexible substrate. Reprinted with permission from [50]. Copyright 2017 American Chemical Society.

Early examples of ISFETs for sweat analysis focused on discrete forms of transistors, where a stand-alone sensor is characterised through a connection to external electronics. The first FET tested specifically for sweat sensing was presented by Cazalé *et al.* [80]. The sensors were each manufactured on a silicon wafer as single devices measuring 2.1 mm $\times$ 1.6 mm and were tested against pH and Na^+, the latter by adding a polymer membrane. Later, the same group used the same sensor alongside ISEs for a follow up study for physiological stress monitoring through sweat analysis [81]. Subsequently, Nakata *et al.* presented an InGaZnO-based ISFET sensor printed on a flexible polyimide substrate for monitoring sweat pH [50], shown in Figure 10.7. The device was not directly integrated with electronics, but was tested on-body with a wired connection. Both of these devices were presented as single sensor prototypes, but are well characterised and prove the viability of using FETs for sweat sensing.

More recently, CMOS ISFET arrays have been shown to be suitable for multi-ion sensing, to detect sodium, potassium, calcium and magnesium on a single microchip [69,78]. Zhang *et al.* presented the first integrated ISFETs specifically for multiple-ion sensing in sweat [24]. This involved etching into the passivation layer in order to reduce non-ideal effects associated with ISFETs, leading to a robust sensing device with good sensitivity and selectivity. A form factor in the style of an NFC patch was also proposed for batteryless operation, but at the time of writing this proposal is only a concept and the system has not been tested with on-body trials. While CMOS ISFET arrays have yet to be demonstrated with on-body measurements, this milestone is fast approaching. Once achieved, the task of integrating further features, including advanced processing and ML.

10.6 Applications of machine learning in wearable electrochemical sensors

Taking a step back to look at the field of wearable electrochemical sensors, and particularly those focused on sweat analysis presented so far, it is not hard to imagine the potential impact of machine learning. In the review of literature carried out in Section 10.4.1, we saw many devices that can detect multiple biomarkers simultaneously [22] and some are beginning to combine these measurements with other physiological signals, such as ECG or EEG.

Consequently, the amount of physiological data obtained by these sensors is increasing, which is ideal for building data-sets for machine learning. From a device perspective, wearables need to process and transmit the information obtained from the body. In this case, the increasing popularity of edge-ML processors could be beneficial to reducing data at the source and saving transmission power. With developments in energy-efficient hardware machine learning, classifications could be carried out on-chip, thus improving the diagnostic capabilities of wearable sweat sensors [5]. At the system-level, ML could be applied off-device (e.g. in the cloud or larger receiver) in order to perform deeper analysis of the collected information to improve the quality of measurements taken as well as increasing the range of inferences that can be made.

10.6.1 Existing research into ML for biosensors

Recently, Cui *et al.* reported that ML has yet to be applied widely to electrochemical sensors [14]. However, due to the complexities of the sensors themselves in terms of non-idealities and degradation over time, there is an opportunity to use ML to improve the sensitivity and reliability of such sensors. They review several ML algorithms and their effectiveness for different applications summarised. A popular option that has been demonstrated for cyclic voltammetry and amperometry is support vector machine (SVM) regression, which has been applied to improve the performance of bench-top instruments [82] and portable sensors [83,84].

There has been further work developing multi-modal wearable sensors with machine learning. Zeng *et al.* presented a epidermal electronic system for measuring electrophysiological signals including respiration rate, electrocardiogram (ECG) and galvanic skin response in order to characterise fatigue [85]. Veeralingam *et al.* [61] presented a wearable device with embedded ML to classify data from sensors measuring skin hydration, sweat glucose and pH. They used a k-nearest neighbour (KNN) algorithm which was chosen to quickly process sensor data for real-time results. The device represents the first wearable with edge-ML for the application of sweat monitoring. Recently, a flexible circuit for machine learning has been demonstrated for classifying results of a gas sensors [86]. Such technology could be implemented in a similar way for electrochemical sensors.

10.6.2 Existing research into ML for ISFETs

There has been some progress in developing ML for ISFETs. Hsu *et al.* utilise ML algorithms to allow ISFETs to detect both pH and photons for light sensitivity [87].

They explore the use of support vector machines and back-propagation neural networks to separate the chemical and optical signals. The ability of the ML-enhanced sensors to partially quantify (through classification) the pH level and light intensity is assessed, showing that An SV classifier is more suitable achieving a higher accuracy for lower computation. In an effort to fully quantify the two signals, regression models are used, with the BPNN shown to be more powerful in dealing with varying conditions and device mismatch. This investigation demonstrates the possibility of using ML to both improve ISFET performance and allow differentiation between multiple coupled input signals.

Sinha *et al.* demonstrate the use of ML for temperature and drift compensation in ISFETs [88]. They first apply regression to create a SPICE that accurately reflects ISFET behaviour under temperature variations and due temporal drift. This model is then used to generate data to train a range of ML algorithms for temperature and drift compensation. The root mean square error of the algorithms is used to compare performance, and it is shown that a random forest algorithm gives best result for this problem. In general, this work shows that a range of machine learning algorithms can be applied to improve ISFET performance, compensating for the challenges described in Section 10.5.3.

Wang *et al.* [89] propose the use of sensor learning in ISFET arrays to cluster pixels with similar sensitivities and selectivities in multi-ion sensing. Offline training is performed using a density based clustering algorithm with noise (DBSCAN) to cluster the nearby ISFETs, and in-pixel memory allows the results of the training to be stored. The current-mode outputs of the pixels within a cluster are then averaged to reduce uncorrelated noise, which improves the output resolution. Finally, the system relaxes bandwidth requirements by reducing data to be transferred.

10.6.3 Integration of analogue classifiers with ISFETs in CMOS

To illustrate potential applications of integrating ML techniques directly with ISFETs, a hardware example which implements analogue classifiers for multi-analyte electrochemical sensing is presented. Analogue computing has long been known to provide advantages of energy efficiency and real-time computation [90]. Furthermore, bio-inspired neural networks are increasingly being used to classify sensor data and several groups have presented analogue-based neural networks in recent years [91,92], and neuromorphic sensor outputs are becoming popular [93].

However, with increasingly advanced CMOS processes, analogue computation is becoming more challenging due to features such as reduced supply voltages. To overcome these challenges, performing analogue operations with signals encoded in the time-domain has been proposed [94], and recently basic neural networks utilising time-domain inputs and outputs have been demonstrated [95,96]. In [95], we proposed a time-domain, analogue multiply-accumulate (MAC) operation circuit which used capacitors charged with weighted currents to produce analogue output. This principle will be described in more detail below. The design was simulated in a flexible electronic process with the intended application of wearable ISFET sensing and showed improvements a reduction in energy consumption when compared to an equivalent digital implementation in the pseudo-CMOS technology. Bavandpour *et al.* [96] presented a time-domain vector–matrix multiplier using the same operating principle. This work was designed in a 55 nm process with embedded NOR flash memory technology, allowing the use of floating-gate (FG) devices to implement the weights. The use of FG devices allows for easier configuration of weights, allowing on-line training.

In this section, this principle of performing multiplication and addition with time-domain inputs in CMOS will be briefly described. Such an architecture could interface to ISFET architectures with time-domain outputs [97,98], which are gaining popularity for robustness and scalability. As these are the underlying operations of many ML algorithms, such circuits bring intelligent classifiers in wearable chemical sensors a step closer.

10.6.3.1 A time-domain MAC engine architecture

The basic block of an artificial neural network (ANN) is the multiply-accumulate (MAC) operation, as shown in Figure 10.8. Each 'neuron' has a number of inputs, each with a corresponding weight. The products of these input-weight pairs are added together, typically along with a bias signal to form the MAC operation. The output of this operation is then typically subjected to a non-linear activation function. By connecting many neurons in a network, the weights can be tuned through training data to allow the network to make classifications.

The time-domain input MAC architecture is based on the concept that the charge stored on a capacitor is proportional to the product of the charging current and integration time. Furthermore, the voltage and the charge are related by a factor of $1/C$, such that, in a discrete form

$$V = \frac{1}{C}(I \times T) \tag{10.7}$$

where C is the capacitance of a capacitor. This relationship allows MAC operations to be performed efficiently using the inherent physics of electronics. In the proposed design, weights are set through the magnitude of currents and size of each input is determined by the length of time each corresponding current source is active. Together these values determine the charge of each pair. Each current signal is directed to a single capacitor which stores the charge from all sources, forming an addition. The input signals could be set by the pulse width output of a time-domain ISFET

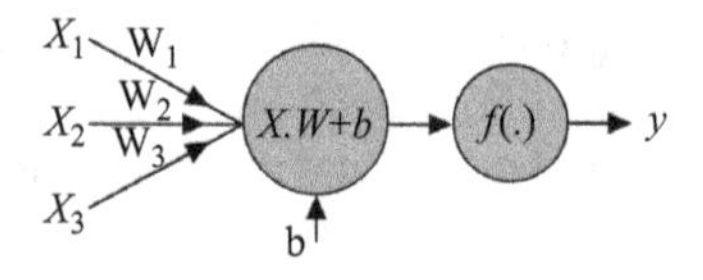

Figure 10.8 A concept image of a neural network on flexible electronics interfacing to a chemical sensing array. The operation of a single neuron is shown, X is a vector of N inputs, W is a vector of N weights, b is a bias term and y is the output value.

sensor [97]. The advantage of using this approach with ISFETs is that since both components can be implemented in CMOS, there is a direct interface between the required input electrochemical sensors and the neural network, reducing the required components and circuits which brings efficiency and fewer sources of noise. The disadvantages are that it becomes more difficult to create adjustable weights and train the neural network, as will be discussed later.

Current can be sourced or sunk from NMOS or PMOS current mirrors, creating positive and negative input-weight products. As a result, the CMOS design utilises a single capacitor per MAC operation, as p-type current mirrors can add current for a positive input-weight products and n-type current mirrors can remove current for negative input-weight products. This architecture is shown in Figure 10.9. In order to cascade stages, the analogue voltage of the capacitor needs to be converted back into a time-domain pulse. Typically, at this stage an non-linear activation function would also be applied for a neural network. Simple rectified linear unit (ReLu) is a popular choice and has been shown to be simply implemented using a single digital logic gate [96].

The networks described in this section only consist of few nodes and perform simple classifications, but they form a proof-of-concept for more complex designs that be implemented in future. Two examples are presented to demonstrate potential uses for this architecture with CMOS ISFETs, one for frequency classification and another for ion identification. The designs were simulated in a 0.18 μm CMOS process with a 1.8 V supply voltage. Inputs and weights were generated in MATLAB® and transferred to the cadence simulation using a Verilog-a block.

10.6.3.2 Frequency classification

The output of some time-domain sensors is a repeating pulse [98], so to identify the magnitude of the signal or measure the rate of drift, it is necessary to measure

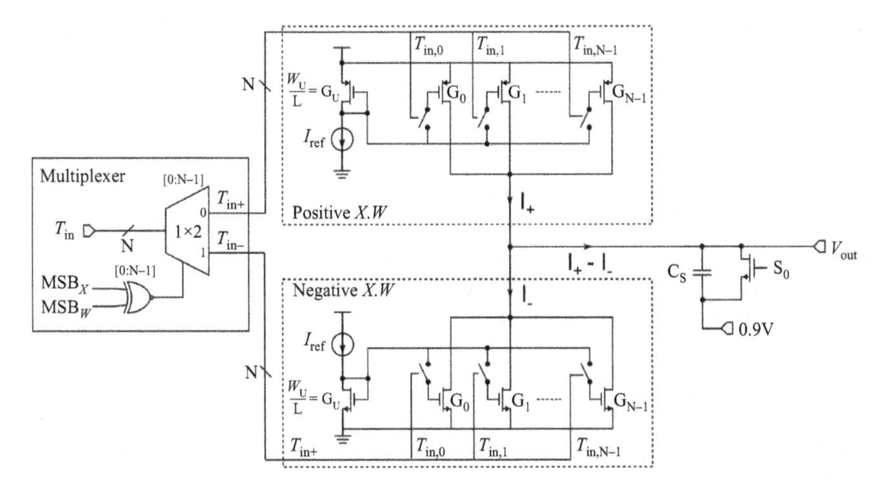

Figure 10.9 An architecture for a time-domain, current-mode MAC engine in CMOS technology

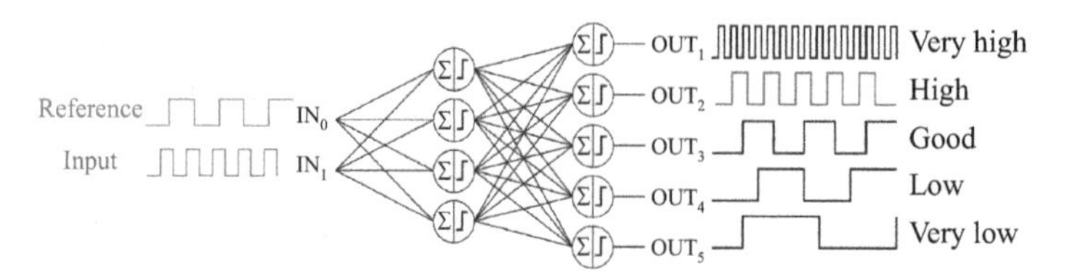

Figure 10.10 Architecture for frequency classification perceptron

frequency in comparison to a reference oscillator. Typically, such measurements are performed with a microcontroller, requiring interrupts, which adds significant complexity for a relatively simple output, and is slow if many sensors need to be read.

To achieve this, a two-layer perceptron was designed, which effectively implements a series of logical comparisons of the two input pulse widths. This is achieved by using a binary step function at the output of each MAC to produce a 1 or 0 result depending on the sign of the accumulation. 100 pairs of pulse widths were randomly generated between 0.5 and 1.5 μs and used as inputs to the topology described in Section 10.6.3.1. These pulse widths would represent half the period of the input frequency of an ISFET architecture with a time-domain output. The perceptron produces one of five outputs depending how far one input frequency deviates from the other, as illustrated in Figure 10.10.

The results of the simulation are shown in Figure 10.11, which shows the classification of each frequency pair and the classification boundaries. As expected the errors, which are circled in red, occur close to the boundaries. The overall accuracy for this dataset is 80%, which could be improved with further optimisation. The average power consumption is approximately 450 μW for a 10 μs compute time. In comparison, a microcontroller when fully active could draw power in the order of milliwatts, making this design an viable alternative for two inputs. However, future development should aim to reduce this power consumption as it is too high to be used for every pixel in an array, which may be desired for parallel read outs.

10.6.3.3 Ion identification

Another application for ISFETs is differentiating between ions in a multi-ion sensing array. Cross-sensitivity is a known issue when using unmodified CMOS sensing surfaces and ion-selective membranes. In [78], Moser *et al.* presented an ISFET array with ion-selective membranes, and showed that they also exhibit cross-sensitivity. If these sensitivities are known, then given an unknown solution it is possible to determine the quantities of ions present by performing an inverse matrix multiplication [78], as shown in (10.8):

$$\Delta a = S^{-1}.\Delta V \tag{10.8}$$

where $\mathbf{\Delta}a$ is the change in ion activity, $\mathbf{\Delta}V$ is the sensor output voltage and S is the cross-sensitivity matrix in mV given by 10.9, with each index denoted as S_{XY} representing the sensitivity of membrane X to ion Y [78]. Note that the activity of ion i, a_i, is related to the ion concentration, x_i, by the formula $a_i = \gamma_i x_i$, where γ_i

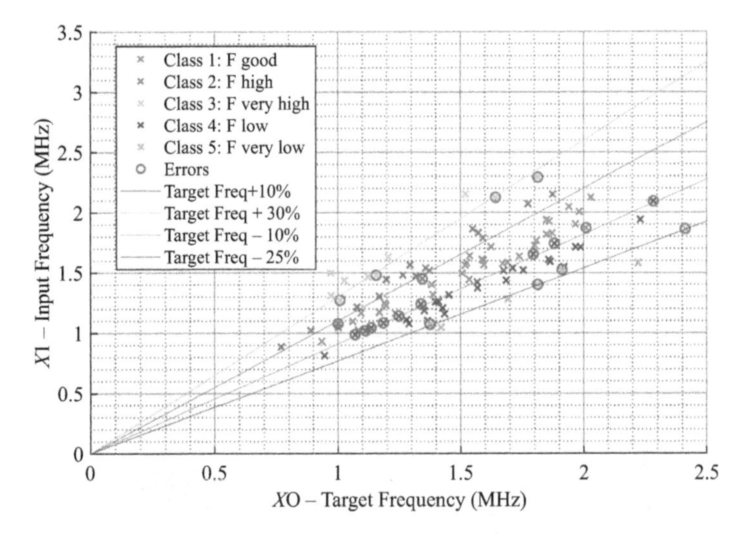

Figure 10.11 Classification of frequency difference

is the activity coefficient which depends on the ionic charge and properties of the solution [78]:

$$S = \begin{bmatrix} S_{K^+K^+} & S_{K^+Na^+} & S_{K^+Ca^{2+}} & S_{K^+H^+} \\ S_{Na^+K^+} & S_{Na^+Na^+} & S_{Na^+Ca^{2+}} & S_{Na^+H^+} \\ S_{Ca^{2+}K^+} & S_{Ca^{2+}Na^+} & S_{Ca^{2+}Ca^{2+}} & S_{Ca^{2+}H^+} \\ S_{H^+K^+} & S_{H^+Na^+} & S_{H^+Ca^{2+}} & S_{H^+H^+} \end{bmatrix} \begin{bmatrix} 58.9 & 8.15 & -1.48 & -1.33 \\ -0.99 & 53.83 & 1.85 & 2.71 \\ -1.71 & 10.73 & 25.31 & -1.96 \\ 14.8 & 24.3 & 2.14 & 19.2 \end{bmatrix} \tag{10.9}$$

Since this is simply a matrix multiplication, a vector–matrix multiplier (VMM) was built using the time-domain MAC topology, which consists of parallel matrix multipliers as shown in Figure 10.12. The output stage of the MACs in this case is a linear function, producing a time-domain output with a pulse width proportional to the final capacitor voltage. To test this design, the sensitivity matrix in 10.9 was used to create a sensor response for a set of randomly generated ion concentration changes (uniform distribution, 0 to 1 decade). These values were then converted into a time-domain signal set as the inputs to the VMM. The output is a pulse width for each ion that is proportional to the recovered concentration change. Figure 10.13 visualises this change between input and output data in the form of a 10 × 10 array, split into four equal sections. The raw sensor outputs indicate that there is a higher Na^+concentration than H^+, however once the inverse cross-sensitivity matrix has been applied it can be seen that the H^+ change is greater.

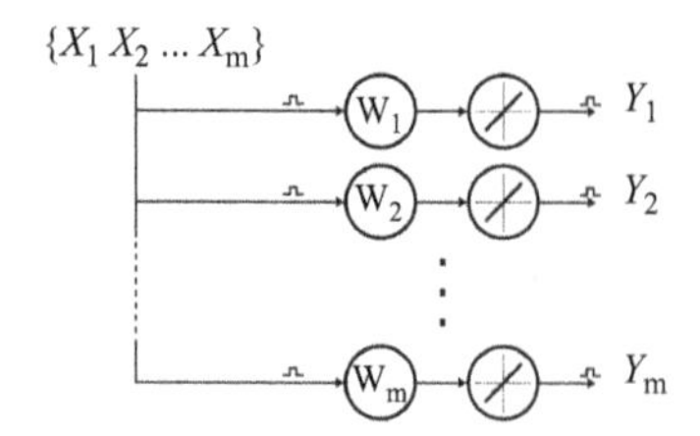

Figure 10.12 Block diagram of a matrix multiplier

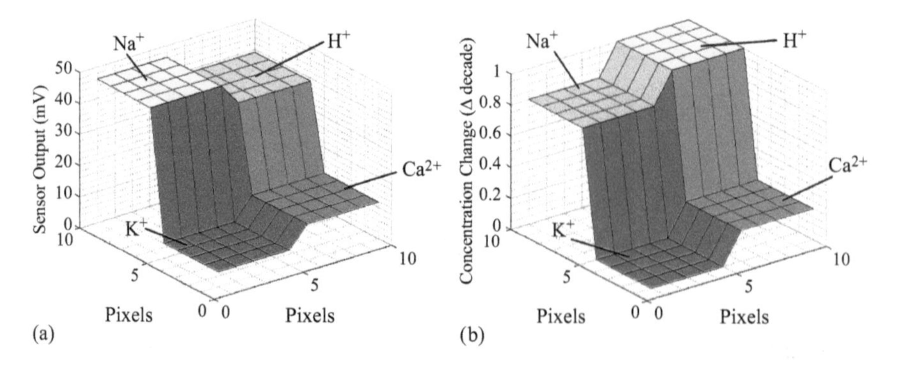

Figure 10.13 Input and output to the vector–matrix multiplier (not-scaled to time domain)

Figure 10.14 shows the output error of the VMM compared to the expected error due to rounding of the ideal weights. The error is found by taking the Euclidean distance between the generated data points and the actual points computed in MATLAB®. It can be seen that there is a significant drop in accuracy due to the errors accumulated in the analogue circuit, but it is possible to determine the output signal. The 4-input, 4-output VMM operates with an average power consumption of 64.7 μW over the 3 μs cycle.

10.6.3.4 Discussion

Sections 10.6.3.2 and 10.6.3.3 present two examples of how analogue classifiers could be used to improve the performance of ISFET-based sensing systems. A key metric for evaluating a neural network architecture is accuracy. The frequency classification problem is carried out by a simple two-layer perceptron and weights were created using the ideal classification boundaries, giving 80% accuracy due to the errors introduced during analogue computation such as deviations in switching threshold or final capacitor value. It is thought that this accuracy could be improved by training the network to overcome these errors if they are consistent across a number of samples. More generally, the impact of mismatch and process variation in analogue neural networks is a legitimate concern, however it has been proposed that by injecting noise during the training process that represents the expected random variations of analogue

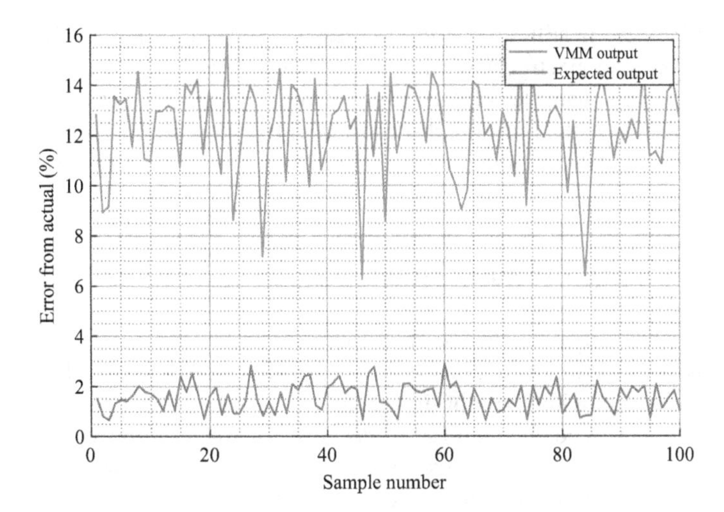

Figure 10.14 Output error of VMM based on euclidean distance between simulated output and ideal points

devices, such potential sources of error can be reduced [99]. In fact, in [100], Chen *et al.* used the random variation in subthreshold current mirrors to their advantage in a crossbar network by implementing an 'extreme learning machine' algorithm that requires normally distributed weights in a hidden layer.

Further consideration should also be given to the balance of current, maximum pulse width and capacitor size, the combination of which affects both power and accuracy. Reducing the reference current of the current mirrors would reduce power consumption, but also voltage swing of the capacitor node, decreasing accuracy. The same effect would be seen by reducing the maximum pulse width. However, since a key limitation of the design is capacitor size, it is beneficial to further reduce this, which would increase the voltage swing but also kT/C noise and the influence of parasitics. As a result, an investigation into the lowest current and pulse width and capacitor size allowable for a given accuracy, and how this scales with network width and depth, would be beneficial to optimise the design.

A potential drawback of the proposed design is the implementation of the weights in the current mirror blocks. This method would work well if training can be done off-line during the design process, taking into account potential process variations as discussed previously, and they do not need to be changed for the application. However, if online training is required, or weights need to be changed, then a large number of transistors and signals for switching them in and out would be required. Additionally, there is a finite precision which introduces quantisation noise, as seen in Figure 10.14. For matching purposes it is typically best to implement current mirror ratios in multiple transistors of unit widths, rather than two different sized transistors, but this makes the weights discrete. A possible alternative solution would be to use a time-domain multiplier, as proposed in [94], which would add an additional signal for every weight, but allow continuous weights encoded in time.

Overall, there is still significant progress to be made to realise full analogue neural networks for electrochemical sensors. However, as neural networks are ideal for handling large quantities of data, using them to integrate directly with the huge number of ISFETs in state-of-the-art arrays and with other sensors, without needing to first convert to digital, offers an exciting prospect for efficient handling of information at the edge.

10.7 Summary and conclusions

This chapter has aimed to provide an introduction to the topics of wearable sweat sensors and ISFETs in CMOS, and provided a review of existing work on integrating machine learning into both of these fields. While these investigations are early stage, there is a lot of potential for further development to help improve sensor performance and identify patterns amongst complex bio-fluids such as sweat. For example, the balance between $Na^{+}$and pH could determine whether an athlete is at risk of hypo/hyper-natremia (a lack or excess of sodium respectively) [8,18]. One challenge is that our understanding of the complexities of sweat components and how they relate to underlying physiology are not fully understood. While ML algorithms could aid with identifying patterns and connections, robust wearable biosensors and carefully planned studies are required to generate the large data sets to make this feasible. There are significant research efforts around the world developing such sensors, and the use of CMOS technology is key to allow the integration of multiple sensors and suitable signal processing to create robust interfaces. We have already seen that ML can be harnessed to further improve the performance. With key trends of wearable electrochemical sensors being both multi-analyte and multi-modal sensing, such as the addition of sensors for electrophysiological signals such as heart rate (ECG) and brain activity (EEG) have been demonstrated [35,48], the quantity of data that will be obtained by these devices will only increase. In this aspect, ISFETs could prove advantageous due to their CMOS compatibility and density, producing the large quantities of data required in a compact form.

Acknowledgements

The authors would like to thank Wesley Gaunt for his contribution to the work described in Section 10.6.3 as part of his Master's project.

References

[1] Yoo HJ, Yoo J, and Yan L. Wireless fabric patch sensors for wearable healthcare. In: 2010 Annual International Conference of the IEEE Engineering in Medicine and Biology Society EMBC'10. Buenos Aires, 2010. p. 5254–5257.
[2] Dunn J, Runge R, and Snyder M. Wearables and the medical revolution. *Per. Med.* 2018;15(5):429–448.

[3] Imani S, Mercier PP, Bandodkar AJ, *et al.* Wearable chemical sensors: opportunities and challenges. In: *Proceeding of IEEE International Symposium on Circuits and Systems*, Montreal, Canada, 2016. p. 1122–1125.
[4] Kim J, Campbell AS, de Ávila BEF, *et al.* Wearable biosensors for healthcare monitoring. *Nat. Biotechnol.* 2019:389–406. Available from: http://www.nature.com/articles/s41587-019-0045-y.
[5] Bariya M, Nyein HYY, and Javey A. Wearable sweat sensors. *Nat. Electron.* 2018;1(3):160–171. Available from: http://www.nature.com/articles/s41928-018-0043-y.
[6] Heikenfeld J, Jajack A, Feldman B, *et al.* Accessing analytes in biofluids for peripheral biochemical monitoring. *Nat. Biotechnol.* 2019;37:407–419. Available from: http://www.nature.com/articles/s41587-019-0040-3.
[7] Sonner Z, Wilder E, Heikenfeld J, *et al.* The microfluidics of the eccrine sweat gland, including biomarker partitioning, transport, and biosensing implications. *Biomicrofluidics.* 2015;9(3):031301. Available from: http://www.pubmedcentral.nih.gov/articlerender.fcgi?artid=4433483&tool=pmcentrez&rendertype=abstract.
[8] Sato K, Kang WH, Saga K, *et al.* Biology of sweat glands and their disorders. I. Normal sweat gland function. *J. Am. Acad. Dermatol.* 1989;20(4):537–563.
[9] Mena-Bravo A and Luque de Castro MD. Sweat: a sample with limited present applications and promising future in metabolomics. *J. Pharm. Biomed. Anal.* 2014;90:139–147. Available from: http://dx.doi.org/10.1016/j.jpba.2013.10.048.
[10] Baker LB. Sweat testing methodology in the field: challenges and best practices. *Sport Sci. Exch.* 2016;28(161):1–6.
[11] Bergveld P. Thirty years of ISFETOLOGY: what happened in the past 30 years and what may happen in the next 30 years. *Sensors Actuat. B Chem.* 2003;88(1):1–20. Available from: http://linkinghub.elsevier.com/retrieve/pii/S0925400502003015.
[12] Rothberg JM, Hinz W, Rearick TM, *et al.* An integrated semiconductor device enabling non-optical genome sequencing. *Nature.* 2011;475(7356):348–352.
[13] Moser N, Lande TS, Toumazou C, *et al.* ISFETs in CMOS and emergent trends in instrumentation: a review. *IEEE Sensors J.* 2016;16(17):6496–6514.
[14] Cui F, Yue Y, Zhang Y, *et al.* Advancing biosensors with machine learning. *ACS Sensors.* 2020;5(11):3346–3364.
[15] Baker LB and Wolfe AS. *Physiological Mechanisms Determining Eccrine Sweat Composition.* New York, NY: Springer Berlin Heidelberg;120:719–752. Available from: https://doi.org/10.1007/s00421-020-04323-7.
[16] Rose D, Prendergast J, Ratterman M, *et al.* Adhesive RFID sensor patch for monitoring of sweat electrolytes. *IEEE Trans. Biomed. Eng.* 2015;62(6):1457–1465. Available from: http://www.ncbi.nlm.nih.gov/pubmed/25398174.
[17] Alizadeh A, Burns A, Lenigk R, *et al.* A wearable patch for continuous monitoring of sweat electrolytes during exertion. *Lab Chip.* 2018;18(17):2632–2641.

[18] Buono MJ, Ball KD, and Kolkhorst FW. Sodium ion concentration vs. sweat rate relationship in humans. *J. Appl. Physiol.* 2007;103(3):990–994. Available from: http://www.physiology.org/doi/10.1152/japplphysiol.00015.2007.

[19] Lynch A, Diamond D, and Leader M. Point-of-need diagnosis of cystic fibrosis using a potentiometric ion-selective electrode array. *Analyst.* 2000;125(12):2264–2267. Available from: http://eutils.ncbi.nlm.nih.gov/entrez/eutils/elink.fcgi?dbfrom=pubmed&id=11219064&retmode=ref&cmd=prlinks%5Cnpapers2://publication/uuid/E019D6DB-68B9-4002-8E45-2ADCF85F3262.

[20] Choi DH, Thaxton A, Cheol Jeong I, *et al.* Sweat test for cystic fibrosis: wearable sweat sensor vs. standard laboratory test. *J. Cyst. Fibros.* 2018;17(4): E35-8.

[21] Schmid-Wendtner MH and Korting HC. The pH of the skin surface and its impact on the barrier function. *Skin Pharmacol. Physiol.* 2006;19(6): 296–302.

[22] Gao W, Emaminejad S, Nyein HYY, *et al.* Fully integrated wearable sensor arrays for multiplexed in situ perspiration analysis. *Nature.* 2016;529(7587):509–514. Available from: http://www.nature.com/doifinder/10.1038/nature16521.

[23] Parrilla M, Ortiz-Gómez I, Canovas R, *et al.* Wearable potentiometric ion patch for on-body electrolyte monitoring in sweat: towards a validation strategy to ensure physiological relevance. *Anal. Chem.* 2019;91(13):8644–8651. Available from: http://pubs.acs.org/doi/10.1021/acs.analchem.9b02126.

[24] Zhang J, Rupakula M, Bellando F, *et al.* Sweat biomarker sensor incorporating picowatt, three-dimensional extended metal gate ion sensitive field effect transistors. *ACS Sensors.* 2019;4(8):2039–2047. Available from: http://pubs.acs.org/doi/10.1021/acssensors.9b00597.

[25] Nyein HYY, Gao W, Shahpar Z, *et al.* A wearable electrochemical platform for noninvasive simultaneous monitoring of Ca2+ and pH. *ACS Nano.* 2016;10(7):7216–7224.

[26] Kim J, de Araujo WR, Samek IA, *et al.* Wearable temporary tattoo sensor for real-time trace metal monitoring in human sweat. *Electrochem. Commun.* 2015;51:41–45. Available from: http://www.sciencedirect.com/science/article/pii/S1388248114003701.

[27] Mondal S and Subramaniam C. Point-of-care, cable-type electrochemical Zn 2+ sensor with ultra-high sensitivity and wide detection range for soil and sweat analysis. *ACS Sustain. Chem. Eng.* 2019;7(17):14569–14579.

[28] Kim J, Sempionatto JR, Imani S, *et al.* Simultaneous monitoring of sweat and interstitial fluid using a single wearable biosensor platform. *Adv. Sci.* 2018;5(10):1800880. Available from: http://doi.wiley.com/10.1002/advs.201800880.

[29] He W, Wang C, Wang H, *et al.* Integrated textile sensor patch for real-time and multiplex sweat analysis. *Sci. Adv.* 2019;5(11):1–9.

[30] Lu Y, Jiang K, Chen D, *et al.* Wearable sweat monitoring system with integrated micro-supercapacitors. *Nano Energy.* 2019;58:624–632. Available from: https://doi.org/10.1016/j.nanoen.2019.01.084.

[31] Zhao J, Lin Y, Wu J, *et al.* A fully integrated and self-powered smartwatch for continuous sweat glucose monitoring. *ACS Sensors*. 2019;4(7):1925–1933.
[32] Yokus MA, Songkakul T, Pozdin VA, *et al.* Wearable multiplexed biosensor system toward continuous monitoring of metabolites. *Biosens. Bioelectron*. 2020;153:112038. Available from: https://doi.org/10.1016/j.bios.2020.112038.
[33] Koh A, Kang D, Xue Y, *et al.* A soft, wearable microfluidic device for the capture, storage, and colorimetric sensing of sweat. *Sci. Transl. Med.* 2016;8(366):366ra165.
[34] Hauke A, Simmers P, Ojha YR, *et al.* Complete validation of a continuous and blood-correlated sweat biosensing device with integrated sweat stimulation. *Lab Chip*. 2018;18(24):3750–3759.
[35] Rosa BG, Anastasova-ivanova S, and Yang GZ. A low-powered and wearable device for monitoring sleep through electrical, chemical and motion signals recorded over the head. In: *IEEE Biomedical Circuits and Systems Conference*. Naja, Kapan: IEEE, 2019. p. 1–4.
[36] Madhu S, Allen JA, Ramasamy S, *et al.* ZnO nanorods integrated flexible carbon fibers for sweat cortisol detection. *ACS Appl. Electron. Mater.* 2020;2(2):499–509.
[37] Taylor NAS and Machado-Moreira CA. Regional variations in transepidermal water loss, eccrine sweat gland density, sweat secretion rates and electrolyte composition in resting and exercising humans. *Extrem. Physiol. Med.* 2013;2(4):1–29.
[38] Barnes KA, Anderson ML, Stofan JR, *et al.* Normative data for sweating rate, sweat sodium concentration, and sweat sodium loss in athletes: an update and analysis by sport. *J. Sports Sci*. 2019;37(20):2356–2366. Available from: https://doi.org/10.1080/02640414.2019.1633159.
[39] Bandodkar AJ, Hung VWS, Jia W, *et al.* Tattoo-based potentiometric ion-selective sensors for epidermal pH monitoring. *Analyst*. 2013;138(1): 123–128. Available from: http://www.ncbi.nlm.nih.gov/pubmed/23113321.
[40] Liu C, Xu T, Wang D, *et al.* The role of sampling in wearable sweat sensors. *Talanta*. 2020;212:120801. Available from: https://doi.org/10.1016/j.talanta.2020.120801.
[41] Coyle S, Lau KT, Moyna N, *et al.* BIOTEX: biosensing textiles for personalised healthcare management. *IEEE Trans. Inf. Technol. Biomed.* 2010;14(2):364–370. Available from: http://ieeexplore.ieee.org/xpl/freeabs_all.jsp?arnumber=5373946.
[42] Boysen TC, Yanagawa S, Sato F, *et al.* A modified anaerobic method of sweat collection. *J. Appl. Physiol.* 1984;56(5):1302–1307. Available from: http://www.ncbi.nlm.nih.gov/pubmed/6327585.
[43] Twine NB, Norton RM, Brothers MC, *et al.* Open nanofluidic films with rapid transport and no analyte exchange for ultra-low sample volumes. *Lab Chip*. 2018;18(18):2816–2825.
[44] Garcia-Cordero E, Wildhaber F, Bellando F, *et al.* Embedded passive nano-liter micropump for sweat collection and analysis. In: *Proceedings of*

the IEEE International Conference on Micro Electro Mechanical Systems, 2018. p. 1217–1220.

[45] Li S, Ma Z, Cao Z, *et al.* Advanced wearable microfluidic sensors for healthcare monitoring. *Small* 2020;16(9):1903822.

[46] Terse-Thakoor T, Punjiya M, Matharu Z, *et al.* Thread-based multiplexed sensor patch for real-time sweat monitoring. *npj Flex Electron.* 2020;4(1): 1–10. Available from: http://dx.doi.org/10.1038/s41528-020-00081-w.

[47] Yu Y, Nassar J, Xu C, *et al.* Biofuel-powered soft electronic skin with multiplexed and wireless sensing for human–machine interfaces. *Sci. Robot.* 2020;5(41):eaaz7946. Available from: https://robotics.sciencemag.org/lookup/doi/10.1126/scirobotics.aaz7946.

[48] Imani S, Bandodkar AJ, Mohan AMV, *et al.* A wearable chemical–electrophysiological hybrid biosensing system for real-time health and fitness monitoring. *Nat. Commun.* 2016;7(1):11650. Available from: http://www.nature.com/doifinder/10.1038/ncomms11650.

[49] Douthwaite M, Koutsos E, Yates DC, *et al.* A thermally powered ISFET array for on-body pH measurement. *IEEE Trans. Biomed. Circuits Syst.* 2017;11(6):1324–1334.

[50] Nakata S, Arie T, Akita S, *et al.* Wearable, flexible, and multifunctional healthcare device with an ISFET chemical sensor for simultaneous sweat pH and skin temperature monitoring. *ACS Sensors.* 2017;2(3):443–448. Available from: http://pubs.acs.org/doi/abs/10.1021/acssensors.7b00047.

[51] Anastasova S, Crewther B, Bembnowicz P, *et al.* A wearable multi-sensing patch for continuous sweat monitoring. *Biosens. Bioelectron.* 2017;93(September 2016):139–145. Available from: http://dx.doi.org/10.1016/j.bios.2016.09.038.

[52] Kim SB, Lee KH, Raj MS, *et al.* Soft, skin-interfaced microfluidic systems with wireless, battery-free electronics for digital, real-time tracking of sweat loss and electrolyte composition. *Small.* 2018;14(45):1–9.

[53] Wang G, Zhang S, Dong S, *et al.* Stretchable optical sensing patch system integrated heart rate, pulse oxygen saturation and sweat pH detection. *IEEE Trans. Biomed. Eng.* 2018;66(4):1000–1005.

[54] Dang W, Manjakkal L, Navaraj WT, *et al.* Stretchable wireless system for sweat pH monitoring. *Biosens. Bioelectron.* 2018;107:192–202. Available from: http://linkinghub.elsevier.com/retrieve/pii/S0956566318301131.

[55] Chen Cm, Anastasova S, Zhang K, *et al.* Towards wearable and flexible sensors and circuits integration for stress monitoring. *IEEE J. Biomed. Heal. Informatics.* 2020 aug;24(8):2208–2215.

[56] Hanitra IN, Criscuolo F, Carrara S, *et al.* Multi-target electrolyte sensing front-end for wearable physical monitoring. In: *2019 15th Conference on Ph.D Research in Microelectronics and Electronics*, Lausanne, Switzerland, 2019. p. 249–252.

[57] Liu X, Zhao L, Miao B, *et al.* Wearable multiparameter platform based on AlGaN/GaN high-electron-mobility transistors for real-time monitoring of pH and potassium ions in sweat. *Electroanalysis.* 2019;32(2):422–428.

[58] Nappi S, Mazzaracchio V, Fiore L, *et al.* Flexible pH sensor for wireless monitoring of the human skin from the medium distances. In: *IEEE International Conference on Flexible and Printable Sensors and Systems*, Glasgow, UK, 2019. p. 1–3. Available from: https://ieeexplore.ieee.org/document/8792291/.

[59] Rosa BMG, Anastasova-Ivanova S, and Yang GZ. NFC-powered flexible chest patch for fast assessment of cardiac, hemodynamic, and endocrine parameters. *IEEE Trans. Biomed. Circuits Syst.* 2019;13(6):1603–1614.

[60] Sempionatto JR, Martin A, García-Carmona L, *et al.* Skin-worn soft microfluidic potentiometric detection system. *Electroanalysis*. 2019;31(2):239–245.

[61] Veeralingam S, Khandelwal S, and Badhulika S. AI/ML-enabled 2-D – RuS2 nanomaterial-based multifunctional, low cost, wearable sensor platform for non-invasive point of care diagnostics. *IEEE Sensors J.* 2020;20(15): 8437–8444.

[62] Bergveld P. Short communications: Development of an ion-sensitive solid-state device for neurophysiological measurements. *IEEE Trans. Biomed. Eng.* 1970;BME-17(1):70–71.

[63] Bausells J, Carrabina J, Errachid A, *et al.* Ion-sensitive field-effect transistors fabricated in a commercial CMOS technology. *Sensors Actuators, B Chem.* 1999;57(1):56–62.

[64] Baker RJ. *CMOS Circuit Design, Layout and Simulation*, 3rd ed. New York, NY: Wiley-IEEE Press, 2010.

[65] Yates DE, Levine S, and Healy TW. Site-binding model of the electrical double layer at the oxide/water interface. *J. Chem. Soc., Faraday Trans 1.* 1974;70:1807–1818.

[66] Bard A and Faulkner LR. Electrochemical Methods: Fundamentals and Applications. Hoboken, NJ: John Wiley & Sons, Inc., 2001.

[67] Bergveld P. ISFET, theory and practice. In: *IEEE Sensors Conference*, October, Toronto, 2003. p. 1–26.

[68] Malpartida-Cardenas K, Miscourides N, Rodriguez-Manzano J, *et al.* Quantitative and rapid *Plasmodium falciparum* malaria diagnosis and artemisinin-resistance detection using a CMOS Lab-on-Chip platform. *Biosens. Bioelectron.* 2019;145:111678. Available from: https://doi.org/10.1016/j.bios.2019.111678.

[69] Moser N, Leong CL, Hu Y, *et al.* An ion imaging ISFET array for potassium and sodium detection. In: *2016 IEEE International Symposium on Circuits and Systems*, Montreal, QC, Canada, 2016. p. 2847–2850.

[70] Zeng J, Miscourides N, and Georgiou P. A 128 × 128 current-mode ultra-high frame rate ISFET array for ion imaging. In: *2018 IEEE International Symposium on Circuits and Systems*, Florence, Italy, 2018. p. 1–5. Available from: https://ieeexplore.ieee.org/document/8351178/.

[71] Martinoia S and Massobrio G. Behavioral macromodel of the ISFET in SPICE. *Sensors Actuators, B Chem.* 2000;62(3):182–189.

[72] Liu Y, Georgiou P, Prodromakis T, *et al.* An extended CMOS ISFET model incorporating the physical design geometry and the effects on performance and offset variation. *IEEE Trans. Electron. Devices*. 2011;58(12):4414–4422.

[73] Georgiou P and Toumazou C. ISFET characteristics in CMOS and their application to weak inversion operation. *Sensors Actuators, B Chem.* 2009;143(1):211–217.

[74] Dei M, Aymerich J, Piotto M, *et al.* CMOS interfaces for Internet-of-wearables electrochemical sensors: trends and challenges. *Electronics*. 2019;8:150.

[75] Milgrew MJ and Cumming DRS. Matching the transconductance characteristics of CMOS ISFET arrays by removing trapped charge. *IEEE Trans. Electron. Devices*. 2008;55(4):1074–1079.

[76] Moser N, Panteli C, Fobelets K, *et al.* Mechanisms for enhancement of sensing performance in CMOS ISFET arrays using reactive ion etching. *Sensors Actuators, B Chem.* 2019;292:297–307. Available from: https://doi.org/10.1016/j.snb.2019.04.031.

[77] Webb P. Temperatures of skin, subcutaneous tissue, muscle and core in resting men in cold, comfortable and hot conditions. *Eur. J. Appl. Physiol. Occup. Physiol.* 1992;64(5):471–476.

[78] Moser N, Leong CL, Hu Y, *et al.* CMOS potentiometric FET array platform using sensor learning for multi-ion imaging. *Anal. Chem.* 2020;92(7): 5276–5285. Available from: https://pubs.acs.org/doi/10.1021/acs.analchem.9b05836.

[79] Bellando F, Mele LJ, Palestri P, *et al.* Sensitivity, noise and resolution in a beol-modified foundry-made isfet with miniaturized reference electrode for wearable point-of-care applications. *Sensors*. 2021;21(5):1–19.

[80] Cazalé A, Sant W, Launay J, *et al.* Study of field effect transistors for the sodium ion detection using fluoropolysiloxane-based sensitive layers. *Sensors Actuators, B Chem.* 2013;177:515–521. Available from: http://dx.doi.org/10.1016/j.snb.2012.11.054.

[81] Cazalé A, Sant W, Ginot F, *et al.* Physiological stress monitoring using sodium ion potentiometric microsensors for sweat analysis. *Sensors Actuators B, Chem.* 2016;225:1–9. Available from: http://linkinghub.elsevier.com/retrieve/pii/S0925400515305827.

[82] Gonzalez-Navarro FF, Stilianova-Stoytcheva M, Renteria-Gutierrez L, *et al.* Glucose oxidase biosensor modeling and predictors optimization by machine learning methods. *Sensors. (Switzerland)* 2016;16(11):1483.

[83] Massah J and Asefpour Vakilian K. An intelligent portable biosensor for fast and accurate nitrate determination using cyclic voltammetry. *Biosyst. Eng.* 2019;177:49–58. Available from: https://doi.org/10.1016/j.biosystemseng.2018.09.007.

[84] Aiassa S, Ros PM, Hanitra MIN, *et al.* Smart portable pen for continuous monitoring of anaesthetics in human serum with machine learning. *IEEE Trans. Biomed. Circuits Syst.* 2021;15(2):294–302.

[85] Zeng Z, Huang Z, Leng K, *et al.* Nonintrusive monitoring of mental fatigue status using epidermal electronic systems and machine-learning algorithms. *ACS Sensors*. 2020;5(5):1305–1313.

[86] Ozer E, Kufel J, Myers J, *et al.* A hardwired machine learning processing engine fabricated with submicron metal-oxide thin-film transistors on

a flexible substrate. *Nat. Electron.* 2020;3:419–425. Available from: http://dx.doi.org/10.1038/s41928-020-0437-5.

[87] Hsu WE, Chang YH, and Lin CT. A machine-learning assisted sensor for chemo-physical dual sensing based on ion-sensitive field-effect transistor architecture. *IEEE Sens. J.* 2019;19(21):9983–9990.

[88] Sinha S, Bhardwaj R, Sahu N, *et al.* Temperature and temporal drift compensation for Al2O3-gate ISFET-based pH sensor using machine learning techniques. *Microelectronics J.* 2020;97:104710. Available from: https://doi.org/10.1016/j.mejo.2020.104710.

[89] Wang H, Moser N, and Georgiou P. An ISFET array for ion multiplexing with an integrated sensor learning algorithm. In: *IEEE International Symposium on Circuits and Systems*, Seville, Spain, 2020. p. 1–5.

[90] Indiveri G, Linares-Barranco B, Hamilton TJ, *et al.* Neuromorphic silicon neuron circuits. *Front. Neurosci.* 2011;5:1–23.

[91] Yakopcic C, Alom MZ, and Taha TM. Extremely parallel memristor crossbar architecture for convolutional neural network implementation. In: *Proceedings of the International Joint Conference on Neural Networks*. Anchorage, AK: IEEE, 2017. p. 1696–1703.

[92] Tsai H, Ambrogio S, Narayanan P, *et al.* Recent progress in analog memory-based accelerators for deep learning. *J. Phys. D. Appl. Phys.* 2018;51:283001.

[93] Tripathi P, Moser N, and Georgiou P. A neuron-based ISFET array architecture with spatial sensor compensation. In: *2019 IEEE International Symposium on Circuits and Systems*. Sapporo, Japan: IEEE, 2019. p. 1–5.

[94] D'Angelo R and Sonkusale S. Precise time mode multiplier using digital primitives and passive components. In: *Proceedings of IEEE International Symposium on Circuits and Systems*. Montreal, QC, Canada: IEEE, 2016. p. 1802–1805.

[95] Douthwaite M, Garcia-Redondo F, Georgiou P, *et al.* A time-domain current-mode MAC engine for analogue neural networks in flexible electronics. In: *Proceedings of IEEE Biomedical Circuits and Systems Conference*, Nara, Japan, 2019. p. 1–4.

[96] Bavandpour M, Mahmoodi MR, and Strukov DB. Energy-efficient time-domain vector-by-matrix multiplier for neurocomputing and beyond. *IEEE Trans. Circuits Syst. II Express Briefs*. 2019;66(9):1512–1516.

[97] Moser N, Lande TS, and Georgiou P. A novel pH-to-time ISFET pixel architecture with offset compensation. In: *Proceedings of IEEE International Symposium on Circuits and Systems*, Lisbon, Portugal, 2015. p. 481–484.

[98] Cacho-Soblechero M and Georgiou P. A programmable , highly linear and PVT-insensitive ISFET array for PoC diagnosis. In: *2019 IEEE International Symposium on Circuits and Systems*, Sapporo, Japan, 2019. p. 1–5.

[99] Öžrenci AS, Dündar G, and Balktr S. Fault-tolerant training of neural networks in the presence of MOS transistor mismatches. *IEEE Trans. Circuits Syst. II Analog Digit. Signal Process.* 2001;48(3):272–281.

[100] Chen Y, Yao E, and Basu A. A 128-channel extreme learning machine-based neural decoder for brain machine interfaces. *IEEE Trans. Biomed. Circuits Syst.* 2016;10(3):679–692.

Chapter 11
Last words

Miguel Hernandez-Silveira[1], *Su-Shin Ang*[1], *and Yeow Khee Lim*[2]

11.1 Introduction

We ventured out to explore the role of machine learning (ML) in the medical technology industry and were rewarded with a fascinating tapestry that reveals the rich and layered nature of these applications. Indeed, even in this relatively short book, we saw ML applied to a large-scale problem – predicting the mortality rate of malaria as a result of policy changes in Chapter 2, to microscopic ISFET based neural networks used to analyse sweat contents directly on the skin surface in Chapter 10, while not forgetting the many exciting applications in between.

It is undisputed that ML is a powerful technique that has been applied in many areas of the industry. In tandem, exciting research is on-going in many areas, which can provide novel and valuable tools for clinicians and improve the level of patient care. In spite of these benefits, the uptake of ML in the industry has been slow partly because of the heavily regulated nature of the medical devices industry as well as the fact that there is considerable hype around ML. Unfortunately, this is accompanied with unrealistic expectations and the subsequent disappointments have affected the standing of ML as a serious tool that can be used to improve health outcomes. Apart from inspiring readers about the possibilities of ML, we hope that this book can help with clarifying what ML means and correct any misconceptions about it.

In this chapter, we will start by reviewing the preceding chapters of this book, beginning with the types of data that are used, followed by the ML frameworks and methods that are used. Clearly, effective implementation and deployment of ML is crucial for its success and there are a wide range of factors involved. We will focus solely on the types of platforms where ML is deployed and highlight some interesting methods that are used to overcome challenging conditions where ML is deployed. Since it is important for manufacturers to demonstrate regulatory compliance, we will discuss some of the issues that they face during the ML design and deployment

[1]Medical Frontier Technology Ltd, UK.
[2]Nanyang Technological University, School of Aerospace and Mechanical Engineering, Singapore

phase. Finally, we will conclude the book with some thoughts about the future of ML in healthcare.

11.2 A review of the state-of-the-art

Table 11.1 depicts the breakdown of this book into the various types of data and methods that were found to be effective in their respective application domains. In the first couple of chapters, reinforcement learning was used to create epidemiological models that can allow health outcomes to be determined in a fine-grained manner e.g. disability adjusted life year (DALY) and years of life with disability (YLD). A real-time guidance system for micro-surgeries is described in Chapter 4. These systems create holograms that allow surgeons to navigate the complex pathways in

Table 11.1 Data types and ML techniques used in various chapters of this book

Chapter/s	Data types	ML techniques	Application/s
2 and 3	Time-series epidemiological data	Reinforcement learning	Epidemiological models
4	Microscopes with Video feeds	Gaussian-mixture Model K-means clustering	Imaging system for microsurgeries
5	EEG, ECG, and PPG	NA	Assessing level of consciousness
6	Brain MRI, CT, and EEG	SVM LeNet5 (transfer learning) DenseNets	Early diagnosis of Alzheimer's disease
7	ECG	SVM Random forest Boosting methods LSTM Autoencoders	Atrial fibrillation detection
8	MRI	Dictionary-based clustering KSRDL classification	HCM fibrosis
9	Temperature Heart rate & HRV Aerobic indicators Strength indicators Nutritional indicators	ANN SVM K-means clustering Random forest	Athletic performance
10	Sweat pH, sodium, potassium	Analogue ANN	Electrolyte, metabolism, and stress monitoring

the patient's body during surgeries. Since these video feeds have to be processed in real-time, deploying computationally demanding software algorithms to do so while fulfilling the real-time requirements is a challenge. As such, the inclusion of ML into the loop is still very much work-in-progress. However, ML methods such as Gaussian mixture model and K-means clustering were found to be useful in image segmentation.

While there was not any in-depth discussion about ML in Chapter 5, the authors noted that there is substantial scope for the application of ML in helping to determine the level of consciousness for perioperative patients. Apart from routine monitoring of a patient's level of consciousness, ML was found to be useful in the early detection of Alzheimer's disease (AD) as well. More specifically, conventional ML methods such as support vector machines (SVMs) were found to be effective in classifying between control normal (CN) and diagnosed AD patient MRIs. Lately, transfer learning, allowing the re-use of well-known networks like the *LeNet5*, as well as large, dense and deep networks such as *denseNets* were found to produce compelling classification results between MCI and AD patient MRIs.

In the area of arrhythmia detection, ML is a well-entrenched technique that was found to be effective at identifying various types of arrhythmias, including atrial fibrillation as well as fibrosis detection in patients with hypertrophic cardiomyopathy (HCM). Apart from health outcomes, ML was found to be effective in related areas such as human performance, which is particularly relevant to athletes and generally for personnel in occupations where they are frequently exposed to extreme conditions. Consequently, they are frequently outfitted with body worn sensors that are designed to monitor their status and performance. The software algorithms and low power sensors used in these applications are described in Chapters 9 and 10 respectively. Based on the contents in these chapters, we made the following observations:

1. Video and image are the most popular types of data used in medical ML applications. In this book, these are the dominant data types used in Chapters 4, 6, and 8.
2. Transfer learning can allow established networks used in other applications to be re-purposed for medical applications such as the early diagnosis of Alzheimer's disease.
3. Modern ML networks tend to be deep and compute intensive. As such, they are typically deployed in centralised and well-resourced cloud-based servers.
4. There is an emerging class of ML applications that are deployed in close proximity to body-worn sensors, such as the sweat sensors described in Chapter 10. Such devices are usually battery powered. Therefore, different design techniques are required in order to negotiate these constrains.

Implementing and deploying a ML solution would be subjected to constraints that are particular to that problem. However, there are common issues that are faced by ML engineers. The training time, execution time, and the power budget (or energy per operation) are among those concerns. Indeed, ML computational requirements far exceeds typical image-based operations such as compression. For these reasons, special techniques and platforms have been employed to allow the efficient operation of ML solutions. This will be discussed in the following section.

11.3 Implementation and deployment

The typical sub-structures in neural networks are the fully connected network (FCN) and the convolutional neural network (CNN), as illustrated in Figure 11.1. Further, the computational operations involved in obtaining the inputs to each hidden layer in the FCN is the product between a matrix and a vector, followed by the addition with a vector, as shown in (11.1). Each element in matrix W_{ij} corresponds to a connection or edge between node i in the previous layer to node j in the following layer (11.2). Vectors I and b are populated by the sum of products from the previous stage and a single bias term. Equation (11.1) refers to the forward propagation step or inference:

$$\mathbf{O} = \mathbf{W}\mathbf{I} + \mathbf{b} \tag{11.1}$$

$$\mathbf{W} = \begin{bmatrix} w_{0,0} & w_{0,1} & \cdots & w_{0,N-1} \\ \vdots & & \ddots & \vdots \\ w_{M-1,0} & w_{M-1,1} & \cdots & w_{M-1,N-1} \end{bmatrix} \tag{11.2}$$

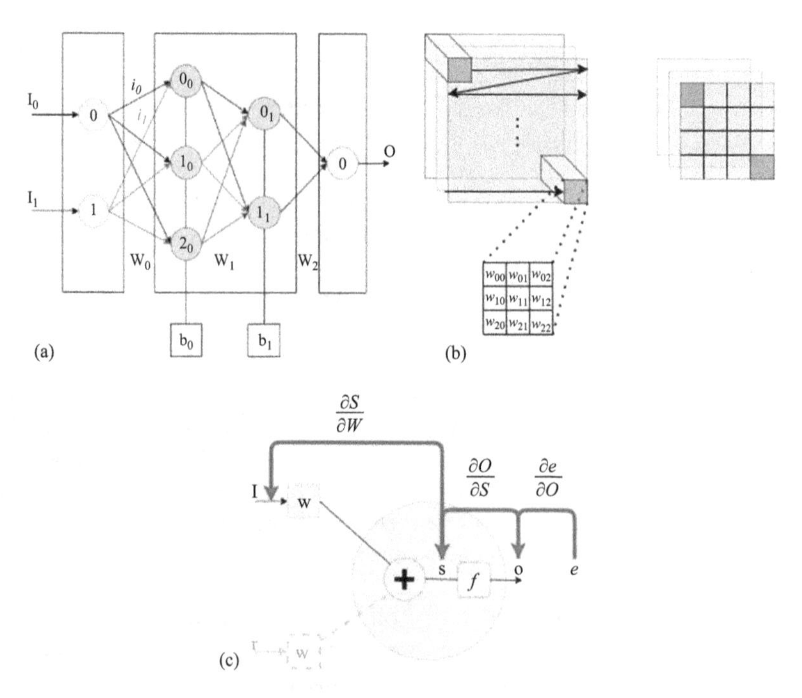

Figure 11.1 (a) Fully connected network (FCN). (b) Convolutional neural network (CNN). (c) Error back-propagation.

In Chapter 1, we derived single weight updates during error back-propagation. In practice, during training, weights belonging to a layer are updated together as a sequence of vector multiplication and matrix addition operations that are described in (11.3). In this case, $\boldsymbol{W_j}$ refers to column j in the weight matrix of a particular layer. Equivalently, each element refers to the weight of the edge that is incident on Node j, from the previous layer. $\boldsymbol{\partial E/\partial W}$ refers to a column vector that is obtained by taking the product of 3 other partial derivatives based on the chain rule, as shown in (11.4). Details of how individual partial derivatives, as well as an additional bias term are provided in Chapter 1:

$$\boldsymbol{W_j} = \boldsymbol{W_j} - \mu \frac{\boldsymbol{\partial E}}{\boldsymbol{\partial W}} \tag{11.3}$$

$$\frac{\boldsymbol{\partial E}}{\boldsymbol{\partial W}} = \frac{\boldsymbol{\partial S}}{\boldsymbol{\partial W}} \frac{\boldsymbol{\partial O}}{\boldsymbol{\partial S}} \frac{\boldsymbol{\partial E}}{\boldsymbol{\partial O}} \tag{11.4}$$

In addition, to implementing each step of the CNN operation in a straightforward manner, this involves multiplying the contents of the filter kernel with corresponding elements in the image or feature map, and subsequently accumulating the results (sum of products). Clearly, the naive method does not correspond to matrix or vector multiplications. We will discuss a technique that converts this operation into a conventional matrix multiplication in a subsequent section so as to take advantage of efficient matrix/vector multiply and accumulate (MAC) modules. Apart from performing matrix computations efficiently, access and movement of data are extremely critical factors to the overall efficiency of the ANN operation. The mitigation methods used to overcome the memory bottleneck will also be discussed.

11.3.1 Traditional computing and the memory hierarchy

Arguably, the first modern computer is conceived by Professor John von Neumann, who designed the Von Neumann architecture [1], as illustrated in Figure 11.2. It shares many similarities with modern computers – at the core of the design is a central processing unit (CPU), that consists of a control unit, arithmetic logic unit (ALU), and some registers that is used to hold intermediate data. Specifically, the Control Unit contains a program counter and instruction registers which are used to store the execution stage of a program and the current instruction that is targeted for execution respectively. The ALU is used to carry out arithmetic operations such as multiplication and addition. Separately, data registers are used to store the data required by the program, as well as to store the computed outputs of the program. Since the register bank has a very limited capacity, the actual instructions and data that are required by the program are stored in the monolithic memory unit, which has a much higher capacity than the register bank. Finally, input (e.g. a keyboard) and output (e.g. a monitor) devices provide the interface to interact with the computer, such as the initiation of programs and obtaining the results from these programs.

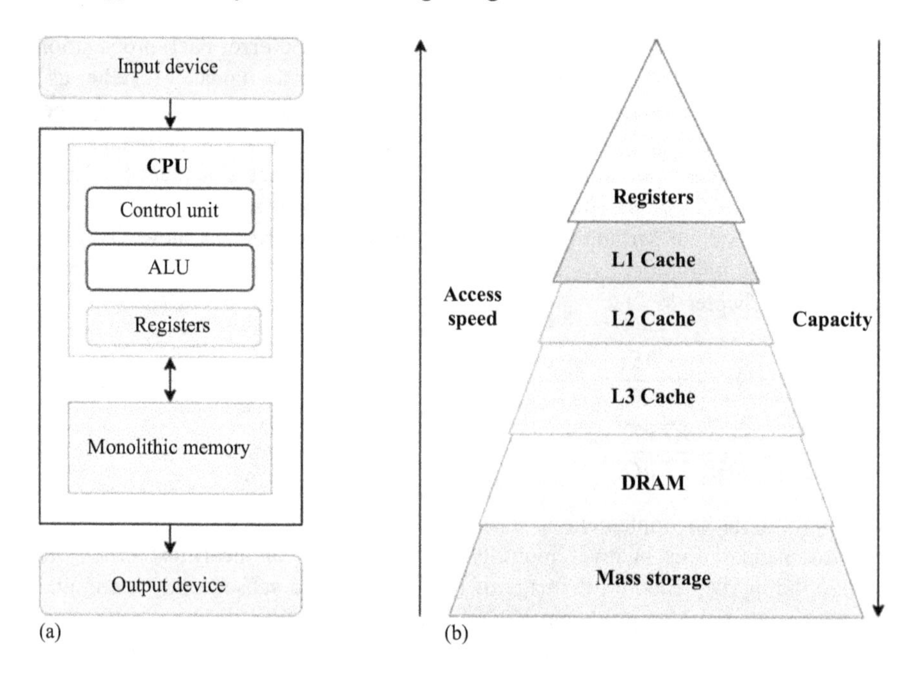

Figure 11.2 (a) The Von Neumann architecture [1]. (b) A modern memory hierarchy – trade-offs between access speed and capacity.

An example of how to create a program to add two numbers a and b, to obtain output c, could be executed is as follows. Note that essential start-up operations have been left out in this case for simplicity:

1. The numbers a and b are stored in the monolithic memory unit.
2. At the start of the program, the program counter points towards the first instruction, which is to fetch Datum a. The fetch instruction is retrieved from the Monolithic memory unit and loaded into the instruction register. Along with the instruction, the memory address for Datum a is also retrieved from the main memory unit.
3. The fetch instruction is executed and Datum a is fetched and loaded into the data Register rA.
4. The preceding two steps are repeated for Datum b, and b is then stored in Register rB.
5. The Add instruction is subsequently fetched from the monolithic memory unit and loaded into the instruction registers. This contains both the instruction to add the contents of Registers rA and rB, and load the result into Register rC. The execution of this instruction involves the ALU.
6. After completion of the Add instruction, the program counter is again incremented, and the next instruction is fetched. This involves storing the contents of Register rC into a specified memory address in the main memory unit.

7. The final instruction involves displaying the contents of Register rC onto the display, to let the user know the result of the computation.

Note that the computer operations are limited to *Load*, *Execute*, and *Store*. Crucially, the instruction and data fetch have to be performed sequentially because both the program and the data are stored in monolithic memory with a single port for access. This limits the speed of program execution and this issue is typically known as the Von Neumann bottleneck. There are many ways to get around this bottleneck – conventional approaches include the separation of program and data memory so that instruction and data fetches can be performed in parallel. This undoubtedly speeds up program execution, as both of the instruction and data can be concurrently fetched. Since only one instruction and datum is involved, this is known as the single instruction and single data (SISD) paradigm.

In the context of this chapter, we are particularly interested in matrix and vector multiplications since they are critical for ML inference and training. For reasons that will be explicit in the subsequent sections, we tend to perform the same instruction set on different data. Therefore, rather than the SISD paradigm, accelerators such as graphics processing units (GPUs) typically make use of the single instruction and multiple data (SIMD) paradigm.

To facilitate the efficient execution of the SIMD paradigm, it is essential that sufficient memory with fast access times is used. However, due to the limitations of memory technology, there is a trade-off between capacity and access time. This memory bottleneck is particularly serious for deep ML networks with a large number of connections (high dimensions for the weight matrix W in (11.1)). To solve this problem, a high-level matrix tiling approach is combined with a memory hierarchy to increase memory bandwidth and reduce execution time.

Static random access memory (SRAM) is a type of volatile memory that is built using CMOS transistors arranged in a bistable circuit. Registers have the fastest access time since they are built using SRAM technology and also because they are nearest to the processing unit. However, Register banks have the lowest storage capacity. For this reason, they are at the top of the memory hierarchy but occupy the least area, as shown in Figure 11.2(b). Caches are typically used to buffer the most frequently used data in fast access memory. The contents of the cache are dependent on digital hardware modules that determine if specific data should be retained or purged based on the spatial and temporal locality of the data itself. They are also built using SRAM technology and frequently divided into three layers – L1, L2, and L3. The L1 cache is usually on-chip (close to the processor), and therefore the fastest as well as smallest in capacity in comparison with the other two layers.

Another type of commonly used memory technology is the dynamic random access memory (DRAM). Similar to SRAM, it is volatile and dependent on capacitors rather than CMOS transistors for the storage of each memory bit. Apart from longer access times, it is also more variable meaning that access to each datum may require a different number of CPU cycles depending on the state of the DRAM. However, DRAM units have higher density because of their comparatively simpler structure and are therefore orders of magnitude higher in capacity compared to caches. The last layer in the memory hierarchy is the mass storage, and it is implemented using technologies like flash memory (which are based on floating gate transistors) or magnetic disks. Mass storage devices have the longest access time. Flash memory has a number of limitations, including a limited number of read/write cycles as well as a coarse granularity during write access.

The efficient use of the memory hierarchy involves maximising data re-use, which implies that the most frequently used data should be buffered at higher levels of the memory hierarchy to avoid accesses to lower levels, which tend to be expensive in execution time and energy. Since matrix and vector products involve very specific access patterns, it is possible to organise the data and computations in a way that will maximise re-use. We will discuss that at length in Section 11.3.2.

11.3.2 Graphics processing unit

Originally, the GPU was designed with graphics applications in mind. However, since they are especially efficient at implementing matrix computations, they have since been re-purposed for general purpose GPU applications (or GPGPU), and this includes ML. At the time of writing, a popular GPU that is well suited for deep learning applications is the Nividia GeForce RTX 3090Ti GPU. Some of the specifications for this GPU are listed below.

- *GPU clock frequency*: 1,560 MHz.
- *Memory capacity*: 24 GB.
- *Number of CUDA cores*: 10,752.
- *Number of Tensor cores*: 336.
- *Cache capacities*: 128 KB (L1), 6 MB (L2).

Obviously, higher clock frequencies and larger onboard memories would translate to higher performance i.e. training and inference can complete more rapidly. Apart from higher clock speeds and memory bandwidth, higher performance can be achieved by increasing the amount of hardware parallelism – performing more MAC operations and/or matrix multiply operations concurrently. This is facilitated by having multiple CUDA and tensor cores respectively. The latter allows products between 4-by-4 matrices to be performed efficiently. Since ML applications primarily involve matrix–vector multiplications, tensor cores are especially effective at optimising these computations. Having this information, we should probably ask two questions:

1. In general, matrix multiplications in deep learning do not involve 4-by-4 matrices. How do we leverage products between sub-matrices when we are computing the product of far larger matrices, while taking into account the dependencies between those computations?
2. Matrix computation rate has to be matched with how rapidly we are able to load new data and store the results in memory. How do we leverage the memory resources in the GPU in order to do so?

We need to understand the nature of matrix multiplications and see how its properties can be exploited in order to answer these questions. First, we look at the recursive nature of matrix multiplication. The product between two 4-by-4 matrices are shown in (11.6). The straight-forward way of computing the product is to do that between each element a_{ij} and b_{ij}. However, each matrix can be grouped into 2-by-2 sub-matrices, which is carried out in (11.7). In this manner, the product can be computed as a function of the individual sub-matrices, as we can observe in (11.8). Computationally, this will lead to the same result as doing this computation between a_{ij} and b_{ij}:

$$C = AB \tag{11.5}$$

$$= \left[\begin{array}{cc|cc} a_{0,0} & a_{0,1} & a_{0,2} & a_{0,3} \\ a_{1,0} & a_{1,1} & a_{1,2} & a_{1,3} \\ \hline a_{2,0} & a_{2,1} & a_{2,2} & a_{2,3} \\ a_{3,0} & a_{3,1} & a_{3,2} & a_{3,3} \end{array}\right] \left[\begin{array}{cc|cc} b_{0,0} & b_{0,1} & b_{0,2} & b_{0,3} \\ b_{1,0} & b_{1,1} & b_{1,2} & b_{1,3} \\ \hline b_{2,0} & b_{2,1} & b_{2,2} & b_{2,3} \\ b_{3,0} & b_{3,1} & b_{3,2} & b_{3,3} \end{array}\right] \tag{11.6}$$

$$= \begin{bmatrix} A_{0,0} & A_{0,1} \\ A_{1,0} & A_{1,1} \end{bmatrix} \begin{bmatrix} B_{0,0} & B_{0,1} \\ B_{1,0} & B_{1,1} \end{bmatrix} \tag{11.7}$$

$$= \begin{bmatrix} A_{0,0}B_{0,0} + A_{0,1}B_{1,0} & A_{0,0}B_{0,1} + A_{0,1}B_{1,1} \\ A_{1,0}B_{0,0} + A_{1,1}B_{1,0} & A_{1,0}B_{0,1} + A_{1,1}B_{1,1} \end{bmatrix} \tag{11.8}$$

The GPU makes use of a memory hierarchy (Figure 11.2) to maximise data re-use. Specifically, it has two levels of caches which are not particularly large. If the matrix calculations are performed in a naive manner described in (11.6), data re-use will not be exploited optimally for high dimensional matrices. For example, the entry $a_{0,0}$ in matrix A is accessed 4 times in the example given in (11.6) in order to compute a row in the final matrix. Ideally, this entry should be kept in the cache until the entire row is computed. On the other hand, this might not be possible if the cache is unable to accommodate an entire row of the matrix A, meaning that the entry $a_{0,0}$ will be repeatedly overwritten by new entries, leading to cache misses which have high latency penalties. By exploiting the recursive nature of matrix multiplications, we would be able to load entire sub-matrices of manageable sizes into the cache and avoid the issue of sub-optimal data re-use. To generalise, the steps involved in maximising data re-use between l_0 and l_1 are as follows. Here, we assume that Matrices A and B are held in memory hierarchy l_0:

1. Load sub-matrix $A_{0,0}$ into l_1.
2. Load sub-matrix $B_{0,0}$ into l_1.
3. Compute the product $A_{0,0}B_{0,0}$ and load the result $C_{0,0}$ into the memory bank at l_1.
4. Load sub-matrix $B_{1,0}$ into l_1, overwriting $B_{0,0}$. Compute the product $A_{0,1}B_{1,0}$.
5. Load $C_{0,0}$ from l_1 and add it to the computed product i.e. $A_{0,1}B_{1,0} + C_{0,0}$.
6. Overwrite the entry $C_{0,0}$ in l_1 with the new final result to obtain an entry in the resulting matrix.
7. Apply a similar procedure to compute other entries of the resulting matrix.

In practice, recursive block segmentation is done at two levels, where the first level occurs between L2 cache and DRAM, and the second level occurs between the register bank and the L1 cache (known as register blocking) [2]. Intuitively, we are able to grasp the improvements in efficiency that results from using smaller sub-matrices as they are better able to fit in the cache. However, this scheme is by no means perfect as it requires the same sub-matrix to be fetched multiple times e.g. $A_{0,0}$ is fetched twice in this example. Therefore, what is the precise benefit from using the matrix blocking method? To quantify the benefit precisely, let us reformulate the problem by considering three matrices: A (dimensions m by k), B (dimensions k by n), and C (dimensions m by n).

- Computing the product $C = AB$ naively will require $2mnk$ fetches since each entry in C requires $2k$ fetches.
- We segment the matrix into new matrices of dimensions M, N, and K, such that $M = m/b_m$, $N = n/b_n$, and $K = k/b_k$, where b_m, b_n, and b_k refer to the segment dimensions for matrices A, B, and C respectively.
- Using the matrix blocking approach, the number of fetches now amounts to $MNKb_mb_k + MNKb_kb_n = mnk(1/b_m + 1/b_n)$.

Consequently, the reduction factor in the number of fetches from the lower memory hierarchy is reduced to expression in (11.9). If we restrict $b_m = b_n = 4$, there is a 4-fold reduction in the number of fetches in order to perform the same computation. In addition, this analysis underestimates the benefit from matrix blocking as it does not take into account the fact that data is being loaded into l_1 and re-used multiple times, which further lowers the number of accesses to l_0 [2]:

$$\begin{aligned}\text{Reduction factor} &= \frac{2mnk}{mnk(1/b_m + 1/b_n)}\\ &= \frac{2}{1/b_m + 1/b_n}\end{aligned} \tag{11.9}$$

Another implication of performing the multiply-and-accumulate (MAC) operation in this way is to remove the dependencies for calculating each entry into resulting matrix, and allow several sub-matrix products to be performed in parallel. This is highly advantageous from the point of view of the GPU since it has a large number of tensor cores, which allows such computations to be conducted in parallel and subsequently aggregated to obtain the final product between high-dimensional matrices.

An important operation in ML is convolution as it allows features to be automatically extracted from images. Unfortunately, the naive implementation of convolution does not involve matrix multiplications. Therefore, without modifications, it is unable to take advantage of the parallelism afforded by GPUs. Fortunately, a technique known as convolution lowering [3] can be used to convert convolution into a matrix multiplication operation. Figure 11.3 illustrates an example involving input image data with conventional channels – red, green, and blue, with corresponding 2-by-2 filter kernels for each channel. The parameters relating to convolution in this context are defined in Table 11.2.

Essentially, convolution lowering involves duplicating parts of the image and filter kernel data to form two matrices, D_m and F_m, before taking their product to form the convolution output O_m. Specifically, two filters (contains coefficients *Fi* and *Gi*), are applied to the image data to produce two output feature maps per channel. Instead

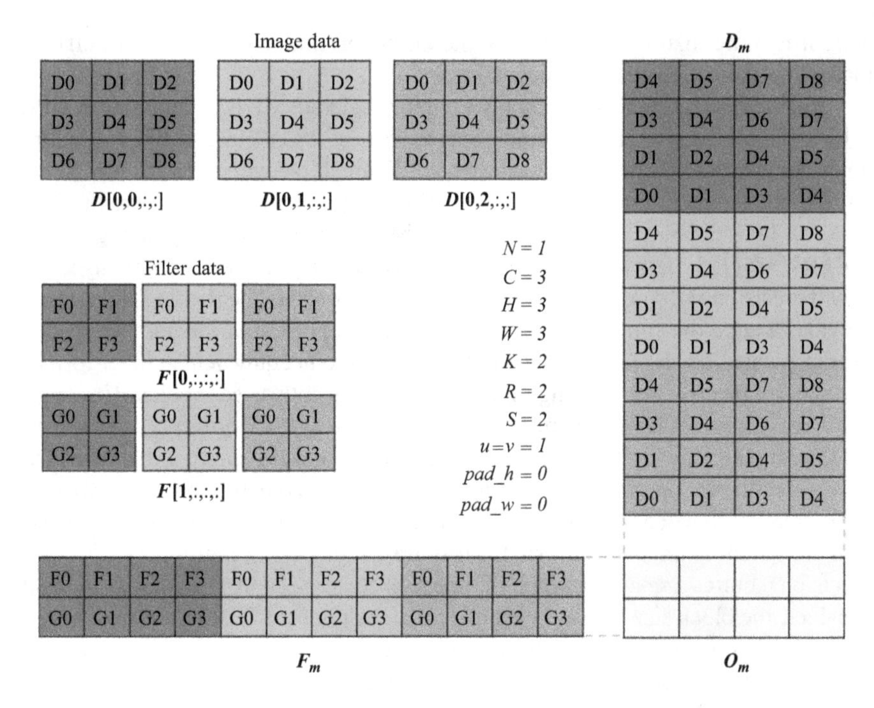

Figure 11.3 Convolution lowering involving three channels – red, green, and blue [3]

Table 11.2 Parameters pertaining to convolution lowering [3]

Parameter	Meaning
N	Number of images in mini-batch
C	Number of input feature maps
H	Height of input image
W	Width of input image
K	Number of output feature maps
R	Height of filter kernel
S	Width of filter kernel
u	Vertical stride
v	Horizontal stride
pad_h	Height of zero-padding
pad_w	Width of zero-padding

of 'sliding' the filter kernels over each channel of the image data and accumulating the results, the same effects are achieved by duplicating filter coefficients to form matrix F_m in a row-major manner and matrix D_m in a column major manner, and taking their product to form output O_m as shown in Figure 11.3. While this approach appears to require a substantial overhead due to the requirement to store duplicated data, it is more than compensated by the execution time reductions from efficient matrix multiplications that can now take place.

Another method that is commonly used to accelerate ML computations in GPUs is by using mixed precision floating-point operations e.g. FP16 and FP32. Floating-point numbers are represented according to Figure 11.4, with three fields: Signed bit, Exponent, and Mantissa. In general, a floating-point number will take the form: $X = (-1)^{\text{Signed bit}} \times (\text{mantissa} \times 2^{exponent})$. The word lengths corresponding to FP16 and FP32 are $(1, 5, 10)$ and $(1, 8, 23)$, respectively, and they are more commonly known as half precision and single precision floating-point formats. The floating-point format provides a large dynamic range, allowing a large proportion of the number space to the represented, as compared to the integer or fixed-point equivalent number formats. As such, single precision floating-point format is frequently used in scientific computations as it provides large dynamic range (determined by the word length of the exponent) with relatively high precision (determined by the word length of the mantissa).

To accelerate operations including floating-point computations, part of those operations could be conducted using half precision floating-point numbers. The reasons for speed-up are two-fold: first, hardware floating-point multipliers with lower precision requires a smaller amount of time to execute. Second, since the word length is halved, the block size of matrix segments can be doubled, leading to smaller number of fetches according to (11.9), as well as increasing the data re-use. However, using lower precision floating-point representations will increase the rounding error in the product, which is not desirable. Often, a compensation term is added to correct for this error [4]. This involves working out the errors between the single and half precision representations, which are shown in (11.12) and (11.13). Subsequently, the compensated term $E_A E_B + A_{half} E_B + E_A B_{half}$ can be derived and added to the original product $A_{half} B_{half}$, as shown in (11.11). This will reduce the error of the product at the expense of three further matrix multiplications and five additions (including the first two subtractions to obtain the errors) for every product:

$$A_{single} B_{single} = (E_A + A_{half})(E_B + B_{half}) \tag{11.10}$$

$$= E_A E_B + A_{half} E_B + E_A B_{half} + A_{half} B_{half} \tag{11.11}$$

$$E_A = A_{single} - A_{half} \tag{11.12}$$

$$E_B = B_{single} - B_{half} \tag{11.13}$$

We looked at a few techniques that are used in the context of GPUs to accelerate matrix arithmetic. However, this is by no means an exhaustive list of methods and there is on-going research in finding other techniques to optimise GPU usage for ML.

Sign bit	Exponent	Mantissa

Figure 11.4 IEEE-754 standard floating-point number representation. Length of each field are represented as (sign bit word length, exponent word length, Mantissa word length).

The reader is reminded that GPUs were not originally designed with ML in mind. For this reason, there are other areas of research, looking into other types of accelerators with architectures that are better suited for ML. We shall look at a ML hardware accelerator in Section 11.3.3.

11.3.3 Hardware accelerators

We saw how GPUs are able to achieve high levels of performance in ML applications, principally because it is capable of performing matrix multiplications efficiently. However, there are operations like convolution that are awkward to perform on GPUs. As we discussed in Section 11.3.2, it is possible to convert the convolution operation into matrix multiplication through duplication, but this is achieved at the expense of substantial memory storage. To achieve further improvements in performance, it is necessary to make use of hardware that are custom designed for operations in ML network layers like convolution, pooling, and fully connected networks, or application-specific integrated circuits (ASICs) that are built to accelerate ML applications.

At the time of writing, there are three well-known hardware accelerators that are deployed in the market.

1. *Neural engine*: used on iPhone mobile devices [5].
2. *HPU CNN co-processor*: used on the Microsoft Hololens, which is a virtual reality (VR) headset [6].
3. *Tensor processing unit (TPU)*: developed by Google as a means of accelerating ML applications in their server farms and more recently, on edge devices such as their pixel mobile devices [7].

Since a limited amount of published information concerning the neural engine and the HPU CNN co-processor architectures exist, we will discuss only about the TPU in this chapter.

The high-level architecture of the TPU is illustrated in Figure 11.5(a). Specifically, the host processor initiates ML operations on the TPU by issuing instructions to it through a PCIe bus interface. Data transfer operations (e.g. ML network feature maps and weights) are facilitated by a DMA controller that allows data transfers without taking up processing cycles of either the CPU or TPU. Also, rather than the use of a micro-controller within the TPU, a specialised control module is used to coordinate operations within the SOC. The core of the TPU is the matrix multiply unit and the accumulators (contains 65,536 MAC units), which carries out essential operations in the FCN and CNN modules. Apart from the *hardware* normalisation and pooling module, a *hardware* activation module is available for downstream computations – users are given a choice of activations, including ReLU and Sigmoid. Several interesting optimisation techniques are employed in this architecture in order to accelerate ML operations. This is summarised below [7]:

- *Use of a memory hierarchy*: Bulk storage of data occurs on the DRAM. Rather than storing intermediate results (i.e. normalisation/pooling and activation modules results) in DRAM or off-chip caches, they are stored in a large on-chip unified buffer (24 MiB), which is implemented using fast SRAM technology.

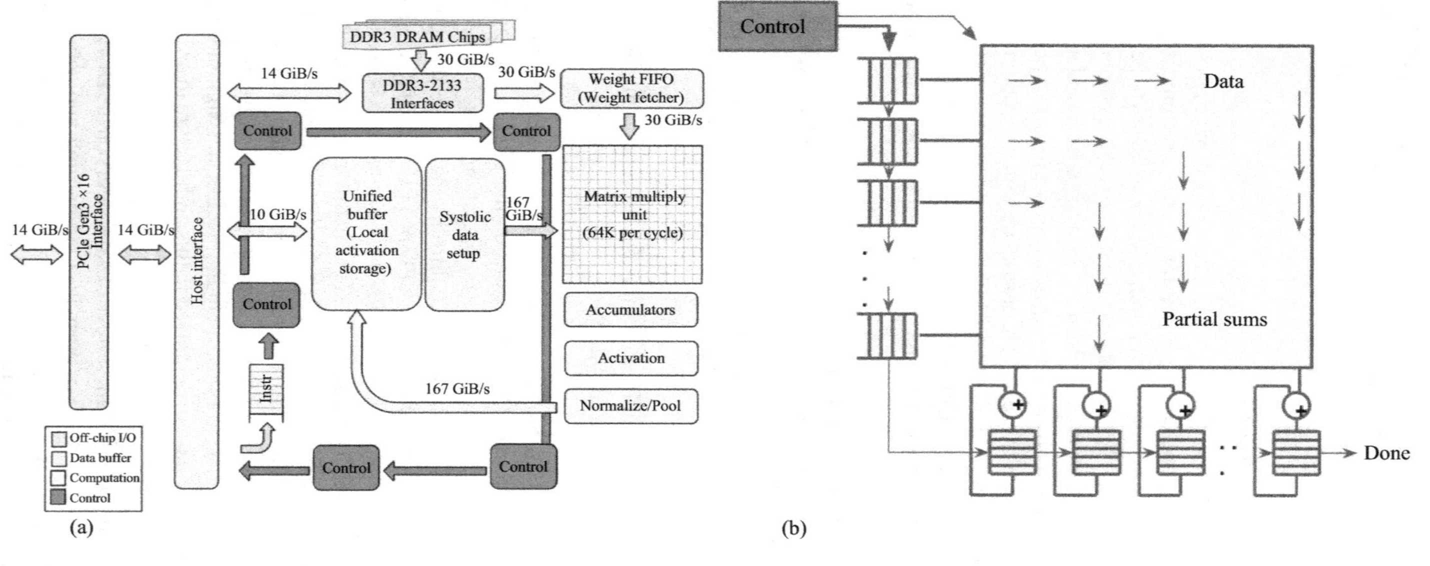

Figure 11.5 (a) High-level architecture of the tensor processing unit. (b) Systolic data flow of the matrix multiply unit [7].

- *Systolic data flow*: A substantial reduction in compute times is achieved by performing hardware MAC operations concurrently (SIMD paradigm). A systolic data flow is used to ensure that the correct data/results are loaded on and off the MAC units at precisely the correct moment. Instead of loading the weights from the unified buffer, this is done through weight FIFOs which in turn feed the MAC units from the top of the array. In tandem, data from the feature maps and the intermediate layers are fed to MAC units from the left to array. Finally, the products from MAC computations are fed downstream to the accumulators and exit at the bottom of the array, as seen in Figure 11.5(b). This data flow relies on a distributed memory sub-system, comprising of FIFOs and intermediate buffers with close proximity to the compute units, rather than a conventional method that is based on a centralised memory sub-system.
- *Reduced precision multiplication*: Conventional single-precision multiplication are reduced to the 8 bits fixed-point format on Edge TPUs (16-bits format is available on server based TPUs). This allows a denser hardware implementation of the MAC array, affording greater parallelism and faster execution times in comparison with single precision-based MAC operations on GPUs.
- *Hardware pipelining*: This is a technique that is commonly used to increase the throughput of an application, while reducing the critical path length of circuit (allowing it to be clocked at a higher rate). Specifically, it allows dependent hardware modules to operate concurrently, and for the results from each module to flow gradually down the pipeline stages. In this case, different stages of the ML layers are executed concurrently, including reading weights from the MAC array, computing the MAC operations, as well as the activation units. In the event that the data is not ready, pipeline stalls would pause all downstream operations until upstream operations have completed successfully.
- *Input/output bottleneck*: The rate at which data/results can be loaded on/off the TPU impacts the operational speed. Inefficiencies due to this bottleneck are avoided by performing as many ML-related operations on the TPU as far as possible. As mentioned earlier, a DMA controller is used to get data on and off chip without involving the host processor.

Now that we have a grasp of the high-level architectural techniques that are used to accelerate ML operations, let us try to answer the original question: *How does one take advantage of the TPU architecture to carry out convolution optimally?*

In answer to the above question, we consider an example where a 1 by 4 filter kernel, W, is convolved with a feature map, I, that is also 1 by 4 in dimensions. We assume that there are 4 channels/filters, and this will result in a 4-dimensional output feature map O. A simplified implementation of the convolution operation is shown in Figure 11.6(a), which describes a loop nest that can be implemented sequentially. Assuming that there are 4 hardware MAC units available, we can unroll the loop to take advantage of the hardware parallelism as shown in Figure 11.6(b). In this case, the computations that have been unrolled are now executed in parallel. A more succinct representation is shown in Figure 11.6(c) with the *par* construct [8], which represents parallel implementation of all the operations described in its body. The equivalent

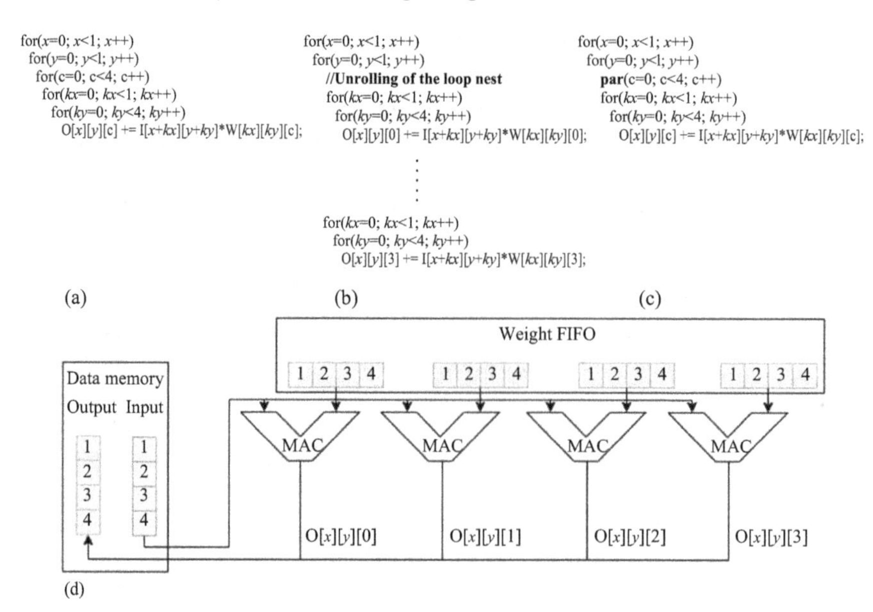

Figure 11.6 *(a) Non-parallel implementation of convolution. (b) Parallel implementation with loop unrolling. (c) Succinct representation of parallel MAC operations. (d) Hardware abstraction of the MAC operations.*

topology is seen in Figure 11.6(d), and the sequence of operations that take place is described as follows:

1. The weights from all of the all of the channels are loaded at the MAC inputs (position one in the green input arrays).
2. Concurrently, the feature map data $I[x + kx][y + ky]$ is loaded and replicated at the other MAC inputs (position one in the blue input array).
3. the MAC operation is computed for each channel and stored to the output array at Position 1.
4. If necessary, repeat the above operations to compute the other elements in the output feature map.

The example above has been deliberately simplified, but extending it to a larger feature map and/or a filter with a larger kernel size should be straightforward. The *par* construct can be applied at different levels of the loop nest as long as there are no loop dependencies between different loop iterations. In this way, the hardware parallelism of the MAC array can be exploited to obtain the convolution sum without the duplication overheads that we have seen in the GPU. In addition to loop unrolling, other compiler-based optimisation techniques such as loop interchange and tiling can be applied to improve the locality of data, to make it more amenable to the memory hierarchy in terms of the buffer capacities at different levels of the memory hierarchy.

ML applications have more demanding computational requirements compared with conventional applications such as real-time video data compression [9], and will continually grow [6]. As well as the increasing sophistication of individual ML applications, there is an increasing demand for ML voice recognition-based services which will lead to scaling issues if the problem is un-addressed [7]. Consequently, the recent increase in efforts to accelerate ML applications arises from a genuine need to do so. Hardware acceleration is by no means the only avenue to improve ML application in both data centres and edge devices. In [6], three other critical areas are identified as areas of research and development:

1. *Accelerator interface*: The interface between the hardware accelerators and the host processor should be carefully designed to ensure that data throughput is not compromised by the IO bottleneck.
2. *Software application programmer interface (API)*: An API that can be used with ease by developers, and allow the capabilities of the accelerator to be easily exploited is important, particularly for open platforms that offer software development kits (SDKs) to third party developers.
3. *System-level optimisations*: Rather than considering optimisations at the modular level only, it is potentially beneficial for designers to consider them at the system level as well. As an example, it is expensive to execute an ML application used for object recognition if it has to be done at the arrival of every video frame. However, since motion vector estimation is routinely done as part of the image pre-processing pipeline, the derived motion vectors could be re-used to track the location of the object that have been found by the ML application in the previous frame. As such, the computational and energy footprint of the application can be substantially reduced at the system level [6].

Clearly, there is still a gap between the deployment of ML at data centers and implementation in edge devices. Promising devices such as the edge TPU, the apple neural engine, and Microsoft HPU CNNs are certainly encouraging. However, these stacks/hardware are usually used for inference rather than for learning, which hinders the widespread usage of online ML algorithms. Even then, at a power supply voltage of 3.3 V, the current draw requirements of the Edge TPU is estimated to be about 500 mA, with a peak current draw of 3A [10]. These requirements put the TPU out of reach of many embedded applications that operate at the micro and sub milli-amperes range, confirming the fact that flexible, power efficient ML accelerators, that can be readily applied at scale are still some way off.

11.4 Regulatory landscape

Previously, we attempted to answer the question about how to implement and deploy compute hungry ML applications at scale. In this section, let us consider the following question:

> *Would you rather subject yourself to a two-hours hip replacement surgery or a 13-hours flight between London and Singapore?*

This question is deliberately provocative and would reasonably trigger an indignant response from the reader. This is tantamount to comparing apples with oranges with no meaningful conclusion. While the question appears foolish, it puts into stark perspective the concept of *risk* – how do we perceive it, and more importantly, how do we manage risk in the context of medical software development, with the objectives of delivering the intended benefits of the product without compromising the level of patient safety?

11.4.1 A brief interlude

Instead of marching into a discussion about medical software risk management, let us take a break and take a look at the airlines industry. Specifically, the aircraft has to operate in exceedingly harsh and unforgiving environments, where the external temperature is in the range of −50 °C to 50 °C, and air pressures as high as 100 kPA are common for a modern jet aircraft flying at 40,000 ft. Further, it is at the mercy of the weather so it has to be designed to be robust against thunderstorms and lightning strikes. Therefore, the design of the aircraft is necessarily complex and nothing short of an engineering marvel. For example, the Airbus A380 is famously known to consist of 530 km of electrical wiring – the cabin wiring alone requires 100,000 wires and 40,300 connectors [11].

Due to the complexity of the aircraft, teams of engineers, air crew, ground crew, and other staff are required to operate each aircraft. Apart from safety concerns, the level of utilisation of an aircraft is an important KPI for any airline i.e. short maintenance and turn-around times, with minimal delays between take-off are necessary to maintain the profitability of the airline. Yet, in spite of these challenging demands, the industry is able to maintain an average accident rate of about 0.0004% between 1983 and 2003 [12]. In comparison, the surgical mortality rate is found to be about 1.89% in one study [13]. Since an aircraft accident does not necessarily lead to casualties, the difference between actual mortality rates for both industries is likely to be even greater. How is the airline industry able to maintain such low mortality rates for a sustained period of time?

The aircraft and airline industry are subjected to risks with varying levels of predictability. Examples of risks that affect safety and/or the bottom line of the airline include the weather, late passengers, aircraft traffic, and aircraft structural and component failures. Advanced weather prediction and warning systems go a long way in helping pilots to plan safe flight paths. Similarly, excellent logistical planning is an important factor that keeps airlines and airports afloat and running like clockwork. In this section, we will focus on the mitigation steps that are taken to prevent structural and component failures, and how they can be rapidly identified to prevent any accident from occurring.

- *Redundancies and fail-safe design*: Critical systems are often duplicated to provide back-up in case of total failure of a system. Navigation equipment are sometimes triplicated to ensure reliable performance. For systems that cannot be duplicated because of bulk and weight, fail-safe design is incorporated. For example, it is better for landing gears to fail extended than retracted and brakes to fail in the applied position.

- *Built-in monitoring*: Traditional flight instruments are always designed with internal failure monitors so that the pilot will not be fooled by an instrument which was not working. Normally, a red flag or a flashing light will warn the pilot that the instrument has failed or degraded performance. With increasing complexity in modern aircraft, electronic units are designed with built-in test equipment (BITE) for pilots and engineers to test the box on ground. The indication is normally a Go : No-Go light for release of aircraft. More failure mode information can be retrieved by the engineer in the workshop. These are useful tools for the engineers to make good diagnostic decisions. Airbus came up with the electronic centralised aircraft monitoring (ECAM) system and Boeing has the Engine Indication and Crew Alert System (EICAS) [14]. Both provide quick display of aircraft systems operations and fault monitoring.
- *Component fault categorisation/prioritisation*: The MSG-2 program categorized all components as hard life, on-condition or condition monitoring (Figure 11.7). Hard life faults are for components whose failure can affect the safety of the aircraft and where deterioration cannot be detected by checks. Many flight control and landing gear components come under this category. On-condition items are periodically checked and released while condition monitoring items are allowed to operate to failure. Most avionics boxes come under this category as critical

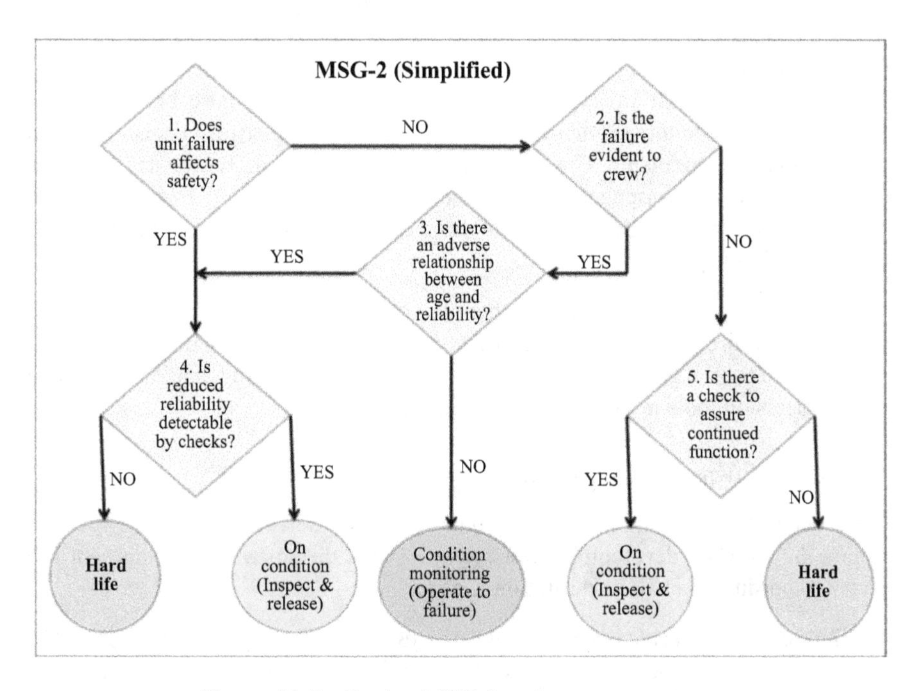

Figure 11.7 Boeing MSG-2 maintenance program

functions are often duplicated for redundancy. Crucially, this method of fault categorisation helps the pilot and engineers to identify and prioritise faults rapidly.
- *Effective maintenance checks*: The integrity and reliability of each system are continuously monitored by maintenance tasks and analyses of failures and unscheduled maintenance tasks. With this method, maintenance check intervals have been extended by more than 20%. This represented significant saving for the airline.
- *Modularity in design*: Modular design with line replaceable units (LRU) is almost universally used to reduce ground time for repair. A faulty component is quickly replaced with a serviceable unit from the store, and the aircraft is released for flight very quickly.

Note that the points described above are not an exhaustive list describing why the airline industry has such a good safety record. However, they are important engineering design principles that are transferable between different industries. A couple of interesting observations can also be made:

- *Fault detection and diagnosis*: Maintenance engineers use several methods to diagnose defects on aircraft. A new engineer will use the aircraft manual diagnostic chart to try to determine the faulty component. A more experienced engineer will look at the past records of the aircraft to check for similar defects. An engineer with many years' experience working the aircraft may already have a feel of the problem component. He/She will do a few more tests to confirm the faulty component. In most cases, he/she will have got the right component with the first or second try.
- *Complexity of instrument panels*: The original flight instrument panels in the Boeing 747 aircraft are extremely complex, as shown in Figure 11.8, requiring the pilots to monitor several systems at once, and making critical decisions based on these measurements. Clearly, these decisions would affect the safety of the aircraft and the passengers and are therefore very important. Digitisation and automation in more advanced aircraft have eased the load on the pilots but further automation is possible [14].

These observations suggest that there is an over-reliance on human judgement in the area of aircraft diagnostics, and there is certainly scope for the development of ML-based tools e.g. anomaly detection, to aid decision making, similar to clinical support decision tools in the healthcare industry.

11.4.2 Software development life cycle

When one is designing, writing and producing medical software, everything centers around the software development life cycle. What is the software development life cycle? According to [17], its definition is as follows:

> *The software life cycle refers to all the phases of a software product throughout its planning, development, and use, all the way through to its eventual obsolescence or retirement. This process has many variable parts, but it can often be segmented*

(a) 

(b) 

Figure 11.8 *(a) Boeing 747 Pilot's instrument panel, Credit Capt. David O. Hill [15]. (b) Flight Engineer's instrument panel, Seaboard Archives courtesy of Capt. David O. Hill [16].*

into several main pieces. This helps developers and others to understand how a product is created, implemented and used.

The steps that are absolutely necessary in software development are firstly, decide what you intend for the software to accomplish. Second, program the software accordingly to achieve the objective. However, the objectives for the software are often far more contrived, and the execution and deployment are subsequently more complex. Further, apart from achieving these goals, the software application should operate in a manner that will not compromise patient safety. A well thought through, systematic approach to the entire development process is therefore necessary to ensure the delivery of safe and effective medical software.

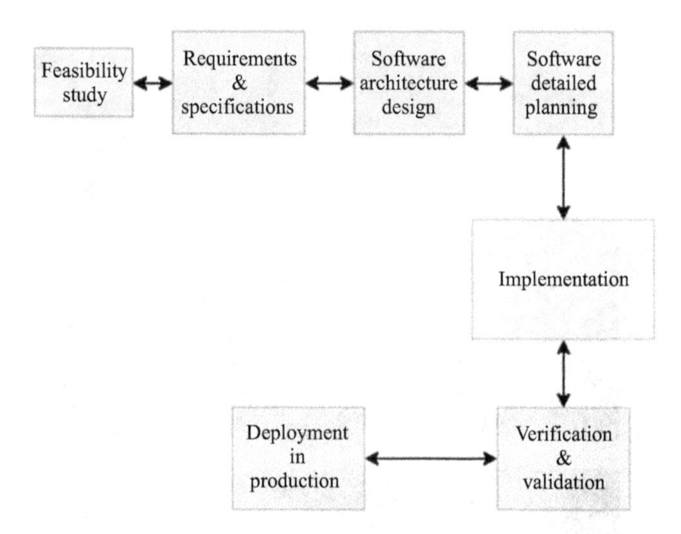

Figure 11.9 The Waterfall framework

The traditional framework for managing the software life cycle is the Waterfall model, which is illustrated in Figure 11.9. Individual stages in the flow diagram are fairly self-explanatory. The only stage that might be unfamiliar to the reader is the verification and validation stage – verification and validation involve answering the respective questions 'Does the implemented software comply with the requirements and specifications?', and 'Does the implemented software satisfy the User needs?'. Going back and forth between consecutive stages is 'allowed' e.g. requirements are drafted from the findings of the feasibility study, but they might trigger a second feasibility study to widen the scope of the investigation as important information that were omitted might be discovered after discussions about the results of the first feasibility study. However, further studies are unlikely to be done at later stages of the project as any changes would affect downstream stages and would therefore be prohibitively expensive in time and resource to implement. There are a few weaknesses associated with the Waterfall model:

1. The outcome of the project has a high level of dependence on how effectively the requirements capture is done at the initial stages of the project. Quite often, the implemented software might not meet user needs either due to omission of requirements, wrong assumptions made during requirements capture, or misinterpretation during the implementation phase.
2. User needs might change after the requirements and specification definition stage of the project. The implemented software might therefore be useless.
3. Verification involves testing and this is carried out after implementation is completed. Writing tests for fully implemented software with a large code base is challenging, and test coverage might not be substantial enough due to hidden complexities.

4. Non-essential software features might be added during the implementation phase. This is a waste of development time and resources. In addition, there might be unintended consequences such as the introduction of bugs and vulnerabilities as a result of these features.

The Agile framework is created to combat these weaknesses, and, in recent years, it has achieved widespread adoption in many software projects. There are 12 principles associated with the Agile manifesto [18]. We will only discuss eight points here (some of them are an amalgamation of these principles):

1. *Customer needs first*: The emphasis is on writing software that will satisfy customer needs. This principle necessitates frequent customer consultations and derivation of customer feedback from intermediate prototypes.
2. *Embrace change*: Changes to software requirements and specifications are welcome and they can be introduced at all stages of the software development cycle.
3. *Team autonomy*: Substantial autonomy is given to the software engineers and other technical staff in the way they carry out the development. Further, they are heavily involved during requirements capture as well as regulatory compliance activities.
4. *Sustainable pace*: The focus of Agile is on the people rather than the processes involved. As such, development should take place at a sustainable pace. Doing so should result in designs and implementations that are well considered, and aligned with customer needs.
5. *Minimal software*: This principle can be broken into three aspects. First, within a product, unnecessary software features should not be implemented. Second, develop only one product that is requested by the customer. Third, the code base and the tests written for it are the most important deliverables. Unnecessary documentation and tasks are to be avoided.
6. *Iterative development*: Going back and forth between various stages of the development cycle is expected to make the process more nimble to changing user needs. Changing and refactoring the code base as a result of changes in the requirements and/or to make the code more organised, robust, and suited for testing are deemed to be necessary and useful exercises.
7. *Tests as a key resource*: Tests at various levels are designated as important markers of progress for the project and should be performed on a regular basis. Substantial effort should be put into developing a comprehensive and effective test infrastructure, which should pay off in the long term.
8. *User stories*: Rather than formal requirements which require specialist knowledge to interpret, user stories should be used to capture requirements in an informal way so that they can be easily understood and extended later on in the project.

At the heart of the Agile method is the *Sprint* – the building blocks of a sprint are shown in Figure 11.10. Unlike the linear nature of the Waterfall framework, the Agile framework is cyclical and iterative. From an initial product vision, user

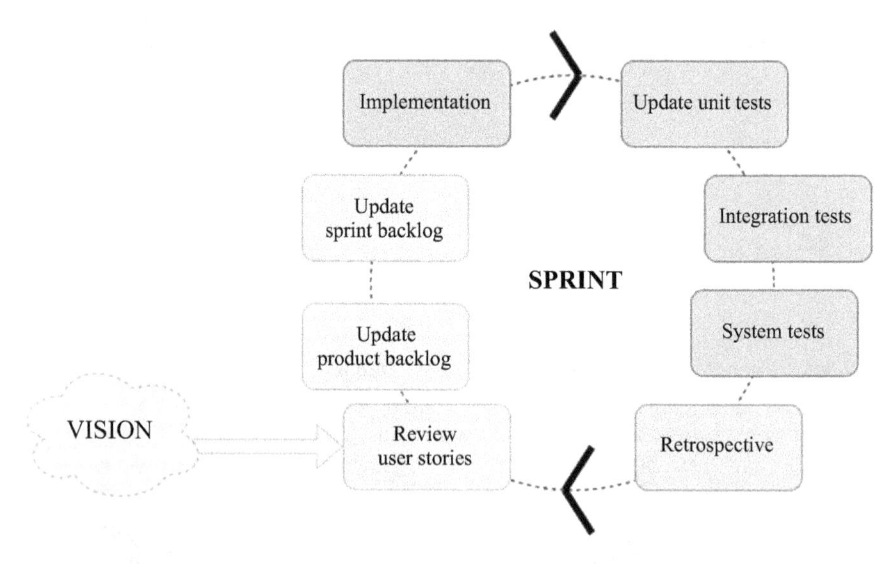

Figure 11.10 An Agile sprint. Green, red, and purple blocks correspond to sprint planning, implementation and the retrospective phases of the cycle, respectively.

requirements are captured in the form of user stories, which specifies user needs in simple language. From these user stories, a product backlog is maintained, containing a list of user features that would satisfy those needs. Subsequently, a subset of these tasks are extracted to form the sprint backlog, which contains features that needs to be implemented in a single sprint cycle. This would form the major tasks involved in sprint planning and would involve all members of the managerial and technical team.

Sprint development tasks could involve implementing new features, resolving bugs in the current feature set, and refactoring code to make it more robust. In addition, it would also include writing the various tests to prove the robustness of the actual code that has been written. As far as possible, these tests should be automated so that they may be re-used as part of regression testing in subsequent sprint cycles without burdening the development team (we will discuss tools and automation later on). This is necessary and useful as it prevents newly implemented features from breaking existing functionalities. Finally, during the retrospective phase of the sprint, the level of success achieved during the sprint will not only be assessed based on the number of features implemented but also the results of those tests. In addition, the team will determine how to further improve their effectiveness and how to avoid mistakes in the future. Typically, the implementation phase of the sprint cycle takes between two and four weeks, with the rest of the time being spent planning the sprint and evaluating user needs, as well as the needs of the development team.

Thus far, we have considered two frameworks that can be used to manage the software development life cycle. For manufacturers who intend to write software for medical devices, they have to ensure that they comply with additional regulatory requirements that are listed below.

1. *IEC62304*: This includes principles relating to the software development processes and the documents that need to be generated as part of the design process.
2. *ISO14971*: This contains the guidelines about how to quantify and manage risk to patient safety through the use of the product.
3. *ISO13485*: This contains information about the quality management system (QMS) that needs to be put in place to ensure the efficacy and safety of the product.
4. *IEC60601-1*: This concerns the design processes that need to be put in place to ensure the electrical safety of the product. Obviously, this is relevant for software products with a physical embodiment that are part of the medical device e.g. firmware in embedded devices.

These regulations are created with a view that software testing cannot possibly provide sufficient coverage for medical software. Therefore, design controls have to be embedded in the processes involved in designing and implementing the software, which should then translate into software that delivers the intended benefits without endangering the patient. On the other hand, manufacturers often run into problems when they implement the Agile method in software development, while trying to comply with the standards described above, due to conflicts between the principles espoused by the standards and Agile. As an embedded firmware engineer working in a medical devices company, the observations below are a combination of my review of existing literature and my personal experiences.

1. *Insufficient thought given to requirements capture*: An important Agile characteristic is the light weight nature of the feasibility study and requirements capture stage, and a reliance on the flexibility of the framework to make changes to the requirements later if necessary. In practice, the rush into development without giving enough thought to requirements or user stories will lead teams into avoidable pitfalls and confusion, just like any other management framework. Crucially, enough time has to be allocated to the creation of these user stories, following the guidelines of *IEC62304*, and particularly to make sure that these user stories are verifiable. As a rule of thumb, proper Sprint planning could take about two to three weeks, especially near the beginning of the project.
2. *When and how to carry out verification*: As we have seen above, Agile champions continuous testing and verification in order to spot bugs and issues early. Ideally, this should be carried out at the end of every sprint. In many instances, this is an onerous burden as verification has to be accompanied with reviews and documentation, as required by *IEC62304*. To feasibly comply with both Agile and the standard, the verification process has be streamlined and automated with good usage of tools as far as possible. We will discuss about some of these tools later on.

3. *Regular consultation with customers*: Agile encourages frequent interactions with customers as part of a continuous validation process, so that the developed product meets customer needs. In practice, direct interaction between the product and the patients is possible only through clinical trials, which requires months, if not years of work with a notified body to demonstrate compliance with the relevant standards before they can submit an application to a competent authority (in the UK, this would be the MHRA) for a letter of no objection to proceed with the clinical trial. Frequently, early validation attempts have to be carried out by proxy i.e. with the principal investigator, other clinical staff who understands patient needs, and members of the public who are not patients. However, this is an important and useful Agile principle and medical devices companies should take every opportunity to obtain realistic feedback about the product that they are developing, and to be creative in bringing about such opportunities. One option is to obtain feedback in simulated hospitals [19].
4. *Frequency of shipping and software releases*: Since medical software is safety critical, shipping or deploying the software in production happens at the very late stages of the project. This is at odds with Agile which encourages frequent releases. On the other hand, it is often possible and important to frequently release intermediate development builds (comprising of partially finished software) for the purpose of internal testing.

The intent of this section is to outline the regulatory landscape where medical devices manufacturers operate, and how they go about complying with these regulations in order to write safe software that will improve patient care. It also describes some of the challenges that manufacturers face in developing and deploying the product. In Section 11.4.3, we will focus on the topic of risk management and some of the tools that are used to streamline development and testing, both as risk mitigation measures as well as to demonstrate compliance with the competent authority.

11.4.3 Risk management in medical software development

The relevant guidelines concerning the assessment and management of risks for medical devices are described in the *ISO14971* standard. To make a start, we ask the following questions:

- What is *risk*?
- How do we quantify *risk*?
- How do we assess if *risk* is acceptable?
- How can we mitigate risks, drawing lessons from the aviation industry?

The terms that are useful in defining risk is explained in Table 11.3, and many of these terms are derived from the *ISO14971* standard [20]. In principle, risk can be computed as a product of the probability of occurrence of harm and its severity, as shown in (11.14). While it is relatively simpler to quantify the impact caused by a hazard, the apparent simplicity of (11.14) hides the complexity in estimating the probability of occurrence. Indeed, during the design and development phase of the

Table 11.3 Risk terminologies [20]

Term	Definition
Hazard	Probable source of harm
Hazardous Situation	Situation when people/property are subjected to one/more hazards
Risk	Risk associated with a hazard
P_o	Probability of occurrence of harm
S	Impact of hazard on patient safety
Harm	Injury/damage to the health of people Damage to property/environment [20]
Sequence of events	Reasonably foreseeable events/combination of events that will lead to hazardous events [20]
P_1	Probability of the hazard occurring
P_2	Probability of harm occurring as a result of a hazardous situation

product, such information is non-existent. Also, while software code itself cannot directly cause injury in people, property, or the environment, it can certain cause indirect harm. Examples include failure in control of the external components such as mechanical actuators leading to injury or causing a misdiagnosis due to software errors and/or providing misleading information. Because of the indirect nature of software hazards, it is useful to estimate the probability of its occurrence based on how probable the hazard will occur in the first place as well as the probability that the hazard will lead to harm, as shown in (11.15):

$$Risk = P_o \times S \tag{11.14}$$

$$P_o = P_1 \times P_2 \tag{11.15}$$

Due to a wide range of medical devices available, it is counterproductive for the health authorities to be over-prescriptive in the methods used to assess risk levels as each medical product is different. It is therefore the responsibility of the manufacturers to demonstrate that they have an objective method to identify hazards, assess the risks associated with each hazard, and determine if the amount of risk is acceptable or not. A common scale that is used among medical device manufacturers is shown in Figure 11.11. Specifically, after estimating the probabilities P_1, P_2, and computing P_o as a product of these terms, a qualitative (numerical) tag can be attached to P_o, ranging from 'Improbable' (Score 1) to 'Frequent' (Score 5), as shown in Figure 11.11(a). The severity of the hazard can be ranked on a simple scale from 1 to 5, which refer to no impact on patient safety to severe impact (potentially fatal) respectively (Figure 11.11(b)). Since the acceptability of risk is contingent on both the probability of occurrence as well as the severity of the hazard, risk acceptability can be visualised as regions in the table.

P_0 (Probability of Occurrence)		
Definition	**Probability**	**Value**
Improbable	$< 1e^{-7}$	1
Remote	$< 1e^{-6}$	2
Occasional	$< 1e^{-5}$	3
Probable	$< 1e^{-4}$	4
Frequent	$< 1e^{-3}$	5

(a)

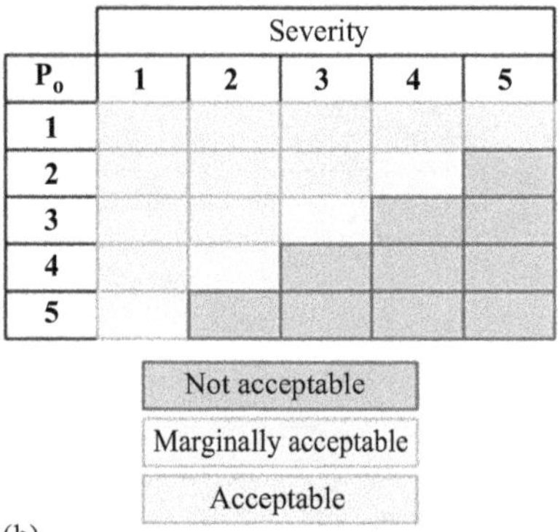

	Severity				
P_0	**1**	**2**	**3**	**4**	**5**
1					
2					
3					
4					
5					

Not acceptable
Marginally acceptable
Acceptable

(b)

Figure 11.11 Risk assessment: (a) categorisation for P_O. (b) Grading risk levels.

While this method is rather subjective and coarse, it is rather useful in practice as it provides manufacturers with a means to compare hazards in relative terms and to prioritise development efforts and resources. According to the *ISO14971* standard, manufacturers are required to maintain a risk register, which contains information about the hazards that have been identified, assessed risk level, the risk mitigation steps that are implemented, the amount of residual risk as a result of mitigation, and whether the residual risks are acceptable based on the scale described in Figure 11.11. An example of a risk register for a wearable, battery operated arrhythmia monitor is shown in Figure 11.12. In this case, a hazard is introduced by the onboard firmware because it fails to notify a caregiver that a patient is in distress. The results are potentially catastrophic as the patient might die as a result of the incident – it is therefore given a severity of 5.

The potential cause of the event is a firmware update, when the manufacturer automatically patches the device firmware to resolve existing bugs and/or add new product features. If this happens while the device is in use, both the caregiver and the patient might be unaware of the situation and assume wrongly that the device is functioning in its normal state. Firmware update events are fairly frequent, particularly in the early stages of deployment. Therefore, it is assigned a value of 0.0001. For a hazardous situation to result in actual harm to the patient, a firmware update event has to coincide with the patient having an arrhythmia episode. Since the product is meant for patients suffering from arrhythmia, this is highly probable and therefore assigned a value of 0.001. The resulting level of risk places this hazard in an unacceptable region, which triggers the need for additional risk control measures. Two risk mitigation examples are shown in Figure 11.12 – firmware updates are allowed only when the patient is not wearing the device, and when the firmware detects a physical connection with the charging port. Further, explicit indications are displayed on the device LCD screen indicating that firmware updates are ongoing. By implementing these risk control measures, there is a large reduction in P_1, and the residual risk level is now in the acceptable region.

Hazard	Sequence of Events	Hazardous Event	Harm
Caregiver not informed when patient is in distress	System performs scheduled firmware update Arrhythmia detection functionality inactive during firmware update	Patient did not receive medical attention as the device did not notify caregiver.	Serious. Potentially fatal

Hazard	P1	P2	Po	S	Acceptable
Caregiver not informed when patient is in distress	0.0001	0.001	3	5	No

Risk Control Measure	P1'	P2'	Po'	S'	Acceptable
1. Activate Firmware update only when device is not worn, while it is charging. 2. Display 'Software Update' in the LCD display during Firmware update.	0.0000001	0.001	1	5	Yes

Figure 11.12 Risk register for a wearable arrhythmia detection device

We saw an example of how one is able to introduce risk control measures and reduce the risk level to an acceptable level. Broadly, we can reduce the level of risk through our development process as well through the use of certain software design principles. More specifically, we suggest some measures below that can streamline the development process and enable efficient and effective verification activities in each sprint and crucially, allow traceability between the requirements and parts of the software application that are linked to them. It is important to note that apart from merely using these tools, it is critical that they are used together in the right way so that the maximum value can be extracted from them.

- *Software version control*: This should be used to track changes throughout the development process. Branching and tagging strategies should be implemented so that new commits are linked to specific software requirements or user stories, as well as the relevant risk control measures. Examples of software version control tools are Git [21] and SVN [22].
- *Issue tracking software*: A requirements traceability matrix is maintained for the whole software development life cycle. Issue tracking software can be used to maintain the product and sprint backlog, and to automatically track changes by linking the tasks in each sprint (linked user stories should be entered in those tasks) to specific software commits. This is achievable through the use of software like Jira [23].
- *Automatic document generation*: Several unit test frameworks have automatic document generation capabilities. Also, there are tools like Doxygen [24] that allows associations between different parts of the code to be established, and for explanations of the code base at a functional level to be possible. These tools allow software documentation to reside in close proximity to the actual code base, which makes it easier to trace code changes through documentation, and for it to version controlled together with the code base. In addition, developers are more likely to keep the documentation up to date.
- *Digital rather than paper-based documentation*: Paper-based documentation tend to be onerous to maintain and changes are difficult to track. As such, most software documentation can be kept in digital form. In addition, tools like Wiki [25] and even the README pages in software version control repositories can be used to hold information about the software such as the release notes. These documentation facilities are usually tightly coupled with other aspects of Software version control, which makes traceability an easier job, particularly at late stages of the project.
- *DevOps CI/CD pipelines*: Continuous integration (CI) and continuous development (CD) facilities involve setting up a pipeline that automates a series of checks before actual software deployment [26]. Such checks would involve unit tests (as part of the regression testing framework), static code analysis, integration and system level tests. Downstream operations are stopped if any of these checks fails. It is also possible to include documentation generation as part of the pipeline. Again, this tool is usually tightly coupled with the software version control and can be activated when the developer checks in new code. In practice, the CI/CD

pipeline takes substantial time and effort to set up and it is usually done when the project reaches maturity, but they can result in substantial time savings if they are properly designed and executed.

- *Code reviews*: Old fashion code reviews are very useful at identifying bugs and improving code quality among peers. However, instead of holding a long and formal review, code reviews can be kicked off with a pull request so that peers have an opportunity to study the code base, look at the tracked changes, and to add comments and suggested changes within the pull request. The pull request can be accompanied by a face-to-face meeting to discuss and clarify any remaining issues within the code. Regular and short code reviews should be done as opposed to occasional but long reviews (ideally at the end of every sprint. This could be considered as part of the retrospective phase of the cycle), so that code base changes linked to the feature set relevant to that particular sprint can be reviewed.

The tools and processes described above concern the development environment that a developer or a software engineer operates in. While they certainly help in terms of complying with the Agile principles and regulatory compliance, and streamlines the development process, they do not relate to how the engineer ought to write his/her code. Rather than quibble over the semantics and various styles of coding, we will borrow the design principles that we described in Section 11.4.1 and provide anecdotal examples of how these principles might apply to medical software development in order to reduce risks to patient safety.

- *Redundancies*: It is important to build in redundancies within the software. As an example, when we are applying firmware updates in the context of embedded devices, the original firmware can be retained alongside the new firmware. Subsequently, if the new firmware fails to load/execute properly, the bootloader should have a mechanism to allow it to revert back to the original working firmware. Ideally, if both application firmwares fail, the bootloader should transit to diagnostic mode and send failure messages to the user and/or manufacturer. This should in turn trigger automatic or manual remedial actions to resolve the issue. In this case, we introduce two levels of redundancies to prevent 'bricking' the device.
- *Fail-safe design*: If the medical device fails, it would be better for it to fail in an acceptable state. For example, if the exposure switch in an X-ray machine fails, the onboard software should not allow the operator to power on the device fully, so that it is improbable that operators or patients will be subjected to harmful doses of radiation. Further, error handling within the code is an important aspect of fail-safe design, and should be subjected to a rigorous review process in order to determine if they allow the device to transit to an acceptable state and also to determine if there are any unintended consequences from the actions that are triggered.
- *Built-in monitoring*: Frequently, electronic components such as accelerometers and audio amplifiers have built-in mechanisms that would detect failure. Ideally, the failure mechanism should be accompanied with software interrupts to rapidly inform the MCU of failure. In the absence of an interrupt mechanism, a polling mechanism should be implemented for the MCU to check the component status on

demand. Crucially, the detection mechanism should be implemented in the driver layer of the firmware code (rather than the application layer) with a portable, hardware abstracted API, so that incorporation into different applications can be done with ease. In addition to these detection mechanisms, informing the user of failure in an effective and intuitive manner is extremely important as well. Indeed, a well-designed software interface e.g. different patterns of LED flashes in combination with error prompts on the LCD screen, will go a long way in helping to diagnose and resolve issues in the field, particularly if remote support by telephone or video conference is part of the support package.

- *Component fault categorisation/prioritisation*: In many ways, the risk register as well as the FMEA are helpful in the categorisation and prioritisation of component faults. This can be done in a more fine-grained and reactive manner in the software. For example, the severity of each component failure can be graded on an arbitrary scale between 1 and 5. These types of failures can then be explicitly written in the header files of the application firmware. By ranking failures in such a way, different error handlers and actions can be executed depending on it's priority. In addition, alignment of these actions with the frontend interface is an important task that needs to be done in order to guarantee a good user experience.
- *Effective maintenance checks*: While build-in monitoring is a reactive mechanism, where the components (particularly sensor and actuator subsystems) report faults on their own initiative, maintenance checks might be triggered by different parts of the system including the MCU, the backend, or even manual checks that manufacturers might ask users to perform. If applied judiciously, overlaps and duplications between checks are not wasteful as they often reveal problems early rather than near the point of use. Corrective action can then be effected to resolve the issue quickly.
- *Modularity in design*: Unlike the case made in aviation, modularity is not implemented for the purpose of rapid replacement. Rather, modularity in the software design makes it easier to identify bugs or issues, and to determine the impact that they have on the rest of the system. Another important reason for modularity is the necessity for unit tests (and subsequent incorporation into a regression testing framework). As such, the organisation of the code base has to be carried out in a very clear, concise and rational way to allow it to be easily understood and maintained.
- *Post-market surveillance*: Similar to air crash investigations at the National Transport Safety Committee, any incidents should be taken seriously and measures should be put in place to ensure that patient safety is not compromised, even after the delivery and deployment of the medical software. Issues could still occur in the field and it is the responsibility of the manufacturer to continue monitoring the product, and include any newly found hazards into the risk register. Specifically, after critical metrics relating to software performance, errors and malfunctions are defined, systems would usually be put in place to collect statistical information concerning these metrics. The resulting empirical data would then be used to improve risk assessments on a regular basis. If unacceptable risks are identified, new risk mitigation measures would be put in place to reduce them to acceptable

levels. If these mitigation measures involve software changes, they should be subject to *change management processes*, that should have been put in place during production. In drastic cases where risks cannot be acceptably reduced, or when immediate threats to patient safety are found, taking the software offline and/or executing a product recall may be deemed necessary.

Unlike the aviation industry, where it is possible to specify and standardise risk management processes due to similarities between aircraft structures and components, it is more difficult to do so in the medical devices industry due to the vast differences between different products. As a result, manufacturers are given a lot of autonomy in the area of risk management and they might make use of different frameworks and methods to create safe and effective medical software. Having said that, we are convinced that there are enough similarities between them for the above design principles to be useful and applicable.

11.4.4 Challenges specific to ML

Given that ML algorithms are inherently software, most of the topics described in the previous sections would apply to deployment of ML for use with patients. However, there are important differences and challenges that are unique to ML, which are highlighted below.

1. *Data availability*: Unlike traditional software development, it is crucially important to possess good quality data that can be effectively used to train, test, and to validate ML models. Due to the need to maintain patient privacy, and to comply with regulations such as GDPR, access to medical data is usually difficult and requires a time consuming and involved process [27]. Even when approval is granted, actual access is limited to a few individuals within an organisation, which can hamper development efforts since fairly large teams are involved in the development and deployment processes.
2. *Data preparation*: The data that is provided by the source e.g. health records cannot be immediately used. Pertinent parts of the data have to be extracted and organised in a manner that can be applied to the ML models. In addition, when feature engineering is required, the feature set has to be determined and the functions used to compute those features have to be written. Subsequently, the data (either raw or in the form of features) might be organised as a multi-dimensional matrix and aligned with the labels. Also, the original dataset is noisy and a preprocessing step is often required to ensure that spurious data points are not included – this is an important step as spurious data might have implications on downstream operations. Frequently, many of the steps above can only be partially automated, making data preparation an exceedingly time consuming and laborious step [28]. Another related challenge is the version control of the dataset. Traditional tools such as Git and SVN tend to have limits imposed on the memory capacity of the repository. Consequently, large datasets (Gigabytes or Terabytes of data for a dataset, as well as different revisions of those datasets are

typically involved in this case) cannot be version controlled in the usual manner, similar to the rest of the code base.

3. *Friction between different specialist groups*: Developing and deploying ML as a product requires data scientists, data engineers, and software developers. The data scientists design and perform experiments on the ML models, while the data engineers organise, process, and make the data more accessible for the models. Finally, the software developers are concerned with packaging the model and the dataset to make them more easily deployable in production, as well as to ensure that the model is smoothly integrated with the rest of the system (this could comprise of a backend and a frontend where ML results are made available to users). Getting these three groups of engineers (or more, depending on the nature of the project) to work effectively together is challenging due to the lack of harmonized processes [28]. The fact that access to the original dataset is usually exclusive to some of the data scientists and/or engineers does not help, and this often means they operate in 'silos' and they hand over the models as assembled binaries to the software developers for integration. Strictly, this is contrary to Agile principles, and one implication is integration issues that are discovered only at late stages of the project, leading to unnecessary delays.
4. *Opaqueness of ML models*: Modern ML models are exceedingly complex and difficult to understand. Increasingly, there are research efforts to make models explainable in a manner that humans can understand. This is important as manufacturers have to understand the model well enough to carry out risk assessments, and demonstrate to regulators that their product will not compromise patient safety. In addition, they are expected to provide a description of the technology and provide evidence relating to the product performance and safety. The opaqueness of ML models hampers the efforts to do so.
5. *Different levels of performance in production*: Quite often, the ML model that is derived from the initial development phase performs differently from when it is deployed. This is because ML models are sensitive to demographic differences, which would alter the distribution of the input data and subsequently its results. Integration and usage flaws might also contribute to substantial differences in it's performance. As such, there is likely to be a need to retrain the models and re-deploy them even in production. This might create considerable problems from the regulatory compliance point of view as any substantial changes to the software would necessitate a re-submission to the competent authority, and introduce unacceptable delays to the project.

The solution to many of the challenges described above is to make use of an MLOps CI/CD pipeline to streamline the processes involved in creating a viable model, checking that it is fit for purpose, and deploying it rapidly, while automating the generation of documentation as far as possible, as illustrated in Figure 11.13. To be more specific, the first part of the pipeline involves the exploratory phase when a set of model candidates are built and evaluated with a dataset. The set of metrics should be sensibly chosen and stored in a way that can be easily executed and extended. After the evaluation stage, the chosen candidate would be packaged in a form that can be

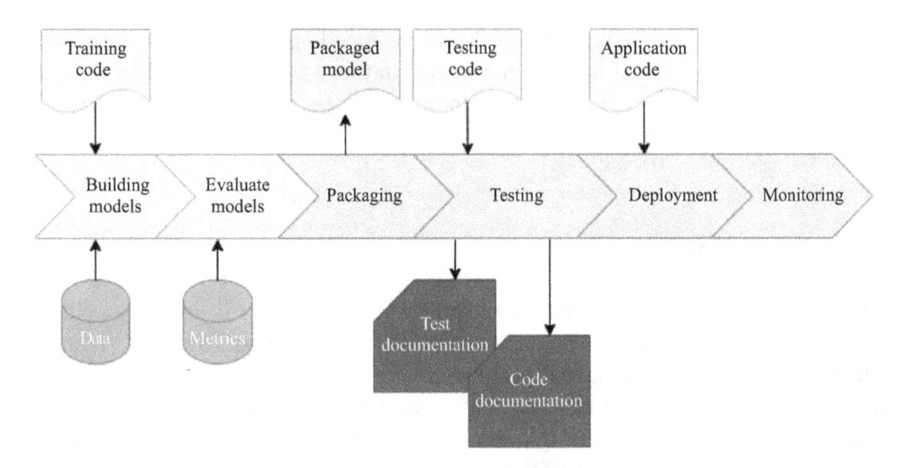

Figure 11.13 A simplified view of the MLOps CI/CD pipeline [28]. Blue and purple stages correspond to MLOps and DevOps stages of the pipeline respectively.

properly integrated within the application code – unit and integration tests should be automatically done to ensure this is the case. The model would then be deployed in the production environment if it passes all of the upstream checks. Accompanying test and code documentation can be automated and generated as part of the testing stage as far as possible. Finally, monitoring of model performance e.g. automatic parsing of server logs is carried out to check for anomalies in the final stage of the pipeline.

There are a number of tools that enable pipeline deployment that are specific to MLOps such as DVC [29]. Apart from pipelines, this tool enables ML data to be version controlled by storing the metadata associated with these datasets and using standard version control tools to maintain the metadata files. Conveniently, it is a layer that is built on top of traditional version control tools as well as the DevOps pipeline facility that is provided together with them. Therefore, the integration of MLOps and DevOps frameworks can be done with relative ease.

Changes to ML models can be managed in a similar way to traditional software. As part of software change management, versioning of software is done at three levels – major, minor, and patch revisions [30]. The impact of changes can be assessed by the team as part of its risk management process. The work involved in making these changes as well as the time taken for them to be effected would then depend on the type of revision that is considered. For example, small and benign changes that are necessary to patch current bugs should be implemented quickly and documented as part of the release notes. Minor changes that are beneficial for the system, and imposes minimal risks can be implemented and signed off only after code reviews to properly assess its impact. Finally, major changes would necessitate a major review and re-design, as well as a re-submission to the competent authority. Regardless of these changes, the MLOps pipeline should minimise the amount of time required to effect changes. In fact, rather than spending time on repetitive and laborious tasks that can be automated, the technical team should be spending most of their time fixing issues or failures at any of the pipeline stages. In this way, necessary changes made during the production stage can be carried out in a safe, resource and time effective manner, while complying with the necessary regulations.

11.5 Conclusion

Clearly, ML is an extremely versatile and powerful tool and its role in being a clinical support aide to improve health outcomes will continue to expand. To realise ML applications more ubiquitously in healthcare, there are a number of hurdles to overcome, including its application in edge devices as well as complying with the regulatory requirements. While these are important topics, there is limited discussion about them in the preceding chapters of this book. Therefore, we hope that the information provided in the concluding chapter will be useful in providing a more rounded view of ML in healthcare.

From Section 11.3.3, substantial leaps in the form of efficient hardware accelerators have occurred in recent years. On the other hand, current draw requirements from these SOCs would still put them out of reach for many embedded applications, indicating that much work remains to be done in this particular area. An interesting area of development is power efficient analogue neural networks, which can be realised through existing manufacturing techniques in massive memory storage (an excellent discussion of this topic can be found in Chapter 10). Apart from platform constraints, it is challenging to deploy ML in medical applications while complying with regulatory constraints. We demonstrated that development efforts can be organised more efficiently and effectively through the judicious use of tools in DevOps and MLOps, as well as through sensible application of management frameworks like Agile.

Overall, we are very optimistic about the outlook of ML in healthcare. Indeed, given the gap between mortality rates in the healthcare industry and other safety critical industries e.g. aviation industry, we believe that apart from the need to draw lessons from other industries to design and deploy safe ML applications, the use of ML itself in healthcare will play a significant role in closing this gap and improve the level of care for patients in the long term.

References

[1] Godfrey MD. Introduction to "the first draft report on the EDVAC" by John von Neumann. *Annals of the History of Computing*. 1993;15(1):11–21.
[2] Tan G, Li L, Triechle S, *et al.* Fast implementation of DGEMM on Fermi GPU. In: *Proceedings of 2011 International Conference for High Performance Computing, Networking, Storage and Analysis*, 2011. p. 1–11.
[3] Chetlur S, Woolley C, Vandermersch P, *et al.* cudnn: Efficient Primitives for Deep Learning, 2014. arXiv preprint arXiv:14100759.

[4] Markidis S, Der Chien SW, Laure E, *et al.* Nvidia tensor core programmability, performance & precision. In: *2018 IEEE International Parallel and Distributed Processing Symposium Workshops (IPDPSW)*. IEEE, 2018. p. 522–531.

[5] The Apple Neural Engine. Accessed: 2022-10-30. https://machinelearning.apple.com/research/neural-engine-transformers.

[6] Zhu Y, Mattina M, and Whatmough P. Mobile Machine Learning Hardware at Arm: A Systems-On-Chip (soc) Perspective, 2018. arXiv preprint arXiv:180106274.

[7] Jouppi NP, Young C, Patil N, *et al.* In-datacenter performance analysis of a tensor processing unit. In: *Proceedings of the 44th Annual International Symposium on Computer Architecture*, 2017. p. 1–12.

[8] Antonín H, Jiří K, Rudolf M, *et al.* Pipelined logarithmic 32bit ALU for Celoxica DK1. ÚTIA AV ČR, (Praha 2001) Research Report; 2034.

[9] Sze V, Chen YH, Emer J, *et al.* Hardware for machine learning: challenges and opportunities. In: *2017 IEEE Custom Integrated Circuits Conference (CICC)*, IEEE, 2017. p. 1–8.

[10] Coral Development Board Datasheet. Accessed: 2022-10-30. https://coral.ai/docs/dev-board/datasheet/#features.

[11] Akkermans HA and Van Wassenhove LN. A dynamic model of managerial response to grey swan events in supply networks. *International Journal of Production Research*. 2018;56(1–2):10–21.

[12] Cowan T, Acar E, and Francolin C. *Analysis of Causes and Statistics of Commercial Jet Plane Accidents Between 1983 and 2003*. Gainesville, FL: Mechanical and Aerospace Engineering Department University of Florida, 2006.

[13] Cutti S, Klersy C, Favalli V, *et al.* A multidimensional approach of surgical mortality assessment and stratification (SMATT score). *Scientific Reports*. 2020;10(1):1–10.

[14] Wright P, Pocock S, and Fields B. The prescription and practice of work on the flight deck. In: *ECCE9, Ninth European Conference on Cognitive Ergonomics*, 1998. p. 37–42.

[15] Boeing 747 Pilot's Instrument Panel. Accessed: 2022-11-6. https://www.seaboardairlines.org/aircraft/b747_c1.jpg.

[16] Boeing 747 Flight Engineer's Instrument Panel. Accessed: 2022-11-6. https://www.seaboardairlines.org/aircraft/b747_c2.jpg.

[17] Software Life Cycle Description. Accessed: 2022-10-27. https://www.techopedia.com/definition/20387/software-life-cycle.

[18] Principles of the Agile Methodology. Accessed: 2022-10-27. https://www.agilealliance.org/agile101/12-principles-behind-the-agile-manifesto/.

[19] Kushniruk A, Nohr C, Jensen S, *et al.* From usability testing to clinical simulations: bringing context into the design and evaluation of usable and safe health information technologies. *Yearbook of Medical Informatics*. 2013;22(01):78–85.

[20] ISO14971 Standard: Application of Risk Management to Medical Devices. Accessed: 2022-10-28. https://www.iso.org/standard/72704.html.

[21] Git Version Control Software. Accessed: 2022-10-29. https://git-scm.com/.
[22] Apache Subversion Software. Accessed: 2022-10-29. https://subversion.apache.org/.
[23] Issue Tracking Software. Accessed: 2022-10-29. https://www.atlassian.com/software/jira.
[24] Automatic Document Generation for Source Code. Accessed: 2022-10-29. https://doxygen.nl/.
[25] Wiki Page in Software Repositories. Accessed: 2022-10-29. https://docs.gitlab.com/ee/user/project/wiki/.
[26] The CI/CD Pipeline. Accessed: 2022-10-29. https://docs.gitlab.com/ee/ci/pipelines/.
[27] NHS Data Access Request Service. Accessed: 2022-10-29. https://digital.nhs.uk/services/data-access-request-service-dars.
[28] Granlund T, Stirbu V, and Mikkonen T. Towards regulatory-compliant MLOps: oravizio's journey from a machine learning experiment to a deployed certified medical product. *SN Computer Science*. 2021;2(5):1–14.
[29] Version Control for Machine Learning Projects. Accessed: 2022-10-30. https://dvc.org/.
[30] Semantic Versioning. Accessed: 2022-10-30. https://semver.org/.

Index

CPSIA information can be obtained
at www.ICGtesting.com
Printed in the USA
JSHW061935270523
42323JS00001B/2

9 781839 533358